T0143266

The Courtiers' Anatomists

The Courtiers' Anatomists

Animals and Humans
in Louis XIV's Paris

ANITA GUERRINI

The University of Chicago Press
Chicago and London

Anita Guerrini is Horning Professor in the Humanities and professor of history in the School of History, Philosophy, and Religion at Oregon State University. She is the author of *Experimenting with Humans and Animals: From Galen to Animal Rights* and *Obesity and Depression in the Enlightenment: The Life and Times of George Cheyne.*

The University of Chicago Press, Chicago 60637
The University of Chicago Press, Ltd., London
© 2015 by The University of Chicago
All rights reserved. Published 2015.
Printed in the United States of America

24 23 22 21 20 19 18 17 16 15   1 2 3 4 5

ISBN-13: 978-0-226-24766-3 (cloth)
ISBN-13: 978-0-226-24833-2 (e-book)
DOI: 10.7208/chicago/9780226248332.001.0001

Library of Congress Cataloging-in-Publication Data

Guerrini, Anita, 1953– author.
    The courtiers' anatomists : animals and humans in Louis XIV's Paris / Anita Guerrini.
        pages cm
    Includes bibliographical references and index.
    ISBN 978-0-226-24766-3 (cloth : alk. paper) — ISBN 978-0-226-24833-2 (e-book) 1. Zoology—Study and teaching—France—Paris—History.
    2. Anatomy—Study and teaching—France—Paris—History. 3. Académie des sciences (France)—History. 4. Histoire des animaux. 5. Perrault, Claude, 1613-1688. 6. Du Verney, M., 1648-1730. I. Title.
        QL51.2.F8G84 2015
        590.760944'361—dc23
                        2014039631

To the memory of my parents,

Rita Lillian Greco Guerrini
1923–1981

Armando Severino Guerrini
1922–2014

CONTENTS

In general I have given proper names in the vernacular rather than in Latin—so "du Laurens" rather than "Laurentius," "Stensen" rather than "Steno," and "Johnstone" rather than "Jonsonius." There are a few exceptions, including Johannes Walaeus. There is little agreement about the latter's name in Dutch ("de Wale"? "de Waal"?), and he is universally referred to in secondary literature as "Walaeus."

In referring to animals, when the gender of the animal is known, I refer to it as "he" or "she." Otherwise, "it" is used.

The year is assumed to begin on 1 January, as it did in France. Dates on correspondence are given as their authors gave them.

On quotations from French and Latin: If there is a contemporary English translation, I have quoted from that (sometimes silently modified for clarity), citing both the original and the translation. Most of the quotations from the *Histoire des animaux*, therefore, rely on Alexander Pitfeild's 1687 translation. Any translations without such citation are my own. While I have retained the original spelling in seventeenth-century French quotations, I have silently regularized the use of accents. I have, however, retained original punctuation and other seventeenth-century diacriticals such as umlauts and circumflexes.

On money: The French *livre*, or pound, in use until 1795, consisted of twenty *sols* or *sous*, each of which was worth twelve *deniers*. It is notoriously difficult to assess the value of money in the past. Suffice it to say that 1,500 livres, the amount that many original members of the Paris Academy received as their pensions from the crown, would have been a reasonable but not princely middle-class income in the 1660s. Most academicians had other income sources, so they were not totally dependent on their pensions (although some were). In comparison, Duverney's *garçon*, or lab assistant, received 600 livres. But two things happened in the 1680s: the amounts of pensions for new members of the academy declined (Jean Méry, who entered the academy in 1684, received 600 livres, the same as Duverney's *garçon*); and as Alice Stroup has documented, the value of the livre itself declined quite drastically from the 1690s onward while the amounts of pensions remained the same. Fifteen hundred livres were worth something like 40 percent less in 1700 than in 1660.

The term "vivisection" is anachronistic for the seventeenth century; in English, it only came into common use in the eighteenth century, and it appeared even later in French. In addition, it has the specific meaning of surgical intervention (literally "live cutting"). I therefore use the term sparingly in this book, preferring the more accurate if somewhat more cumbersome "dissection of living animals."

ABBREVIATIONS USED IN THE NOTES

Libraries and Archives

AdS   Archives, Académie des sciences, Paris.

AN   Archives nationales, Paris.

BIUS   Bibliothèque interuniversitaire de santé, Paris.

BL   British Library, London.

BN   Bibliothèque nationale, Paris.

MNHN   Bibliothèque, Muséum nationale de l'histoire naturelle, Paris.

Wellcome   Wellcome Library, London.

Manuscripts and Books

AT   *Oeuvres de Descartes*, ed. Charles Adam and Paul Tannery, new ed., 11 vols. (1974–86).

CdB   *Comptes des bâtiments du roi*, ed. Jules Guiffrey, 5 vols. (1881–1901).

Colbert, *Lettres*   *Lettres, instructions, et mémoires de Colbert*, ed. Pierre Clément, 10 vols. (1861–82).

DSB   *Complete Dictionary of Scientific Biography* (Gale, Cengage, online).

Eloy   Nicolas Francis Joseph Eloy, *Dictionnaire historique de la médecine ancienne et moderne, ou mémoires disposés en ordre alphabétique pour servir à l'histoire de cette science et à celle des médecins, anatomistes, botanistes, chirurgiens et chymistes de toutes nations*, 4 vols. (1778).

HARS   *Histoire de l'Académie royale des sciences*, 2 vols. (vol. 1, 1666–86; vol. 2, 1686–99), published 1733.

HdA   *Mémoires pour servir à l'histoire naturelle des animaux* (1671, 1676, 1733; 1733 in *MARS*, vol. 3).

Huygens, *Oeuvres*   *Œuvres complètes*, 22 vols. (1888–1950).

JdS — *Journal des sçavans.*

MARS — *Mémoires de l'Académie royale des sciences, depuis 1666, jusqu'à 1699,* vols. 3–11 (continuing *Histoire*, above), published 1729–34.

Mersenne, *Corres.* — *Correspondance de P. Marin Mersenne,* ed. P. Tannery et al., 17 vols. (1933–88).

ODNB — *Oxford Dictionary of National Biography* (2004–).

Oldenburg, *Corres.* — *The Correspondence of Henry Oldenburg,* ed. A. Rupert Hall and Marie Boas Hall, 10 vols. (1965–75).

PdS — Pochettes de séance, Académie des sciences, yearly until 1723, monthly 1724–54, by séance from 1755.

Peiresc, *Corres.* — *Lettres de Peiresc,* ed. P. Tamizey de Larroque, 7 vols. (1888–98).

Perrault, *Essais* — Claude Perrault, *Essais de physique,* 4 vols. (1680–88).

Phil. Trans. — *Philosophical Transactions of the Royal Society of London.*

Pitfeild — [Claude Perrault, ed.], *Memoir's [sic] for a Natural History of Animals* (1688).

PV — Procès-verbaux, Académie des sciences.

PWD — *The Philosophical Writings of Descartes,* trans. John Cottingham, Robert Stoothoff, and Dugald Murdoch, 3 vols. (Cambridge: Cambridge University Press, 1985–91).

ILLUSTRATIONS

Chapter 1

Chapter 2

Chapter 3

Introduction

This book is about bodies, human and animal, dead and living: particular bodies in a particular place and time. The place and time is the city of Paris during the reign of Louis XIV, between 1643 and 1715. In this context, *The Courtiers' Anatomists* explores the role of dissection in the development of experimental methods in seventeenth-century science[1] and the central role of animals in those methods and that science, as well as in the wider cultural field of the second half of the seventeenth century and into the eighteenth.

This is a book about seventeenth-century science, but it is not a book about the scientific revolution. The medieval historian Caroline Walker Bynum recently commented that "ideas, events, and practices always change in a complex mutual relationship."[2] My goal in this book is to add another level of complexity to the ongoing historical discussion of ideas, events, and practices in that era commonly known as the scientific revolution. I don't use that phrase in this book, not so much because I believe "there was no such thing as the Scientific Revolution,"[3] as because the phrase has been taken by many historians (and nonhistorians) to mean a particular set of ideas, events, and practices focused on mathematics, physics, and astronomy, with, more recently, the addition of alchemy and chemistry. This book is not about those things.[4]

1. Paris as a Site of Science

Why Paris? In the seventeenth century, Paris was the most populous city in Europe, and Louis XIV the most powerful monarch. As a center of cultural production, Paris contributed to the birth of many of the cultural markers of modernity, including the novel, modern contrapuntal music, and the arts of printing, engraving, and illustrating (and painting), as well as science.

Yet, compared to London, seventeenth-century Paris has been much less explored as a site of science or these interrelated activities. This was a period of vast changes in Paris and in science, and while Parisians certainly participated in the wider world of early modern science, they also developed ideas and practices unique to their particular place and moment. Scientific Paris under Louis XIV included royal institutions such as the Paris Academy of Sciences and university institutions such as the Faculty of Medicine as well as an array of salons, private academies, and independent lecturers that both complemented and competed with more official bodies.

Science was embedded in other cultural pursuits, not only because Louis and his chief minister, Jean-Baptiste Colbert, aimed to control cultural production, but also because the same people practiced science, architecture, art, music, and literature simultaneously. The court of Louis exercised control by means of its patronage, but that control developed gradually, and despite the best efforts of Colbert and his successors, it was never all-encompassing. Much has been written about the way Louis reined in the aristocracy via the court life of Versailles.[5] Much less has been written about the government functionaries—some aristocratic, some bourgeois—and other professionals in Paris, including physicians, lawyers, and a variety of religious men, who attended salons and academies (and the opera), published works of literary criticism alongside works of science, and acted at various times as both patrons and clients in Parisian cultural life. These men cannot easily be labeled as "ancients" or "moderns," although by the end of the seventeenth century the moderns had definitely come to dominate.[6]

This book concerns the courtiers' anatomists, not the court's anatomists, and I define "courtier" broadly to encompass those whose position depended on the crown, whether or not they were in attendance at the court. This category includes such men as Henri-Louis Habert de Montmor, master of requests; Chancellor Pierre Séguier; Finance Minister Nicolas Fouquet; the prince de Condé; and Charles Perrault, controller of the king's buildings; as well as Colbert. These courtiers sponsored practitioners of literature, music, art, and science from the 1630s onward; clients such as Marin Cureau de la Chambre often pursued several of these activities at once. Close networks of families, institutions, and professions intertwined and overlapped. The Paris Academy of Sciences, founded in 1666, was the king's, and Colbert's, answer to the proliferation of patrons and projects in the 1640s and 1650s, but it did not end them.

Because of the place of science among other cultural pursuits in Louis XIV's Paris, I talk about a lot of things in this book that at first glance may seem to have little to do with science: not only printing, publishing, and illustration, whose connections with natural knowledge are now well established, but also music, architecture, literature, rhetoric, and aesthetics. Historians of early modern Europe know that modern disciplines did not then exist, and that the quest for knowledge mapped a myriad of paths through its broad terrain. But too often historians (including me) have nonetheless focused on a disciplinary corner of that terrain. The people I consider in this book have made a narrowly focused disciplinary study impossible. Such men as Claude and Charles Perrault, Marin Cureau de la Chambre, Louis Gayant, Jean Pecquet, and Joseph-Guichard Duverney are not household names, even among historians of the era. But in Paris, they were the most important investigators of the human and animal body.

2. Animal Bodies and Human Bodies

The other individuals I consider in this book are animals, and they too have resisted disciplinary categories. The uses of animals of all kinds in the development of modern life science is well known but largely taken for granted. While the development of the field of "animal studies" has certainly helped to insert animals into scholarly discourse, I believe it has also defined the field in a way that separates it from other disciplines, so that ignoring animals as historical actors remains the norm outside of that small subdiscipline. Little work in animal studies as it is currently practiced, for example, looks at the use of animals in science in the past other than to condemn it.[7] By placing animals at the center of my study, I document and acknowledge the enormous role they played in the birth of the experimental method as well as in natural history and the reconfiguration of the human and animal body. The metaphorical and symbolic meanings of animals in this period are moreover intimately entwined with their scientific meanings.

Historians, even historians of science, have been willing to leave consideration of animals to animal studies in part because the dissection of living animals in this period is, to current sensibilities, a distasteful and even shocking phenomenon. There was no anesthesia or reliable means of reducing pain, and experimenters seem, to modern eyes, stunningly cavalier in their use of animals. Most historians have taken the work of René Descartes, particularly his comments on animal consciousness in *Discours de la méthode* (1637) and more detailed comments in *Traité de l'homme* (written in the 1630s but not published until the 1660s), as establishing the philosophical basis for animal experimentation in the seventeenth century. As this book will show, this is simply not true. Descartes's ideas on animal consciousness were one opinion among many and not the most prominent. Most anatomists believed animals could indeed feel pain; Niels Stensen, who had no illusions on that issue, nonetheless expressed the consensus among men of science when he wrote, "We can treat animals as we please."[8]

The dissection of dead animals and humans was also largely taken as a matter of course by its practitioners and witnesses. This is even more surprising in an era when, as some have argued, the new science was a polite and gentlemanly activity. Yet the court and courtiers such as Nicolas Fouquet authorized cutting up animal and human bodies, and these activities took place regularly at scientific salons such as that of Montmor, another man of the court. Even at Versailles, at the same time that Louis schooled his noblemen on proper behavior, his son the Dauphin followed anatomical lessons. The violence and violation of dissection mirrored a violent society, a time of public execution and widespread and public corporal punishment. The journal of the Swiss medical student Felix Platter, who attended the University of Montpellier in the 1550s, detailed several horrific executions as well as several illicit dissections of stolen corpses. He witnessed one execution from the windows of a doctor's house, alongside the other dinner guests. Not much changed in the subsequent two centuries; the descriptions of the execution of the attempted regicide Damiens in 1757 are not for the faint of heart.[9] As my opening pages indicate, the juxtaposition of physicians, surgeons, dead bodies, and living and dead animals was an everyday occurrence in Paris, and courtiers were not isolated from this. Even the king witnessed the dissection of the elephant at Versailles in 1681. Hunting and seventeenth-century battles often involved killing at close quarters. Seventeenth-century people were not unfamiliar with blood and death.

Animals had been accepted as stand-ins for humans in anatomical study since at least the time of Aristotle. Many acknowledged that animals were not perfect models for humans; Vesalius had criticized Galen on precisely this point, and indeed Galen knew quite well that animals were not always ideal. But there was widespread acceptance of the general principle that animal bodies—some more than others—were similar enough to humans to act as stand-ins. The use of animals was neither controversial nor unusual, and the practice of dissection, even of live animals, and resulting doctrines such as the circulation of the blood, were not topics of concern to the Catholic Church, or to Protestant churches, in the way that astronomy was in this period. No works of anatomy appeared on the Catholic *Index of Prohibited Books*, and ecclesiastical censors across Europe readily approved them. Harvey's Protestant beliefs were not used to argue against the circulation, and in the 1630s and 1640s Pierre Gassendi, a priest, dissected with the Protestant physician Abraham Du Prat. Anatomical textbooks indeed claimed anatomy was first a moral practice, whose primary purpose was self-knowledge rather than medical or surgical uses. Knowledge of the human body, even if mainly gained from animals, would lead to that recognition of human uniqueness that in turn led to moral behavior.

Only humans had souls. It had long been understood that animals did not. When Descartes proposed his radical division of mind from body, he believed it would uphold this essential distinction between animals and humans and even strengthen it. But the idea that the human (and animal) body was merely a machine might also, it turned out, lead to speculation about the role of the soul: was it even necessary to human functioning? In *Traité de l'homme*, Descartes recognized this difficulty, presenting his machine-man as merely an ingenious construct, recognizing that a clockwork human would not have the rational soul that could lead to salvation. Even with this precaution, Descartes did not publish *Traité de l'homme* in his lifetime. When it finally did appear in 1662, it was quickly placed on the *Index*. Niels Stensen acknowledged that some of Descartes's followers were less cautious than their master on this topic. But most anatomists agreed with Stensen that while Descartes had some interesting ideas, his anatomical work was seriously flawed. Cartesian ideas about animal mechanism competed with many others. When Stensen dissected a human brain in Paris in 1665, he acknowledged that it contained the seat of the soul; but he did not attempt to find it.

3. Dissection and Natural History

Dissection has a justifiable claim to be the most widespread and significant scientific activity of the seventeenth century, and Paris increasingly became its epicenter over the course of the century. By the late 1680s, the anatomical courses and demonstrations of Joseph-Guichard Duverney at the Jardin du roi had become an essential stop in the medical peregrinations of ambitious European physicians and naturalists. It should be noted that "anatomy" in this period referred both to what we now call anatomy—that is, the study of structure—and to what we now call physiology, the study of function. Indeed, disentangling the relationship of structure to function was a major goal of anatomical study. The critical figures in the development of anatomical practice in the seventeenth century were William Harvey and the Parisian physician Jean Riolan, best known as antagonists over the theory of the circulation. Riolan has definitely received the short end of the historical stick, but even Harvey acknowledged him as "the prince of anatomists" and his emphasis on human and animal dissection unleashed forces he could not control. Harvey in particular developed research methods that depended heavily on the use of living and dead animals.

The explanations of Aristotle and Galen that had formed the basis of natural history and medicine for close to fifteen hundred years were no longer adequate by the seventeenth century to account for how nature works. New observations, new plants and animals, and a new curiosity about the natural world had begun to explode long-held beliefs. In the second half of the century, a number of new theories of human and animal function emerged. Most of them assumed some form of the new mechanical philosophy, and argued that the body could be analyzed and explained in terms of new concepts of mechanics that had been developed most prominently by Galileo and Descartes.

All of these new theories of human and animal function were based first on dissection and only second on concepts imported from mechanics. Dissection was not a new technique. Its practice had revived in European medical schools beginning in the fourteenth century. But in medical schools it was used like the illustrations in anatomy texts: to show what was already known. Indeed, since most anatomy texts were not illustrated, this was the main function of dissection. It was not taught as a technique or as a way to find new knowledge until the sixteenth century, when Berengario da Carpi and Vesalius began to challenge the passive anatomy of the past. Some sixteenth-century anatomists, including Berengario and Vesalius, also dissected live animals. But this did not become a method until the seventeenth century.

Most scientific practitioners of the seventeenth century, from Galileo to Newton, dissected at some point in their careers. It is well known that Descartes dissected, but less known that his philosophical rival Gassendi did too. Those who did not themselves dissect witnessed it. Christiaan Huygens, for example, looked through telescopes and witnessed anatomical demonstrations, sometimes on the same day, during his visits to Paris in the early 1660s and later at the Paris Academy of Sciences.

This book shows that dissection played an essential role in the development of experimental methods in seventeenth-century science. It can no longer be claimed that dissection, as a nonexperimental practice, did not take part in the new science of the seventeenth century except in a peripheral way as an expression of the mechanical philosophy, or that its only value was in its applicability to medical practice.[10] The story of dissection as an experimental discipline must begin with William Harvey and Gaspare Aselli in the 1620s, both of whom made significant discoveries by means of dissection of live animals. Aselli discovered a new structure, the lacteal veins, while

Harvey demonstrated a new function, the circulation of the blood. In the eyes of its practitioners, this story concluded with the publication of Le Clerc and Manget's *Bibliotheca anatomica* in 1685, which declared the triumph of mechanistic, experimental anatomy.[11] However, that book minimized the role of animals and the connections between dissection and natural history. To complete this story requires inserting these missing elements and looking forward to the eighteenth century and particularly to Buffon's *Histoire naturelle*, the major work on natural history of its era.

The nexus of activities usually considered under the rubric of "natural history" constitutes the other half of the story of animals and the new science in seventeenth-century Paris. Natural history encompassed collecting, describing, classifying, and, since I am concerned with animals and not plants, dissecting. Books and their illustration were critical to this practice, as they were for anatomy. Anatomy and natural history formed two sides of the same coin: one could not take place without the other. The role of animals in anatomy and natural history changed over the course of the sixteenth and early seventeenth centuries from passive objects of observation to active subjects in demonstration and experimentation. Before 1640, anatomists and naturalists—more often than not, the same individuals—valued experience and occupied a liminal philosophical space between descriptive practice and causal knowledge, between *historia* and *scientia*. However, in the second half of the seventeenth century, for reasons that this book will explain, this space closed.

The dissection of living and dead animals contributed to ongoing debates on living function as well as debates on the structure and function of nature as a whole, and on the proper methods to attain knowledge. The collecting of previously unknown and exotic (both in the modern sense of not native, and in the older sense of simply foreign) animals contributed to that crisis of classification that dominated seventeenth-century natural history as well as to methodological debates about particulars and universals.[12] The activities under the category of *"anatomie"* of the Paris Academy of Sciences and elsewhere in Paris included both dissection and collecting, and produced new knowledge that was not always the knowledge of causes that constituted natural philosophy but was more than mere description.

Over the course of the sixteenth and early seventeenth centuries, anatomists and naturalists developed techniques of observation, description, and dissection of animals. Dissection evolved into a practice distinct both from medicine and from ancient philosophies, and natural history increasingly emphasized direct observation, while both maintained their humanist ties to textual knowledge. Both were driven, moreover, by a curiosity that would not easily be satisfied until everything possible was known about the human and animal body. Aided by the microscope, this curiosity—for example, in Duverney's work on the ear of 1683—penetrated more and more deeply, revealing the amazing craftsmanship of God, but also opening yet more vistas for curiosity to conquer.

4. Chapters

Chapter 1 begins with the dead body's journey to the places of dissection. There were several of these in Paris, most of them within the area of the Left Bank demarcated by the Pont Neuf and the Pont de la Tournelle (see map, fig. 1.1). Dissections took place in a variety of settings, both inside and outside schools of medicine and surgery, and the long tradition of the natural history of animals that dated back to Aristotle had little connection to medical practice. Living and dead animals were as important to these anatomical uses as the human body.

This chapter asks who the anatomist was in this period and how he gained his knowledge, framed by the experiences of the four anatomists who were charter members of the Paris Academy of Sciences: Marin Cureau de la Chambre, Jean Pecquet, Louis Gayant, and Claude Perrault. Their varied experiences chart diverse paths toward anatomical knowledge, including books. Contemporaries as well as modern historians portray doing rather than reading as the primary activity of seventeenth-century science. But less than two centuries after the introduction of printing to Europe, books retained central roles not only in recording and modeling knowledge but also in producing it; reading could constitute doing.[13] For example, physician André du Laurens advised that one could learn anatomy from reading textbooks and then practicing on one's own. Publication, and in the case of anatomy and natural history, the accompanying illustrations, made concrete the discoveries of the age and allowed for their replication and expansion. In France the *"honnête homme"* courted by both Descartes and Molière sought in the French language a vernacular to replace Latin, and anatomical works in the seventeenth century increasingly were published in French. Books and later journals situated anatomy within the broader cultural framework of the Republic of Letters. At the same time, a flourishing manuscript economy of student lecture notes disseminated anatomical knowledge across time and space.

Chapter 2 explores the implications in France of two discoveries of the 1620s. The English physician William Harvey reconceived the relationship between natural history and anatomy and between *historia* and *scientia* in two important ways. His discovery of the circulation of the blood validated a descriptive methodology that explicitly did not include final causes. In addition, his dissection techniques constituted an experimental method that, although not entirely new, compelled the generations that followed toward an experimental science based on animals that events of the seventeenth century would repeatedly confirm. The publication of Harvey's 1628 *De motu cordis* is therefore a watershed moment, and the key to Harvey's innovation was his experimental use of animals, especially live animals. The Milanese physician Gaspare Aselli also used live animals in his discovery of the so-called lacteal vessels, published in 1627. Aselli's discovery caused almost as much comment as Harvey's, and underscored the importance of the dissection of live animals in the production of new knowledge. Passionate discussion of these issues accompanied by much more dissection over the next two decades led to a second watershed, Jean Pecquet's discovery of the thoracic duct, published in 1651.

Anatomical facts were not like facts in the mathematized sciences. They were mutable, subject to changes in light, setting, time, temperature, and the skill of the dissector. Each individual human and animal body had particular characteristics. Although, as Harvey recognized, a heart was a heart, hearts differed not only across species but within them. Certain phenomena, like the lacteal veins (or the pores in the septum) could only be seen under particular circumstances. Anatomical phenomena were therefore singular, and evoked the wonder and admiration that natural phenomena of all kinds evoked, reminding observers of God's fecundity and invention. Although particular, anatomical events were not strange: they were neither miraculous nor preternatural, nor were they rare.[14] But the kind of induction and generalization that led to causal explanations was much more difficult to achieve in anatomy than it was in mechanics. The experimental method developed by Harvey, Pecquet, and others established anatomical facts but did not lead to a common theory of anatomical causes. Instead, many different causal theories, most of them in some way based on the mechanical philosophy, were advanced in the second half of the seventeenth century.

Most anatomists and naturalists were physicians or surgeons, and the medical frame of mind that valued empirical evidence and a multiplicity of causes is evident in the practice of anatomy and natural history. Yet connections between those fields and medical practice became increasingly tenuous over the course of the century. Thus, French men of science, including many physicians, accepted Harvey's theory of the circulation long before its official acceptance by the Paris Faculty of Medicine. Medical practice, even among those who accepted the circulation, continued to rely both on

Galenic ideas of humors and individual uniqueness and on bleeding as a remedy.

With Harvey's work, dissection of live animals moved beyond demonstration to become a quintessentially experimental act, an act that revealed new knowledge. In the last third of the seventeenth century, the members of the Paris Academy of Sciences pursued two intersecting animal projects, dissection of exotic animals and dissection of living and dead domestic animals, the latter coupled with the dissection of human cadavers. Out of these projects emerged a new experimental comparative anatomy that valued animals both as models for humans and as legitimate objects of knowledge in themselves.

Chapter 3 looks at the anatomical projects of the Paris Academy of Sciences, returning to the four anatomists introduced in chapter 1. Claude Perrault, previously an unassuming member of the conservative Paris Faculty of Medicine, improbably became a leader in the academy. But if we think of the academy as it thought of itself, as a branch of the Republic of Letters, Perrault's rise is more understandable. The original plan for the academy put forward by Claude's brother, Colbert's *commis* Charles Perrault, called for a general academy of arts and sciences, and Claude's combination of literary, architectural, and medical knowledge made him well suited for such a group. Claude's plans for the *physique* side of the academy, presented at the end of 1666, included an extensive program of dissection.[15] Beginning with blood transfusion experiments in January 1667, the academy established a style of experimenting, collaboration, and publication that set it apart from other European academies, particularly the Royal Society of London, founded in 1662. Collaboration meant that all contributed to projects, but it also meant that any publication was in the name of the academy rather than any individual author. Consensus could be elusive, and transfusion was curtailed when all could not agree to pursue it. But new subjects for anatomical exploration appeared when exotic animals that had died at the royal menageries at Vincennes and Versailles were sent to the academy for dissection, beginning with a lion in June 1667. Publications soon followed.

Underlying the academy's animal work was a mechanistic philosophy, but it was not the "animal-machine" philosophy of René Descartes. Following instead the writings on animal cognition and soul of academician Marin Cureau de la Chambre, Perrault and his colleagues believed that animals had consciousness and feeling and some degree of rationality. This did not prevent the academicians from dissecting living animals with impunity, but no one claimed that these animals felt no pain. Perrault articulated in his *Essais de physique* (1680–88) a distinctly non-Cartesian theory of animal mechanism based on the idea of an incorporeal and self-moving soul, a Christian vitalism that permeated the academy's animal work. In addition, his Galenist medical training at the Paris Medical Faculty had imbued Perrault with the notion that each animal, like each patient, was unique and that generalization concerning either diseases or species was impossible. Any attempt at classification, the great debate of seventeenth-century naturalists, would be misguided.

Artists and illustrations were part of the academy's work from the outset; the royal printing house was next door to the academy's rooms on the rue Vivienne in central Paris, and artists and engravers, particularly Sébastien Leclerc, engraved images of plants and animals along with images of Versailles, the towns of France, and items from the royal collections. To Colbert at least, these different images held similar value. But along with the images—of two exotic animals dissected in 1667, and five more by 1669—Perrault and the academy had written detailed anatomical descriptions of the animals. These were published with the images in the form of pamphlets. The imperative to contribute to the glory of the king led to the publication of several larger and more elaborate volumes of the academy's work in the 1670s, chief among them the elephant folio *Mémoires pour servir à l'histoire naturelle des animaux* (*Histoire des animaux*), published in 1671 with an augmented edition in 1676 and discussed in chapter 4. The *Histoire des animaux* described over thirty animals from the king's menageries that academy members had dissected. Edited and largely written by Claude Perrault, the volumes displayed the academy's distinctive collective style and its mechanistic philosophy. Full-page engravings by Leclerc and others displayed the animal in life as well as its dissected parts. As a work of natural history and comparative anatomy, it far surpassed any previous efforts.

But the beauty and size that made it an effective tool in promoting the *gloire* of Louis, its images echoed in paintings and tapestries, made it less useful to fellow men of science. Few copies were produced, and most of them were not sold but given away to deserving clients of the crown. An English translation a decade later expanded its readership, but not until its 1733 reprinting as part of the academy's *Mémoires* did it become widely available, sixty years after its first publication. Moreover, although Perrault's name appeared as a "compiler" of the 1676 volume, the academy's policy of anonymity stands out in an era when credit for other cultural productions—art, music, literature—was readily assigned.

Pecquet, who had made a splash two decades earlier with his discovery of the thoracic duct, died in 1674. Originally a client of Colbert's rival Nicolas Fouquet, who had spectacularly fallen from power in 1661, Pecquet's scientific renown overcame his political liabilities to allow his appointment to the academy. Cureau de la Chambre died in 1669, and Louis Gayant, who had taught Pecquet to dissect in the 1640s, died on the battlefield in 1673. Filling the void left by these deaths was the precocious Joseph-Guichard Duverney, appointed in 1674. Duverney had arrived in Paris from the provinces while a teenager and immediately gained a reputation for his dazzling dissections at the private academies of the *abbé* Bourdelot and the royal physician Jean-Baptiste Denis. Known as the "anatomiste des courtisans" (the courtiers' anatomist) for his instruction of the Dauphin, Duverney embarked on a collaborative relationship with Perrault that lasted until Perrault's death fourteen years later.

Duverney, unlike other members of the academy, pursued a single task: dissection. Chapter 5 examines Duverney's role in the second volume of the *Histoire des animaux* and in the development of Perrault's ideas about animal mechanism. While the discovery of the circulation made some ideas about the body untenable, the rapid collapse of ancient understanding of the body opened the door for a multiplicity of new ideas. Owing to the magisterial historical work of Robert Frank, we know quite a lot about ideas of animal mechanism in Britain, but we know much less about other ideas and other places.[16]

Perrault's essay "De la méchanique des animaux" in volume 3 of *Essais de physique* (1680) employed evidence from hundreds of dissections performed at the academy to elaborate a theory of the animal body that was quite unlike others. He traced all actions of the animal body to peristaltic motion. This motion, the "cause of all the operations of life," derived from the cohesion and elasticity of the particles of matter, such that the living body was in a constant state of alternate contraction and relaxation, an innate vibration of its every fiber. It was a throbbing, pulsating machine in continual motion, governed by an innate soul. Duverney employed this notion of *péristaltique* in his work on the respiratory system of the tortoise, published in the 1676 *Histoire des animaux*.

By the mid-1670s, the menagerie at Versailles was in full flower, and the 1676 *Histoire des animaux* reflected its emphasis on decorative birds. Animal themes pervaded Versailles, particularly in the new Labyrinth, designed by André Le Nôtre with Charles Perrault, which included thirty-nine fountains with verses from Aesop's fables and over three hundred animal sculptures. But the *Histoire des animaux* project proved to be an imperfect mirror of the universe of Versailles as scientific and political goals battled for prominence in the volumes. Royal sponsorship gave the *compagnie*, as the academy called itself, pensions and resources, including the use of royal artists such as Leclerc and access to animals such as the elephant dissected with great fanfare in 1681. But it also imposed constraints, including policies of anonymity and secrecy, which would not be resolved until the reorganization of 1699.

Perrault's *Essais de physique*, composed in the 1670s, focused particularly on two senses: vision and hearing. Vision took up much of the academy's time between 1668 and 1676 with Edmé Mariotte's much-debated theory about its seat. But Perrault devoted an entire volume to sound, hearing, and music, and Duverney's one published monograph was his work on the ear (*Traité de la organe de l'ouïe*, 1683). Duverney's book, grounded in meticulous anatomical investigations, was scientifically more sophisticated, but Perrault's work connected the theory of sound to ongoing debates about the value and legitimacy of modern music, particularly opera. His brother Charles was deeply involved in these debates, which soon extended beyond music to the entire Republic of Letters. Claude Perrault's volume on sound placed him firmly with the moderns in this debate.

Duverney did not directly address these debates but, as chapter 6 explains, became a symbol of the moderns. Even before he entered the academy, his anatomical skills had made him well known in the private academies and salons of Paris. In the late 1670s, he gained a more public arena with his appointment as anatomy lecturer at the Jardin du roi, the King's Garden. The garden had been founded in the 1630s as a counterpoint to the Paris Faculty of Medicine, and the faculty had done its best to keep it under its control. Marin Cureau de la Chambre, known for his philosophical works but not for his dissecting skills, had been named at the garden's foundation to make "demonstration of all operations of surgery," but he appears never to have delivered a lecture. Anatomy only began to be taught at the garden in 1673, when the surgeon Pierre Dionis began to lecture, assisted by royal edicts that made the lectures free and open to the public and granted the garden first rights to the bodies of the executed. Dionis's audience rapidly grew, and when he retired in 1679 to become the Dauphine's surgeon, Duverney was his obvious successor.

Duverney taught at the garden for nearly forty years to large audiences of medical and surgical students and the public. For a time, anatomy became fashionable: Molière's Dr. Diafoirus took his fiancée to a dissection, and Duverney was singled out by the critic (and supporter of the ancients) Nicolas Boileau as an example of all that was bad about modernity. Duverney's lectures, as we learn from student notes, were resolutely modern but did not accept all modern theories equally. A particular specialty was osteology, the study of the bones, and he used to his advantage the skeletons of the numerous animals the academy had dissected, which were housed in the skeleton room of the garden.

Duverney was widely praised for his eloquence, and language and its uses were at the heart of the ancients and moderns debate. While Duverney's lectures conformed to classical standards of oratory, his subject matter placed them in the same category as novels and the opera, appealing, said critics, to the emotions rather than to reason. The riots that often followed his lectures at the garden seemed to confirm this opinion. But Charles Perrault, in his 1687 poem *Le siècle de Louis XIV*, defended modern science along with art, literature, and music as evidence of the superiority of the moderns over the ancients: Louis XIV was indeed the new Alexander, as numerous works claimed, including the *Histoire des animaux*.

The *Histoire des animaux* project continued after the publication of the 1676 volume, with plans for a third volume continuing to Perrault's death in 1688 and beyond. Despite attempts in 1688 and 1695, with new dissections, descriptions, and illustrations, the third volume did not appear. The financial problems of the academy following Colbert's death in 1683 are partly to blame, but Duverney also bore responsibility. His increasingly fractious relationship with a new academy anatomist, Jean Méry, led to lengthy disputes in the 1690s and Duverney's departure from the academy by 1706, taking the *Histoire des animaux* manuscripts with him. Duverney's departure marked the end of an era for the academy, and although he continued to lecture at the garden and the Hôtel-Dieu hospital, comparative anatomy did not revive in Paris for many years.

A conclusion assesses the many comparative anatomies of this period and the role of French anatomists in the wider European scene. The French used comparative anatomy in distinctive ways, joining it closely to natural history and seeking differences in specific organs and their uses rather than assuming uniformity in nature as most mechanical philosophers did. Perrault and his colleagues rethought a number of the prominent topics of seventeenth-century life science, ranging from the pineal gland to the function of the lungs, and reached new conclusions that demonstrated once more the diversity of seventeenth-century research on these topics.

An epilogue explains the afterlife of *Histoire des animaux*. Following Duverney's death in 1730, a complete revised edition of the project was undertaken under the guidance of the academy's secretary Fontenelle, which appeared in three volumes in 1733–34. Five years later, Buffon took control of the King's Garden, and his appointment of Daubenton in the 1740s signaled the revival of comparative anatomy in Paris, made evident in the volumes they jointly undertook of *Histoire naturelle* between 1749 and 1767. Buffon self-consciously took up where Perrault and Duverney left off.

In the historical moment between the Renaissance and the Enlightenment, the courtiers' anatomists practiced their craft and their science, and participated in the ideas and events that produced the new science. The animals they used, in anatomy and in metaphor, did not participate willingly, but they provided evidence and authenticity that have continued to shape our ideas of life.

ONE

Anatomists and Courtiers

1. A Geography of Paris Anatomy in the Seventeenth Century

Under cover of night, the dead of Paris made their journey from the burial grounds to the places of dissection. In this era of recurrent plagues, their numbers never dwindled, and for three centuries from the 1530s, they did not lie quiet in their graves. The cemetery of Saints-Innocents, between the rue de la Ferronerie and the rue St. Denis, was one of the few places with streetlamps until Louis XIV ordered the installation of thousands of candle-lanterns across Paris. But its dim beacon did not deter the trade in the dead, as physicians and surgeons exhumed bodies. Shadows from their torches made the danse macabre carved into the wall of one of the bone-houses, the *charniers*, seem to move. Saints-Innocents had been filled many times over, and as new bodies came in, the bones of the old were disinterred and placed in the *charniers*. The famous Flemish anatomist Andreas Wesel, known as Vesalius, fondly remembered the piles of bones at Saints-Innocents during his days as a medical student; he and his friends blindfolded themselves and took bets as to who could identify the most bones by touch.[1] The smell of decomposing flesh permeated the *quartier*. The medical men and their apprentices trundled the bodies from Saints-Innocents south down the rue St. Denis, the road of kings. At the river they came to the fortress of the Grand Châtelet, which housed the office of the *prévôt de Paris* that administered the king's justice, as well as a court and prisons, and the second streetlamp in Paris.

By the end of the sixteenth century, anatomy teachers offered a variety of courses and demonstrations. The content, setting, and style of anatomical lectures varied widely, from perfunctory demonstrations of surgical techniques to detailed expositions of structure and function according to Galenic theory. They took place mainly in university settings, but also before corpo-

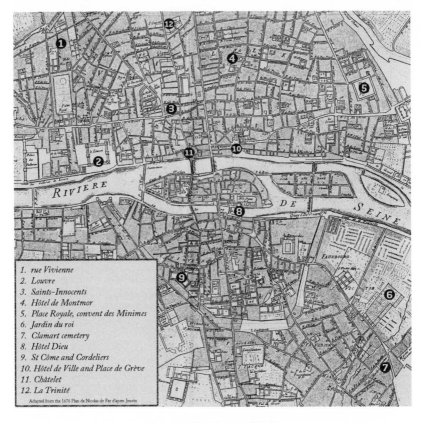

1. rue Vivienne
2. Louvre
3. Saints-Innocents
4. Hôtel de Montmor
5. Place Royale, convent des Minimes
6. Jardin du roi
7. Clamart cemetery
8. Hôtel Dieu
9. St Côme and Cordeliers
10. Hôtel de Ville and Place de Grève
11. Châtelet
12. La Trinité

Adapted from the 1676 Plan de Nicolas de Fer d'après Jouvin

Figure 1.1 Detail, map of Paris.
(Albert Jouvin de Rochefort, printed by Nicolas de Fer, 1676. Wikimedia.)

rate bodies of medical men and in private rooms to fee-paying audiences.[2] In his 1610 dissection manual *La semaine ou pratique anatomique*, the surgeon Nicolas Habicot (ca. 1550–1624) described the ideal anatomical subject. He would be a young man between twenty-five and thirty (older subjects were too "cold and dry"), well fleshed. Drowning victims were best, although the anatomist had to hang the subject upside down by his heels and press the water out before cutting; victims of strangling or decapitation were less desirable. Although many anatomists advised starting with animals, Habicot advised against it, but he admitted that animals had their uses.[3]

However, this ideal candidate, or indeed any human corpse at all, would have been difficult to come by for the many who wished to dissect. This was not because the dead body held any special religious meaning. Catholic doctrine was quite clear that the body would be resurrected whole no matter what indignities it may have suffered before or after death. Pope Boniface VIII had forbidden the division of the body after death in his bull *Detestate feritatis*, but he directed his ire at the trade in relics, not at anatomical demonstration, and the bull did not claim a special value for the dead body. The urge to protect the body after death was personal and cultural rather than legal or religious. Few were willing to allow their bodies or those of their loved ones to be dissected, and municipalities grudgingly granted a few bodies of executed criminals a year to medical faculties for their yearly public demonstrations. Many anatomists were not content with what they could learn in these demonstrations, and to find out more about the human body, they sought additional corpses and animals to dissect. In addition, anatomical preparations and skeletons supplemented these dissections, and their construction also required human corpses or body parts.[4]

Dead and alive, animals far outnumbered human corpses on seventeenth-century dissection tables. Animals became as important to this new practice of anatomy as the human body for a number of reasons, both practical and symbolic. Anatomists could easily acquire the dogs, cats, and pigs that constituted the majority of animals used in anatomical demonstration and research. Animals had no inconvenient relatives to protest their dissection. Unlike "resurrected" corpses, they were fresh—so fresh that they were often still alive. And in many important respects, they resembled humans enough to demonstrate human function. This assumption was neither new nor controversial. The humanist physicians of the sixteenth century, steeped in classical philosophy, knew that both Aristotle and Galen had relied upon the structure of animals in order to talk about the human body.

Even before they reached the dissection room, the human dead commingled with animals. The stalls of Les Halles, the biggest market in Paris, backed up to the walls of Saints-Innocents, and even encroached on the space inside the walls. The stench of Saints-Innocents soon mingled with the stink of the rue de la Boucherie, Butchers' Row, up against the wall of the Châtelet. In the dark, the medical men could not see the butchers or the offal they threw into the Seine that stained its banks red with blood and according to the chronicler Mercier, marked with blood the streets surrounding the fortress as well. The Tour Saint-Jacques is all that now remains of the butchers' church, Saint-Jacques de la Boucherie.[5] Here during the day the anatomists could get the pigs and sheep they preferred to dissect alive, as well as animal parts for practice.

But night was the time to round up the stray cats and dogs that anatomists used by the dozen. The Paris Academy of Sciences employed a man to "find" cats and dogs, but they were not difficult to find. Cats haunted the rooftops and cellars of the city and, complained a contemporary, "celebrate a witches' Sabbath all night long" with their howls.[6] Once a year, one of the *commissaires de la ville* rounded up a sack of cats to be burned alive in the bonfire celebrating the feast of Saint John the Baptist. Cats generally did not follow the sharply defined social hierarchy afforded to dogs, although it was well known that Louis XIII's minister Cardinal Richelieu (1585–1642) loved his cats, as did Louis XIV's minister Jean-Baptiste Colbert (1619–83).[7] Later in the century, the garden of the *salonnière* Mme. de Lesdiguières featured an elaborate marble tomb to her cat Menine.[8] The lap dogs depicted in portraits, known as "chiens de manchons" (literally "muff-dogs," small enough to be carried in a lady's muff), led pampered lives. Louis XIV built elaborate velour-draped niches for his favorites. Hunting dogs formed another category, which overlapped with other working dogs such as herd dogs and guard dogs. In yet another class were the trained dogs who participated in staged fights with other dogs and with wild animals.[9] But many dogs simply wandered the streets; scenes of urban life often depicted one or more dogs at their margins.[10]

The small interior courtyard of the Châtelet that served as a morgue held "animaux urbains" of another kind, corpses of the drowned and the anonymous found dead in the streets, who were exposed to await identification.[11] Those that remained unidentified were conveyed to the nearby convent of Sainte-Catherine, formerly the hospital of Sainte-Opportune, whose sisters washed the bodies and prepared them for burial at Saints-Innocents, conveying them to the burial ground after dusk.[12] But these dead often did not reach their destination. The English physician Martin Lister (1639–1712) reported disapprovingly in 1698 that the anatomist Joseph-Guichard Duverney (1648–1730) obtained some of the many bodies he used for anatomical demonstrations from "the Chatelet, (where those are exposed who are found murthered in the Streets, which is a very common business at *Paris*)."[13]

From the Châtelet, the doctors crossed the Pont au Change to the Ile de la Cité, passing the stalls of the lap-dog sellers on their way to yet another source of corpses, the medieval foundation of the Hôtel-Dieu, the largest hospital in Paris, which occupied several buildings next to the Cathedral of Notre Dame. During the day, medical men could witness many autopsies at the hospital on the corpses of the poor. Duverney held court here from the 1680s to the 1710s, witnessed by physicians, surgeons, residents, apprentices, and medical students. Earlier in the century, the great anatomist Jean Riolan the younger (1580–1657) gave a private anatomy course in a house on the square in front of Notre Dame, between the hospital and the

Figure 1.2 Detail, map of Paris, showing Hôtel-Dieu and the Paris Faculty of Medicine.
(Turgot map, 1739. Wellcome Library, London.)

bishop's residence.[14] The Hôtel-Dieu treated thousands of patients per year, and in Mercier's words, "vomit journellement" the dead.[15] Until the early 1670s, the hospital buried its dead not at Saints-Innocents but farther to the north along the rue St. Denis at the cemetery of the medieval Hôpital de la Trinité, a public lodging house for pilgrims traveling to the Cathedral of Saint-Denis, several more miles north of the city walls. La Trinité had large pits for mass burials in time of plague, and as at Saints-Innocents (indeed probably more often, since la Trinité was smaller), bones were regularly exhumed and placed in *charniers* to make room for more corpses.[16]

Fewer bodies reached la Trinité than had left the Hôtel-Dieu. Because of its distance from the Hôtel-Dieu, corpses destined for burial at la Trinité were particularly vulnerable to thieves. Eager surgeons importuned the *emballeurs* who shrouded the bodies (these dead were too poor for coffins) and piled them on carts for their evening journey to the cemetery, and those who were caught in the act and fired in 1626 for selling bodies were likely not the only ones who attempted this.[17]

From the Hôtel-Dieu the doctors could look back across the river to the

Hôtel de Ville and the wide-open area in front of it known as the Place de Grève, where most of the executions in Paris took place. Surgeons in particular came to executions in the hope of gaining an illicit corpse. From the gates of the Hôtel-Dieu, the medical men could cross the Seine again at the recently opened Pont au Double or the Pont de l'Hôtel-Dieu. Only a few yards away on the Left Bank stood the Paris Faculty of Medicine. Anatomists of all kinds worked in the medieval warren of Left Bank streets that spread out from the Sorbonne known as the Latin Quarter. Before it built an anatomical theater in 1620, the Paris Faculty of Medicine had established a site for dissection in the early sixteenth century—the first in Paris—at the Hôtel de Nesle, where the Institut de France now sits, directly across from the Louvre.[18] The tower of the Hôtel de Nesle contained the third streetlight in Paris, to guide boatmen along the Seine. But when, a century later, the faculty finally built a lecture theater, it was located next to its ancient headquarters in the "house of the Iron Crown" at the corner of the rue des Rats and the rue de la Bûcherie.

The guild of surgeons of Saint-Côme was only a few hundred yards away at the Eglise Saint-Côme, on the corner of the rue de la Harpe and the rue des Cordeliers. The Paris Faculty of Medicine could, by law, claim the bodies of executed felons for dissection, and a 1552 ordinance had granted the faculty absolute power over the distribution of cadavers to the various teaching bodies, including the surgical school at Saint-Côme and those barber-surgeons who took apprentices. Unlike the physicians, the surgeons had no legal right to corpses and obtained them where they could. According to a complaint to the Parlement of Paris in 1615, surgeons absconded with bodies "either by violence or force, or by virtue of some license that they obtain by surprise." Sometimes they stole them from under the noses of the physicians, as in 1622, when a body from the faculty's new amphitheater was transported to Saint-Côme for a public demonstration.[19] Fifty years later, the faculty called on the *huissier* of the Parlement of Paris to enforce the law and seize a corpse that had been transported to the Saint-Côme anatomy theater without its permission. After several unsuccessful attempts to retrieve the corpse, the *huissier* returned with several archers, and following a pitched battle with the surgical students, which entailed calling in reinforcements, the cadaver was transported to the medical school. However, this struggle took two weeks to resolve, by which time the cadaver was of little use.[20]

Thus, surgeons and medical students haunted executions as well as the cemeteries. Gangs of barbers and surgeons assembled at the Place de Grève, the execution ground in front of the Hôtel de Ville, to seize the body of an executed criminal, although the faculty's *huissier* sometimes followed them and wrested back the corpse.[21] Rumors abounded that medical and surgical students incited "vagabonds" to violence in order to obtain their bodies after execution—or simply murdered them. Physicians usually paid three livres each for legally obtained cadavers, but intense competition for dead bodies among surgeons led to prices as high as one hundred livres for a single corpse.[22]

As well as teaching at the Hôtel-Dieu, from the early 1680s Duverney dissected humans and animals at the King's Garden, in the Faubourg Saint-Victor. Farther east along the Seine and outside the medieval city limits, but within the new area of development opened by the building of the Pont de la Tournelle in 1620, the garden was a short walk from the Latin Quarter. Duverney drew large crowds for his public anatomy lessons, and also did the anatomical work of the Paris Academy of Sciences. At the academy's foundation in 1666, dissections and other activities took place at the Royal Library on the Right Bank, not far from the palaces of the royal ministers Richelieu and Mazarin. But the library's limited space as well as the inherent messiness and noise of dissection, especially of live animals, soon drove the academy's dissections back to the Left Bank: before 1673, to the rooms of its dissector, the surgeon Louis Gayant, and from 1682 until the early eighteenth century to the King's Garden with Duverney. A 1673 royal order shifted the authority to obtain bodies from the Faculty of Medicine to the garden.[23]

According to Martin Lister, the cemetery of Saints-Innocents was "in ruins" by the time he saw it in 1698, although it did not fully close until 1780, after which its *charniers* were cleared and the bones placed in what came to be known as the Catacombs beneath the city.[24] La Trinité closed in 1678, and its place as the burial ground for the Hôtel-Dieu was taken by the new cemetery of Clamart, which opened in 1673 on the Left Bank south of the garden in the Faubourg Saint-Marcel. The corpses of the executed and those from the Châtelet morgue also now traveled south and east to what was then a far corner of the city, allowing further opportunities for plunder along the way.[25]

Clamart also replaced the older cemeteries as the favored site for stealing corpses. Although the King's Garden had a guaranteed supply of corpses, this was never enough for Duverney, who continued to obtain bodies illicitly. The Hôtel-Dieu's records detail a number of violations. In 1682, the minutes of the hospital's governors record that the gravedigger at Clamart sold cadavers to surgeons and to the King's Garden. In August 1717, it was reported that Duverney's assistant "often" took "entire dead bodies" from the cemetery, hacking off limbs and cutting out internal organs on site, "to the great scandal of the people, who cannot view such a spectacle without horror." The Hôtel-Dieu decided in 1717 to forbid Duverney from obtaining corpses altogether, but his "suborning" of the gravedigger continued until the latter's dismissal in 1725.[26]

2. Learning the Body

Paris, said Nicolas Habicot, was the best place to learn to dissect, because there were so many opportunities to dissect and to view dissection. When the royal minister Jean-Baptiste Colbert organized the Paris Academy of Sciences in 1666, the four men he appointed as anatomists had all gained their knowledge of anatomy in Paris. Claude Perrault (1613–88), Marin Cureau de la Chambre (1596–1669), Jean Pecquet (1622–74), and Louis Gayant (d. 1673) were all medical men, but they came to the study of animal and human bodies from very different perspectives. One of them may never have dissected. Their varying roads to the Paris Academy reveal that books, language, and patrons could be as important to an anatomical career as dead bodies, stray dogs, and sharp knives.

Back in the 1590s, the Montpellier professor André du Laurens (1558–1609) exhorted readers of his anatomy textbook that physicians, surgeons, and pharmacists alike needed to know anatomy (echoing Galen's insistence on the unity of medical knowledge), both for practical reasons and as a means to know the self. But when he wrote, and for some time thereafter, most medical men rarely witnessed a dissection. The Swiss physician Caspar Bauhin (1560–1624) saw seven bodies dissected during the eighteen months he spent in Padua, probably the best medical school in Europe, in the 1570s.[27] When in 1579 the Paris surgeon Séverin Pineau (d. 1619) dissected the corpse of a woman who had recently given birth, the entire company of surgeons of Saint-Côme came to witness the event, as well as a professor and a surgeon from Montpellier, who came to Paris to witness "by order of the king."[28] In their magisterial work on early modern French medicine, Laurence Brockliss and Colin Jones assert, "Certainly until 1650 it was considered possible for a physician to profess a practical medical science with little or no hands-on experience."[29] It was possible, but by 1650 it was increasingly unlikely. How and where did our four anatomists, born between 1596 and 1622, acquire their knowledge of the human and animal body, and by what means did they gain entrance to the elite circle of the academy?

2.1. The Paris Faculty

When Jean Pecquet crossed the threshold of the Paris Faculty of Medicine on the rue de la Bûcherie in 1647, aged twenty-four, he already knew how to dissect. Most historians view the faculty by the 1640s as a moribund institution mired in the doctrines of Galen, which under the leadership of Jean Riolan the younger refused to accept William Harvey's (1578–1657) theory of the circulation of the blood. As both a regulatory body and a teaching institution, the Paris Faculty continued to focus on that medical humanism that had made it so successful in the sixteenth century. After 1550, it had increasingly turned to Hippocrates and his emphasis on observation and diagnosis, although Galenic ideas of anatomy and physiology still prevailed.[30] One result of this new Hippocratism was a strong suspicion of chemical remedies and Paracelsian philosophy. But there was no necessary contradiction between useful dissection and Galenism.

The Paris Faculty had offered lectures in surgery from the 1490s onward, and regular lectures on anatomy for medical students followed shortly thereafter.[31] The 1552 *arrêt* of the Parlement of Paris that established the faculty's hegemony in the acquisition of bodies also established its dominance in the actual performance of dissection, which could not take place without a faculty physician in attendance. These measures intended to restrain the influence of the surgeons, who, here as elsewhere, formed a separate corporate body—in this case, as we shall see, two, the barber-surgeons and the surgeons of Saint-Côme.

As a center of medical humanism, the faculty revived a number of works of Galen and Hippocrates, including Galen's *On Anatomical Procedures*, translated from Greek into Latin by a Paris professor, Jehan Guinter d'Andernach (1505–74), in 1531. Guinter (or Winter) was one of the teachers of Vesalius during his time in Paris; the other was Jacques DuBois (1478–1555), known as Sylvius, a staunch Galenist and skilled anatomist famous for doing his own demonstrations rather than leaving the manual work to a surgeon. After teaching independently for a time, Sylvius received his medical degree from Montpellier in 1530 and taught outside the faculty at the Collège de Tréguier, one of the colleges of the University of Paris, until he was admitted to teach in the Paris Faculty in 1536 in recognition of his great skills in dissection.[32]

The surgeon Ambroise Paré (1510–90) took the faculty's anatomy course for surgeons in 1533, when the professor of "chirurgie Latine" sat in a chair with a beadle on either side, lecturing on Galen in Latin mingled with French (since his auditors knew no Latin). The demonstrator dissected a corpse on a table below.[33] This scene changed little for a century. The faculty finally named a professorship in "chirurgie en langue française" in 1634, but this professor taught only specific surgical operations, not anatomy in general. Moreover, because all the professorships rotated every two years among the members of the faculty, the level of instruction varied widely.[34] The humanist Pierre de la Ramée, also known as Petrus Ramus (1515–72), criticized the faculty for its lack of practical instruction in his 1562 "Advertissements sur la reformation de l'Université de Paris." The regents of the faculty, he said, should lead students in gathering medicinal herbs, dissecting bodies, and consulting patients, instead of engaging in scholastic wrangling.[35]

Some students, such as the younger Riolan, acquired considerable anatomical skills during their time in the Faculty. But they did not necessarily acquire these skills in the classroom. Faculty lectures on anatomy did not always include an anatomical demonstration, and students received little opportunity to dissect. Anatomical skills gained elsewhere did not advance one toward a degree.[36] Although the faculty jealously guarded its right to obtain corpses and to perform dissections, anatomy was not central to its teaching mission, and until 1598 it formed part of the first-year teaching on "choses naturelles." Statutes from that year ordained two public anatomy demonstrations per year, each using two corpses (similar demonstrations had taken place in Montpellier since 1340).[37] For some students, witnessing these demonstrations was all the knowledge of anatomy they ever acquired. Only in 1707 did anatomy become a compulsory field of study.[38]

A temporary anatomical theater was erected at the rue de la Bûcherie in 1604, and Jean Riolan the younger instigated the building of the permanent theater in 1617, over sixty years after one had been built in Montpellier. When it opened in 1620, he gave the inaugural lecture.[39] But anatomy lectures within the faculty continued to follow an archaic model. Riolan claimed that the faculty regularly received four to six bodies a month, but its narrow interpretive framework confined anatomical teaching to the most basic questions, illustrating what was already known rather than seeking new knowledge. Almost half a century later, the Danish anatomist Niels Stensen (1638–86) bemoaned the continued low state of anatomical knowledge, which he attributed to a lack of curiosity: most who dissected were content to demonstrate what the ancients wrote, and the medical schools were most to blame for this. One attended medical school to learn what the ancients said. While others at the Paris faculty besides Riolan did their own dissecting and demonstrating rather than leaving these tasks to a surgeon, his skills in anatomy were unusual enough to merit frequent comment.[40]

Riolan tacitly acknowledged the deficiencies of the faculty's anatomical instruction when he began to offer his own private anatomy courses beginning around 1614.[41] In this he followed an earlier faculty physician, Germain Courtin (d. 1587), who had lectured on anatomy to surgical students between 1578 and 1587. Étienne Binet, a surgeon of Saint-Côme, issued an edition of Courtin's lectures drawn from student notes in 1612.[42]

Physicians schooled in the Paris Faculty did not learn how to dissect unless they made a special effort to do so. When Claude Perrault, the middle son of a well-off bourgeois family, obtained a degree from the Paris Faculty in 1641, he seemed destined to follow a very conventional path to fortune, if not fame, as a practicing physician. He had made his way through the faculty's course without incident, and the titles of the theses he defended reveal the meaningless propriety on which the faculty prided itself. They included such typically banal topics (for his first set of theses in 1639) as whether the soul ages as the body does, whether it is healthy to put ice in one's wine on a hot day, and whether cautery may be used in the case of "inveterate" trembling of the head or limbs. The next year he defended three more, on whether a physician should travel or marry, whether he should abandon a patient or charge for his services, and whether bleeding or purging is preferable in a quartan fever.[43]

Perrault paid his fees and gained admission as a physician in 1641, with the privilege of signing the minute books each year to affirm the activities of the faculty. He may very well have heard the disputations surrounding theses attacking William Harvey's theory of the circulation of the blood in 1642 and 1645. In due time Perrault took his place as a senior member of the faculty, served as an examiner, and did a two-year stint as professor of physiology in the early 1650s, but no evidence survives that he challenged the status quo in any way. At some point he accepted the theory of the circulation of the blood, but we have no evidence that his physiology lectures, delivered under the deanship of archconservative Gui Patin (1601–72), mentioned it.[44] Yet by 1666, Perrault knew how to dissect, and was an ardent mechanist and supporter of William Harvey and Jean Pecquet, while continuing to sign the faculty's minute books each year. Perrault's case shows that it is possible to overstate the influence and impact of the faculty's teaching. As much a social as an intellectual organization, it could not dictate the opinions of individual members once they left the confines of the rue de la Bûcherie. From the 1640s onward, Perrault frequented private salons and academies where the new science was avidly discussed, as we shall see in chapter 3.

Harvey announced his discovery of the circulation of the blood in 1628, following publication the previous year of Gaspare Aselli's (1581–1626) discovery of the so-called lacteal veins. The lacteals would later prove to be crucial to a new theory of digestion and blood formation brought about by the circulation. Although Riolan the younger was later famous for his opposition to the circulation, he quickly accepted the lacteals, mentioning them in the 1628 edition of his anatomy textbook. When Jacques Mentel (1599–1670), a faculty physician, was appointed its *archdiacre* in 1629—that is, the director of student anatomical studies—he demonstrated the lacteal veins in a dog before the students.[45] But the faculty remained officially opposed to Harvey's theory of the circulation until the 1660s. Its graduates who supported it before that time included Mentel and Perrault as well as Pierre de Mercenne, Jacques Duval, Samuel Sorbière (1615–70), and Abraham du Prat (1616–60).[46] All of them gained their knowledge of Harvey's theory and of the anatomical skills necessary to demonstrate it outside the faculty's walls.

2.2. Surgery and Surgeons

If anatomy was peripheral to the aims of the Paris Faculty, it was central to the city's surgeons, a fact that the faculty recognized in its attempts to monopolize anatomical materials and practice. François Quesnay (1694–1774), in his polemical 1744 history of the Paris surgeons, asserted that "l'Anatomie est un Art qui n'appartient qu'aux Chirurgiens" ("anatomy is an art that belongs only to surgeons").[47] Two distinct companies comprised the Paris surgeons. The barber-surgeons trained by apprenticeship, and a 1506 contract enforced their subordinate status to the physicians, who from that date offered lectures in surgery.[48] As we have seen, these did not impress Ambroise Paré.

When Jean-Baptiste Colbert appointed Louis Gayant to the Paris Academy of Sciences in 1666, he was a member of the confraternity of surgeons of Saint-Côme, a smaller and much more exclusive group whose origins dated from the fourteenth century and who claimed equal status to the faculty physicians as teachers and practitioners. Unlike the barber-surgeons, they were "*gens de lettres*" with university degrees and knowledge of Latin, and supervised the examination of their candidates for licensure without the faculty oversight afforded the barber-surgeons. A royal ordinance of 1533 first referred to the group as a "*collège*," and a decade later the king granted the college the right to "enjoy and use university privileges."[49] The surgeons of Saint-Côme dressed much like the physicians, in long gowns—indeed, the two groups of surgeons were identified as "*robe courte*" or "*robe longue*"—and during the sixteenth century they gained the royal patronage and collegiate status that made them social equals as well.[50] Not surprisingly, competition was particularly fierce between these surgeons and the faculty, but the two groups of surgeons also competed with each other for patients.[51]

Under the influence of humanist translations of classical surgical texts and the subsequent flowering of dissection from the 1530s onward, surgeons of all ranks saw increased status over the course of the sixteenth century. Vernacular translations of classical and modern works recognized their demand for up-to-date texts. Until the early seventeenth century, the surgeons of Saint-Côme performed autopsies and gave occasional courses in the Church of Saint-Côme.[52] In 1615, they built their own headquarters near the church, with a public room for demonstrations and a small amphitheater. The first to dissect there, "portes ouvertes and sans lecture," was Séverin Pineau, known for his discovery of the hymen and his works on the determination of virginity as well as his secret technique of cutting for the stone. In defiance of the faculty, the surgeons advertised their dissections to the public, but they did not offer a regular course of anatomy.[53] Next to Saint-Côme, in the Convent of the Cordeliers, was the Bibliothèque du roi, the royal library, which had moved there in 1603 and remained until 1666.

The surgeons of Saint-Côme from the mid-sixteenth century onward included a number of skilled anatomists. The best known, Ambroise Paré, began his career with the barber-surgeons but passed the examinations for Saint-Côme in 1554. His many books emphasized the value of anatomical knowledge for surgical practice. Others included Jacques Guillemeau (1550–1609), who had published some of Courtin's lectures, and Nicolas Habicot.[54] By the early seventeenth century, various members offered the occasional courses known as "cours particuliers." These were private courses taught to paying audiences. Pecquet was among the students in Gayant's cours particulier in the early 1640s.[55] Although the medical faculty forced the two bodies of surgeons to merge in 1656, the surgeons of Saint-Côme retained their higher status.

Both the physician Riolan and the surgeon Pineau offered cours particuliers, probably during the same years. On the level of anatomical instruction, the disciplinary boundaries between physicians and surgeons were less distinct than on the level of medical practice. Surgeons also served as demonstrators for faculty lectures in anatomy. Professors varied widely in expertise, and surgical assistants provided continuity. Even though Jacques Mentel knew how to dissect, he acquired the services of Gayant to assist him when he was named anatomy professor in 1649.[56]

2.3. The Hôtel-Dieu and the Collège Royal

Physicians and surgeons also mingled at the hospitals, particularly the Hôtel-Dieu, which, said Paré, offered a "nearly infinite" supply of corpses.[57] Its personnel included a master surgeon who directed several journeymen, twelve of whom resided at the hospital, as well as a number of apprentices, "pensionnaires et externes." In addition, medical students in their first two years of training could frequent the hospital and assist the resident physician.[58] The physician made his rounds each morning, accompanied by surgical journeymen and apprentices and medical students. The master surgeon gave instruction to the surgical students, but there does not appear to have been a formal course of study. Hospital surgeons and the attending physician had the right to dissect the bodies of those who died in the hospital to determine the cause of death. The Parlement of Paris issued periodic *arrêts* forbidding the transport of corpses outside the walls of the hospital without the permission of the dean of the Paris Faculty of Medicine. However, the repeated issue of these orders indicates that they were not often followed.[59]

Yet another institution also claimed to offer anatomical instruction apart from the faculty and surgical lecturers. King François I had founded the Collège Royal in 1530 as an independent body of lecturers who instructed the public in a number of topics, including medicine. In 1542, François invited the Florentine surgeon and anatomist Guido Guidi (1508–69), known as Vidus Vidius, to hold the first medical chair at the college. Guidi served as a royal physician and taught medicine at the new college. When he returned to Italy in 1547, the chair was filled by Sylvius, and by the end of the sixteenth century the college included four chairs in medicine.[60] Like other royal offices, the chairs in medicine at the Collège Royal could be bought and sold as stepping-stones to higher office, and by the seventeenth century the *collège* had become another royal sinecure, its teaching disorganized and infrequent, its chairs occupied by little dynasties of faculty physicians. The supposed independence of the college from established doctrine had fallen by the wayside. Riolan the younger held one of its chairs in medicine for fifty years, but it is unknown how much he actually taught.[61] The college did not recover its independence in medical teaching until the eighteenth century.

3. Cureau de la Chambre, the Jardin du roi, and the Republic of Letters

By the 1640s, another potential site for dissection existed, but it was not immediately utilized. The King's Garden (Jardin du roi) opened in 1640 on the

Left Bank. The first physician to King Louis XIII charged Marin Cureau de la Chambre at the time of the garden's founding in 1635 to "make visual and manual demonstrations of all and each of the operations of surgery, whatever they may be."[62] By this time Riolan the younger, the best-known anatomist in Europe, was physician to the queen mother. He had long promoted the idea of a medical botanic garden—a small one existed within the faculty's grounds—but he disapproved of the new garden's independent status and its promotion by Montpellier graduate Gui de la Brosse (1586–1641), well known for his chemical leanings, as were many Montpellier graduates. Founded as a royal institution outside the jurisdiction of the Paris Faculty, the garden depended on the king's physicians, among whom was la Brosse, for its governance. They alone could practice medicine in Paris without the faculty's degree.[63] With its placement on the Left Bank, it represented an inroad of the crown into the independent stronghold of the University of Paris, and an inroad of the generally Gallican intellectuals surrounding the court against the Jesuits who dominated the university.

The faculty viewed the garden as a potential rival and took measures to neutralize its impact. Having just instituted their own professorship in "Chirurgie en langue française" to maintain their hold on surgical instruction against the surgeons of Saint-Côme, the faculty's physicians were not about to compromise their monopoly with another institution. The king's first physician in 1635 was Charles Bouvard, a regent of the Paris Faculty and Riolan's brother-in-law, and in this case his loyalty was clearly to the faculty. He stipulated that the garden's charter require its three "demonstrators" (in chemistry, botany, and anatomy) to be members of the faculty. Two of them were.[64] Cureau de la Chambre was not, and his career displays the power of patronage and the intersections among letters, politics, and the new science in this period.

It is likely that Cureau de la Chambre's appointment to teach anatomy was a strategy to minimize any rivalry between the garden and the faculty, for there is no evidence that he ever delivered lectures. Although Cureau de la Chambre probably obtained medical training, if not a degree, at Montpellier, he owed his fortunes in Paris not to his medical skills but to his literary ones. With a Latin verse, he had gained the attention of Pierre Séguier (1588–1672), the powerful keeper of the king's seals and soon to be chancellor. Séguier, an important literary patron, also (like so many others in this era) had interests in anatomy and in 1633 attended a *cours particulier* of Antoine Charpentier, who became the faculty's new vernacular professor of surgery in the following year. Later, Séguier even presided over a dissection at the faculty.[65]

In this period, Latin poetry provided the most important means of entry to the Republic of Letters, that informal humanist network of correspondents born during the Renaissance that included scholars and writers from across Europe. The ability to compose Latin poetry was a sign of erudition and taste, and as in the case of Cureau de la Chambre, could lead to patronage. Séguier appointed Cureau de la Chambre as his personal physician, who in turn fulfilled his literary obligations to the chancellor with a number of works on natural philosophy dedicated to his patron and written in French rather than in Latin. Thus, Cureau de la Chambre bridged an older erudition based on Latin and a new national Republic of Letters based on French and exemplified in the foundation of the Académie française in 1635.[66]

Cureau de la Chambre's sole publication with any relationship to anatomy consisted of a 1636 book on digestion, which set forth a chemical theory and gave no anatomical evidence.[67] This and other works of the 1630s that displayed a distinctly chemical sensibility may have interested the chemically inclined Gui de la Brosse as well as Bouvard. Although Bouvard was a physician of the faculty, the court favored chemical medicines and therefore so did he.[68] Cureau de la Chambre's particular talents, apart from prose, lay in the arts of physiognomy and chiromancy, the reading of character in the face and the hand. He served the king in this regard by vetting prospective place seekers, and beginning in 1640 he undertook a multivolume work on the passions, which included these topics as well as others. Séguier's library held a copy of la Brosse's own unpublished treatise of physiognomy.[69]

The natural philosopher Marin Mersenne (1588–1648) recommended Cureau de la Chambre's treatise on light in 1634, although René Descartes (1596–1650) seemed to think little of it a few years later.[70] But as important as what he said was how he said it. During a period when the merits of the vernacular and its role in civil society were hotly debated in the Republic of Letters, writing learned works in French, as Descartes also recognized, was a political act.[71] By the seventeenth century, classical rhetoric as filtered through Scholasticism had become synonymous with insincere persuasion: rhetoric and truth were as often contradictory as complimentary. The neoclassical revival, which led to the kinds of poetry (in both Latin and French) that Cureau de la Chambre wrote for Séguier as well as to the foundation of the Académie française in 1635, challenged this view of rhetoric, not with the aim of determining truth, but to discipline and reshape the French language into a worthy vehicle for the greatest nation on earth. "All was thus in the manner, not the matter," as Marc Fumaroli has summarized it.[72] When Louis XIII's minister Richelieu created the Académie française, Cureau de la

Chambre was a charter member. His pamphlet defending Richelieu against his critics earned him letters of nobility.[73]

Cureau de la Chambre's appointment to the garden, then, owed less to his anatomical skills than to his connections to the court and its values. Bouvard filled the other two posts with two faculty physicians: his son-in-law Jacques Cousinot and Urbain Baudinot, a family friend. None of them appears to have delivered lectures. The post of demonstrator, which paid the handsome sum of fifteen hundred livres a year, was a sinecure. At least until the 1660s, others lectured in chemistry and botany. But there is no evidence of any lectures in anatomy at the garden until after Cureau de la Chambre's death in 1669.[74] Fontenelle implied as much when, speaking of the anatomical demonstrations of Joseph-Guichard Duverney in the late 1660s, he said that dissection up to that time was behind the closed doors of the Paris Faculty or the surgeons' school at Saint-Côme.[75]

4. Books and Their Readers

As a visual and descriptive discipline, anatomy was closely intertwined with natural history, and dissection was integral to the natural history of animals.[76] The humanist reexamination of ancient texts that began in the fifteenth century included such works as Pliny's *Naturalis historia* and Aristotle's *Historia animalium* as well as the works of Galen, and the many examples of *historia anatomica* and *historia naturalis* in the sixteenth century emphasized direct observation along with textual knowledge verified by a consensus of authors.[77]

Books, and particularly textbooks of various kinds, constituted an important source of anatomical knowledge. Books could tell the student how to see and what to see; among the meanings the lexicographer Antoine Furetière (1619–88) attributed to the word *"historia"* was "the exposition of things of which we have been the spectators."[78] The textbook tradition in anatomy dated back to Mondino de'Liuzzi's of 1316, which accompanied generations of medical students until the early sixteenth century. André du Laurens recommended reading textbooks as one way to learn about anatomy, and dozens of anatomical textbooks appeared in the century and a half before 1650.[79] They ranged from enormous Latin tomes with few or no illustrations to large-format illustrated atlases to smaller and less expensive surgical texts. Most of the latter texts, such as Paré's *Anatomie universelle du corps humain*, presented learned theory in the vernacular, often alongside instruction in specific surgical operations. Like textbooks in natural philosophy from this era, anatomical textbooks tended to hew to a conservative philosophy. In natural philosophy this continued to be Aristotle, while anatomists referred to Galen.[80]

The first anatomy text to be published in French was that of Charles Estienne (1504–64), who studied anatomy with Sylvius in the mid-1530s. *De dissectione*, composed with the surgeon Estienne de la Rivière, appeared in Latin in 1545 and in French a year later. Although it gave, as the French title page proclaimed, "déclaration des incisions" by la Rivière, both Latin and French editions were lavishly illustrated folios and perhaps therefore both too expensive and too unwieldy to be widely used, although they could have served as reference volumes on-site. Neither version was reprinted.[81]

Three Latin texts from the first half of the seventeenth century enjoyed particular longevity with numerous reprintings, and provide some clues about what medical students were expected to know, and how they learned it. All three reveal a French connection. The *Historia anatomica* of du Laurens, first published in 1593, was translated into French in 1613. Gaspard Bauhin of Basel first issued his *Theatrum anatomicum* in 1605 as an expanded version of earlier texts. Riolan the younger's anatomy text, *Anthropographia*, had numerous incarnations between 1605 and the early 1650s; the best-known edition appeared in 1626 and was translated into French in 1628–29. All of these books summarized current knowledge about the human body, referring both to textual and to anatomical evidence. Much of this anatomical evidence came from animals, even if the human was the ultimate model. The order of dissection, and the slow and careful techniques, continued to follow the model established by Mondino de'Liuzzi in the fourteenth century.[82] Du Laurens, Bauhin, and Riolan also emphasized the moral value of anatomy: its primary purpose was not practical knowledge, but knowing oneself, and this knowledge had no specific religious referent; although du Laurens and Riolan were Catholic, Bauhin was a Protestant. This self-knowledge, as du Laurens explained, came from a recognition of human uniqueness among animals, which included physical characteristics such as an upright stance as well as intelligence. Moreover, only humans have "the spark of heaven": only humans can know God. Acknowledging the special place of humans would inevitably lead to moral behavior.[83]

We know little about how or even if students used such books. Did they purchase copies? How much did they cost? One historian suggests that students carried Bauhin's text to dissections; as a small octavo, about 7.5 inches (19 centimeters) high, the 1605 edition was the smallest book of the three, but it also had thirteen hundred pages.[84] The other two were much larger: the 1599 illustrated edition of du Laurens was a six-hundred-page folio, 15 inches (38 centimeters) high, and the 1626 edition of Riolan was a nine-

hundred-page quarto 9.5 inches (24 centimeters) high; the French translation of 1628–29 did not reduce the page count but bound it in two volumes.

None of these volumes were exactly "how-to" manuals. Bauhin's came
the closest, with his methodical exposition of the course of a human dissection, based on the classical concept of "venters" or sections of the body;
William Harvey used Bauhin's textbook as the basis for his anatomy lectures
before the London College of Physicians in 1616.[85] Du Laurens and Riolan
spent more time explaining the philosophical justification for anatomy. All
of these men were well trained in human and animal anatomy, and all were
firm Galenists, taking seriously Galen's exhortations to dissect (which, in
Galen's case, meant dissecting animals exclusively).

The *Historia anatomica humani corporis* of du Laurens was reprinted at
least a dozen times in French and Latin over the course of the seventeenth
century, and a French edition appeared as late as 1778. The medical faculty
at Montpellier had long advocated dissection. Its anatomy theater dated
from 1556, when Guillaume Rondelet (1507–66) made dissection a priority. Yet it is not clear how much du Laurens actually taught: he was appointed
professor of medicine in 1583, but by the 1590s he was in Paris at the court
of Henri IV, eventually becoming the king's first physician. Du Laurens advised students basically to learn on their own, counseling them to practice
dissection as much as they were able, with human cadavers if possible but
also with living and dead animals. In addition, he recommended looking at
anatomical illustrations, attending demonstrations, and reading books. This
advice also described the practice of natural history, which as we shall see
employed the same combination of textual evidence and direct observation.

Gaspard Bauhin learned to dissect in this way. As an undergraduate,
he studied in Basel with Rondelet's former student Felix Platter. Although
Platter's famous diary of his medical education detailed clandestine expeditions to dig up corpses for dissection and his participation "in every secret
autopsy of corpses," Platter and Rondelet dissected far more animals than
humans.[86] Bauhin left Basel for Padua, where he studied with Girolamo
Fabrici d'Acquapendente (1533–1619) and assisted in dissections outside
the classroom, and then traveled to Bologna and Paris for additional instruction, in the latter city from the surgeon Pineau.[87] Basel highly prized the
skills he thus acquired, and its physicians sponsored his first public anatomy
demonstration soon after his return in 1581.

Unlike du Laurens and Bauhin, Riolan the younger, called by Thomas
Bartholin (1616–80) "the prince of anatomists," was entirely a product of
the Paris Faculty, where he benefited greatly from its inbreeding. Elected to
the faculty's new chair in anatomy as well as a medical chair at the Collège

Royal as soon as he received his degree in 1604, he first published on the topic in the following year with a short treatise attached to the end of the latest edition of his father's works. This publication established his reputation as a solid Galenist and a vigorous defender of the Parisian school against its enemies, real or imagined. He repeated, with embellishments, the claim that Berengario da Carpi and Vesalius vivisected humans to support his overall disdain for Vesalius, who had dared to criticize Galen.

Riolan's *Anthropographia* demonstrated his skills as a dissector and his deep admiration for Galen. "The subject of anatomy," Riolan declared, "is the body of man, the composition and conformation of which is shown by a very precise dissection."[88] Such dissection, he believed, would clarify and add details to Galen's account of the body, but would not fundamentally change it; nonetheless, he did not include illustrations in his book because he believed that one should consult only nature itself. Despite the emphasis on the human body, the title page stated that *Anthropographia* encompassed the anatomy of "men, women, children, and living animals."

Riolan repeated with approval Galen's advice to dissect both the dead and the living, explaining the usefulness of animals in this regard. Writing in 1626, Riolan boasted that he had personally dissected one hundred bodies in the past twenty-four years as a sign of his stature and experience, comparing himself to the "grand Anatomiste" Berengario da Carpi, who had similarly dissected a hundred corpses (even as he castigated Berengario for human vivisection).[89] But this number is what Riolan would have dissected at the twice-yearly public anatomies and does not indicate any experience he must have received outside that realm. As noted above, he later claimed that the Paris Faculty had access to four to six bodies a month "for anatomical use," although it is not clear what these uses were. Even if the hundred corpses were in addition to the public anatomies, this is not many bodies over a quarter of a century. Most students, even most professors, did not have many opportunities to dissect humans. But in this too Riolan followed Galen in advising his readers to dissect humans as much as possible, acknowledging that these opportunities would be few and that animals would be more likely subjects. Ideally, he said, anatomical training should take place solely on human bodies, because one could not find in animals' bodies "the divine composition of man." But a more "expansive" view of anatomy would encompass animals, which were also useful for training the hand in dissection, and if human cadavers were lacking (as they usually were), animals could fill their place.[90]

Du Laurens, following Galen, had described a hierarchy of animals suitable for dissection based on their similarity to humans, ranging from apes to

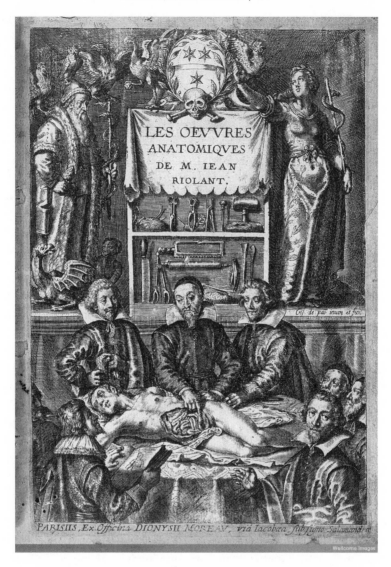

Figure 1.3 Title page, Jean Riolan, *Oeuvres anatomiques*, 1628–29.
(Wellcome Library, London.)

monkeys, bears, and pigs to ruminants. Riolan expressed distaste at Galen's favored animal, the pig, claiming it was too fat and had an unpleasant voice, and asserted that the only animal currently available for this purpose was the dog.[91]

Riolan classified anatomy, the knowledge of "hidden parts," in two categories, theoretical and practical, the first learned by ear and eye through lecture and reading, the second by the hand—what he later called the "seeing hand"—through dissection. He further distinguished between medical and philosophical aims of anatomical knowledge, the former concerned with curing disease, the latter with discovering the craftsmanship of God in the body; again, the moral purpose of anatomy was as important as its practical use. Seeking after useless knowledge, such as "the ramifications of the jugular vein," could corrupt these aims. Knowledge of the body was better acquired "by the sight and the touch, rather than the ear and the reading of books," and the dissection of live animals had a role in this instruction.[92]

A very different sort of anatomical textbook was the surgeon Nicolas Habicot's *La semaine ou pratique anatomique*, first published in 1610 and reprinted as late as 1660. Habicot, a surgeon of Saint-Côme, outlined a course of fourteen lectures to take place over seven days. Like Paré, Habicot was a barber-surgeon who joined the higher-prestige company of Saint-Côme, and his book was based on his public lectures.[93] Habicot began with the basics: choosing a subject, obtaining the proper tools, learning how to cut. Unlike the learned professors of anatomy, Habicot spent little time on theory. The 1660 edition of his manual, printed cheaply on poor paper, made no mention of the lacteals, the circulation, or the discoveries of the 1650s. In this world there are still pores in the septum and Galen rules. He explained that if the reader was dissecting under a physician, such as at the Paris Faculty, he need not know the theory, but only be attentive to the doctor's discourse.

But Habicot also gave readers practical advice, telling them to be sure to have towels to cover the face and the "parties honteuses" of the subject. Whether giving public or private lessons, one should take care with the composition of the audience, avoiding those who would be disruptive, although there will always be someone who wants to show off his knowledge at others' expense. If the anatomist can convey confidence and discretion and choose good assistants, one will be able to acquire "bon renom, & multitude d'amis, qui est une richesse inestimable." Yet even this utilitarian manual claimed that dissection was first "useful and necessary to those who wish to have perfect knowledge of themselves."[94]

By the 1640s, Pecquet and Perrault could have read the Danish physician Thomas Bartholin's new edition of the *Institutiones anatomicae* of his father Caspar (1585–1629), which first appeared in 1641 and in a French translation in 1647. Thomas Bartholin updated the text over a number of editions between 1641 and 1663, first with marginal notes and then with a wholesale rewriting. In 1653, the German physician Michael Lyser (1627–60), Bartholin's assistant, published his *Culter anatomicus* (The anatomical knife), another how-to book for the practicing anatomist, particularly notable for its detailed instruction in the construction of skeletons.[95] Unlike Habicot, Lyser was fully up-to-date with the latest discoveries.

Could students therefore learn to dissect from textbooks or manuals such as these, or a combination of books and observation? Or were books even necessary? In self-exile in Amsterdam at the end of 1629, the young René Descartes wrote to Marin Mersenne, "I wish to begin to study anatomy."[96] He began to teach himself anatomy, first by witnessing human dissections, then by watching the butchering of animals and taking home parts of them to practice dissection, and by 1633, attempting to demonstrate the circulation by dissecting a live rabbit. In his anatomical notes from the 1630s, not published until 1859, he cited Bauhin's text a few times, and in one section he appeared to follow Bauhin's order of dissection.[97] Book 5 of the *Discourse on Method* (1637) displayed how much he had learned. By the end of the 1630s he regularly witnessed dissections in Leiden and practiced on his own with animals, as his anatomical notes detail. Like many of his contemporaries, he was intensely interested in the structure of the eye, and in his *Dioptrique* (1637) he described an ox's eye he had dissected.[98] Samuel Sorbière recounted that when Descartes was asked in 1645 which books he most valued, he led his questioner to the backyard and showed him a calf he was about to dissect.[99]

5. Patrons and Anatomists

Perrault and Pecquet were products of two institutional configurations: the Paris Faculty of Medicine and the Republic of Letters. They gained much of their knowledge of anatomy and the new science from their membership in the republic, which comprised informal salons and academies and their virtual equivalent, correspondence networks. As we have seen with Cureau de la Chambre, the links between men of letters and men of science were close and intertwining. All four of the academy's anatomists relied on these circles and the patronage networks behind them to advance their careers. Colbert attempted to channel the energy of these groups, and not incidentally, their existence outside the university, into the Paris Academy of Sciences. Laurence Brockliss has argued that the Paris Academy gave natural philosophy the "cultural authority" it previously lacked in seventeenth-century France. Before 1666, he claims, the new science failed to "command consent" to its doctrines among university and political elites.[100] But a more expansive view of cultural authority dilutes this claim; the new science held enough cultural purchase to gain powerful political patrons long before the founding of the academy, even if it did not gain the support of the traditional intellectual elites represented in Paris by its university.

In the case of anatomy, the Paris Faculty of Medicine's refusal to accept new doctrines such as the circulation of the blood left it increasingly irrelevant intellectually, even as it continued to dominate the medical profession. The King's Garden and its intended teaching function was, as Riolan recognized, a serious threat to the faculty's hegemony that was temporarily thwarted. Similarly, as we shall see below, Théophraste Renaudot's *Bureau d'adresse* of the 1630s, which Richelieu supported, threatened the faculty's dominance in medical practice as well as in intellectual debate.[101] But at the same time, courtiers such as Séguier recognized that their patronage of intellectuals such as Cureau de la Chambre gained them credit within the court as well as power among fellow courtiers to whom the act of patronage established status.[102]

Historians have identified several characteristics of the seventeenth-century Republic of Letters, a loose, informal and egalitarian community whose boundaries constantly shifted.[103] Membership was voluntary and even self-selecting: if you thought you were a *littérateur*, you were; in the words of April Shelford, the republic was an "intellectual vocation" that its members considered to be "honorable and moral."[104] Members of the republic considered fundamental questions of the nature of knowledge and how we acquire it, and communication was its goal. It was above all social, and its communications took many forms, particularly correspondence networks. Correspondence included writing letters but also exchanging books, pamphlets, and manuscripts, as well as artifacts and other objects. After around 1650 journals were also exchanged. Travel, which was intrinsic to the persona of the denizen of the republic, could be virtually experienced via these means of communication. But perhaps most important were the meetings called variously salons, or academies, or "cabinets," or "*conférences*." The various names indicate the various formats, broad or narrowly focused in topic, informal or structured in organization, with humanist interests in texts and the past merging and overlapping with the new science.[105] In some of these groups, pursuit of the new science eventually overtook humanist admiration for the ancients. But this process took many years, and the same people could appear on either side of what became the ancients and moderns divide.

Informal academies dating back to the sixteenth century had mainly focused on humanist interests in art and literature. Federico Cesi's Accademia dei Lincei in Rome, which ran from 1603 until Cesi's death in 1630, merged humanism with the new science.[106] In Paris, the circle of the historian Jacques-Auguste De Thou (1553–1617) was carried on by the "Cabinet" of Pierre Dupuy (1582–1651) and his brother Jacques (1591–1656) until the 1650s. De Thou was a president of the Parlement of Paris and a councillor of state, and the Dupuy brothers came from a long line of advocates. As royal librarians, they valued books, but their discussions ranged widely and attendees included the literary critic Guez de Balzac, the natural philosopher Pierre Gassendi (1592–1655), the printer Sébastien Cramoisy, and the philosopher Tommaso Campanella, among many others.[107] Participating virtually in these academies and constituting virtual academies on their own were a number of correspondence networks, particularly those of the Minimite friar Marin Mersenne (1588–1648) and the Provençal jurist Nicolas Fabri de Peiresc (1580–1637). Peiresc served as a patron and "broker" of patronage for natural philosophy as well as for the Republic of Letters.[108]

Mersenne's physical "Academia Parisiana" took shape around 1635, but his "académie du papier" had begun many years earlier with Peiresc and his close friend Gassendi. By the 1630s Mersenne was known across Europe for his correspondence network. At a time when the practitioners of the new science were few in number and often scattered geographically, correspondence provided a critical bond.[109] Peiresc's correspondence indicates a brisk commerce in books and scientific instruments as well as in ideas; he reported writing forty-two letters in one day. His correspondents learned of astronomical observations from Aix and dissections he witnessed there.[110] Peiresc supported Mersenne with books and instruments, and in the reciprocal relationship of patronage, Mersenne gave in return the latest intelligence from Paris. Natural knowledge became a form of social currency that could increase one's standing among peers. And unlike the shifting sands of political patronage, the relationships established by Peiresc and Mersenne persisted for decades, in contrast to the fate of Théophraste Renaudot's *conférences*, which perished with his patron Richelieu.[111]

If Riolan was the "prince of anatomists," Peiresc was, in the words of Marc Fumaroli, "the prince of the Republic of Letters." In the networks and academies of the thirty years that preceded the founding of the Paris Academy of Sciences, the Republic of Letters became what Robert Mandrou labeled "the republic of Savants."[112] The relationship between humanism, language, civility, politics, and natural philosophy that Peiresc and Mersenne and later Bourdelot, Thévenot, and Montmor explored came to an official resolution in Colbert's Academy of Sciences, but as we shall see in later chapters, that resolution left some loose ends.[113]

Peiresc, long a member of the Dupuys' network, formed a bridge between their humanist circle and Mersenne's more scientifically inclined group.

Peter Miller argues that Peiresc "worked to expand the varieties of evidence from the textual to the material," and both books and experiments, both civil history and natural history, occupied his days.[114] In earlier years, he had studied anatomy with Fabrici at the same time that he explored classical culture. With Gassendi, Peiresc pursued several anatomical projects, including dissections of humans and animals as well as of a variety of eyes, the latter part of his study of vision, the telescope, and the moons of Jupiter.[115]

Peiresc viewed his practice and method of natural history as a highly social enterprise and also a moral one. In the quest for the *"honnête homme"* that preoccupied the Republic of Letters in the 1630s, Peiresc believed that improvement and not merely entertainment must be the goal of the academies and correspondence networks of the republic. Like his friends the Dupuys and Gabriel Naudé (future librarian to Cardinal Mazarin), he believed learning must be a central activity of the republic. Civility and conversation, while important, were not sufficient. He contrasted his activity with the desire to please of the aristocratic *salonnières* exemplified in the famous blue salon of the Marquise du Rambouillet and well described in Nicolas Faret's 1630 *L'honneste-homme ou l'art de plaire à la court*. The marquise and her followers in turn rejected what they viewed as the ostentation of the royal court.[116] Salons and academies tended to be egalitarian rather than hierarchical; but, as D'Alembert complained a century later, the goals of polite and pleasing behavior at times clashed with a new science that could be less than polite.[117]

Faret's book, dedicated to the king's brother Gaston, Duc d'Orléans, described the *honnête homme* as the ideal courtier: learned but not pedantic, *politique* but Christian, eloquent but not glib. It was, he told the duke, "a portrait of yourself," and one that the rest of the court would be wise to follow. Faret's *honnête homme* had a smattering of knowledge in science and mathematics and deeper knowledge of history and literature. He scorned pedantry, which led to unnecessary quarrels, and "would profit more to study in the great book of the world than in Aristotle."[118] But another vision of this ideal, based on the character of Montaigne, placed more emphasis on learnedness. Nonetheless, this alternative definition of *honnêteté* also involved rationality, sociability, and judgment; according to Blaise Pascal's friend Damien Mitton, who wrote a treatise on the topic, *honnêteté* comprised "the desire to be happy, but in such a way that others are as well."[119] Some, like Cureau de la Chambre and Jean Chapelain (1595–1674), moved easily back and forth between the blue salon and the Dupuys'.

In the Académie française, Richelieu attempted to strike a delicate balance between a true national language and mere courtly eloquence.[120] While its original members included many men of the court, it also included erudite men such as Chapelain, Guez de Balzac, Cureau de la Chambre, and Henri Louis Habert de Montmor (1600–1679). By the end of the seventeenth century, *honnêteté* had become synonymous with aristocratic *politesse*, the outward expression of civility, and the top-down and hierarchical society of Louis XIV's court replaced the egalitarian style of the salons. But the court never entirely displaced salons and academies, which continued to flourish. However, as we shall see, negotiating what Jacques Revel has called the "difficult middle ground" between court and salon required a different kind of person than either Peiresc or Mersenne.[121]

The practice of natural history in the seventeenth century fit perfectly into the Republic of Letters.[122] It required exchanges of specimens and texts, words and images. Like items of ancient art and manuscripts, natural specimens found places in the physical cabinets of Peiresc and later Bourdelot and Fouquet, as did the instruments of natural philosophy: lenses, scales, barometers. Commerce in objects, as well as letter writing, created bonds across boundaries of class and nation. In this period, the Latin language also provided a common intellectual ground for communication; but increasingly, French took that role. Cureau de la Chambre, Claude Perrault and his brothers, and Jean Pecquet all emerged from this milieu.

René Taton justly referred to Mersenne as "le véritable secrétaire de l'Europe savante." Like Peiresc, Mersenne took on the role of patron. His social and financial position meant that his direct patronage was largely intellectual rather than financial, such as when he assisted in the publication of Descartes's *Discours de la méthode* in 1637.[123] Mersenne claimed that his circle, "la plus noble académie du monde," was "toute mathématique," and although his network intersected with Peiresc's, it lacked many of the latter's literary and antiquarian contacts.[124]

But if Mersenne's own interests and his immediate circle centered on the mathematical sciences (which included physics and astronomy), he and Peiresc nonetheless shared a broad view of the new science that included both natural philosophy and natural history, the mathematical sciences and the phenomena of life. The library at the Minimite convent at the new Place Royale (completed in 1612) was one of the best in Europe, and Mersenne received permission from Rome to read prohibited books. The group that met in the porter's lodge at the convent (which, unlike the monks' cells, was heated) included Gassendi as well as the mathematicians Pierre Fermat (1601–65) and Gilles-Personne de Roberval (1602–75, mathematics professor at the Collège Royal), Étienne Pascal and his precocious son Blaise, bureaucrats with interests in mathematics such as Pierre de Carcavi (1603–84) and Bernard Frenicle de Bessy (1605–75), and visitors including Thomas Hobbes. Mersenne was instrumental in publicizing the ideas of Galileo and Descartes in the 1630s. The latter had known Mersenne since their days at the Jesuit college of La Flèche, and was a frequent correspondent.[125]

By 1633 the young medical student Pierre Michon Bourdelot (1610–85) had joined Mersenne's network. Born Pierre Michon, son of a barber-surgeon, Bourdelot adopted the surname of his maternal uncles who took over his education. Edmé Bourdelot (d. 1620?) had been one of Louis XIII's physicians, and his brother Jean (d. 1638) was a jurist and antiquarian and master of requests for the queen. Both brothers were well known in the circles of Peiresc and the Dupuys.[126] Their nephew Pierre entered the Paris Faculty as a student in 1629, and with the help of his uncle Jean, also entered Parisian social and intellectual life. One biographer described him as having "tout le génie d'un courtesan," and Pierre Bourdelot's affability and talent for friendship—not unlike Mersenne's—quickly made him a favored correspondent, a courtier not of the court. The chronicler Le Maire later said of him that "the character [le génie] of Monsieur l'Abbé Bourdelot was cheerful, and his mind well-tuned."[127] Moreover, his talents were not merely personal. The provincial physician Christophe Villiers responded to Mersenne's queries about Bourdelot in September 1633. By November, he was trusted enough to act as a courier to Rome for the circle. In the same month, Villiers described Bourdelot's dissection of a living dog to demonstrate the nerves that control the voice; "Il est bon anatomiste," he added. Gassendi had introduced Bourdelot to Peiresc in the previous year, and when Peiresc witnessed a dissection, he made sure Bourdelot received an account.[128]

Even though he had not yet attained a medical degree, Bourdelot spent the years from 1634 to 1638 in Rome as physician to the French ambassador François, Comte de Noailles while his correspondence network grew. Mersenne had close contacts in Rome via the French Minimite convent of Trinitá dei Monte and its prior Emmanuel Mignon. Peiresc introduced Bourdelot to Cassiano dal Pozzo (1588–1657), secretary to Cardinal Francesco Barberini, nephew of Pope Urban VIII. A physician, collector, and member of the Accademia dei Lincei, Cassiano remained a frequent correspondent after Bourdelot returned to France. Bourdelot no doubt saw Cassiano's famous museums, both the physical collection of antiquities and natural history specimens and the *museo cartaceo*, the "paper museum" of drawings and paintings he commissioned on these same subjects.[129] In Rome, Bourdelot made himself indispensable as a go-between among the *érudits* of Italy and France along with his (and Peiresc's) friend Jean-Jacques Bouchard (1606–41), Barberini's Latin secretary.[130] Upon his return to Paris, Bourdelot continued his medical studies and entered the service of the powerful prince de Condé. When his uncle Jean died in 1638, he inherited his extensive library and cabinet. By 1642 Bourdelot had organized his own academy, which met fortnightly on Tuesdays. After finally gaining his degree from the Paris Faculty, he was named one of the royal physicians and became the physician to the Condé family, settling in the sumptuous Hôtel de Condé. The *hôtel* occupied a large estate on the Left Bank of the Seine, not far from the convent of the Cordeliers and the Church of Saint-Côme.[131]

Bourdelot's meetings may be categorized as *"conférences,"* following Montaigne's definition of rhetoric in his *L'art de conférer* (1580) as *conférence*, something between conversation and the more analytical comparison and implying a delivery of information and not mere small talk. Blaise Pascal referred to this meaning of the term in his *L'esprit géométrique et l'art de persuader* (1658), which explained the three objectives of the pursuit of truth as discovery, demonstration, and "discernment of it from the false." Like the *salonnières*, Bourdelot disliked controversy, and he designed the atmosphere of his *conférences* to avoid confrontations, beginning each session with music and often providing dinner.[132] But unlike the salons, Bourdelot often included dissection among the activities. Civility, it seems, did not exclude the performance of violence in the name of science.

Bourdelot cast his net widely, and his fortnightly meetings included literary and political figures as well as natural philosophers from Mersenne's circle. What little we know of these early meetings is from Bourdelot's correspondence with Mersenne and Cassiano dal Pozzo. The breadth of topics covered resembled the public *conférences* of Théophraste Renaudot, a royal physician and client of Richelieu, which had begun in August 1633. These took place every Monday afternoon from 2:00 until 4:00 at his *Bureau d'adresse*, a general information office he had set up, with Richelieu's backing, in the Maison du Grand Coq on the Ile de la Cité, between the Hôtel-Dieu and the Palais de Justice. Renaudot's conferences ranged widely, from politics to ethics to the occult, and auditors came from all classes of society.[133] Many, then and now, have referred to Renaudot's conferences as a cabinet of curiosities, in the early modern sense of a diverse collection of natural and human-made objects. His audiences remain anonymous; a number of allusions in Mersenne's correspondence to topics Renaudot treated indicate that he and his circle may have frequented the meetings.[134] Bourdelot's broad interests made him a perfect auditor of Renaudot's sessions, but there is no concrete evidence.

Renaudot's conferences ended abruptly in September 1642. His patron Richelieu died in December, and the death of Louis XIII in May 1643 left his widow, Anne of Austria, as regent for the four-year-old Louis XIV. While Richelieu's successor as the power behind the throne, Cardinal Mazarin, was a great collector, he had no interest in institutions such as Renaudot's.[135] Bourdelot's academy, supported by Condé, had begun by the end of 1642. Unlike Renaudot, but like the salons, he limited his audience by social and intellectual standing. However, the breadth of his perspective resembled Renaudot more than Mersenne.[136] René Pintard characterized Bourdelot as consumed with ambition but intellectually little more than a dilettante, and others have followed this characterization. Yet if not genuinely learned himself, Bourdelot nonetheless recognized the significant questions and players. Members of Mersenne's circle, including Roberval, Gassendi, and the Pascals, attended Bourdelot's, as well as newcomers such as Adrien Auzout (1622–91) and Jean Pecquet, both probably introduced by Blaise Pascal.[137] It is likely that Claude Perrault, who had obtained his medical degree from the Paris Faculty at the same time as Bourdelot, also attended.

Henri II, prince de Condé, died in 1646, but unusually, his son Louis II, known as the "Grand Condé," continued the family's patronage of Bourdelot (who continued to serve as the family's physician), and he even attended the academy's meetings from time to time between military campaigns. Medical and anatomical concerns figured prominently among the topics discussed, as well as the Torricellian vacuum. The academies of Mersenne and Bourdelot, with overlapping memberships, flourished through the 1640s. Bourdelot continued to use his connections to do favors for his friends, even after death. Mersenne died in 1648, but in the next year Bourdelot wrote to Cassiano dal Pozzo to introduce Mersenne's nephew Pierre, a recent graduate of the Paris Faculty.[138] Pierre de Mercenne was among those who assisted Pecquet in his dissections.

As with Cureau de la Chambre, the success of Jean Pecquet owed much to his assiduous cultivation of patronage. But Pecquet's talents were definitely not literary. Born in Dieppe in 1622, Pecquet was educated at its Oratorian academy and then in the 1630s at the Jesuit College in Rouen, where he met Adrien Auzout and Blaise Pascal.[139] The three shared interests in mathematics and natural philosophy. Pecquet came to Paris around 1641 with a patron, an elderly marquise, who seems to have paid for his anatomy lessons with Louis Gayant. Between 1641 and his entry into the Paris Faculty in 1647, Pecquet also acted for several years as a *précepteur*, or private tutor, at the Jesuit Collège de Clermont (now the Lycée Louis-le-Grand), part of the University of Paris on the Left Bank. There he earned the master of arts degree required for admission to the Paris Faculty.[140]

By the mid-1640s Pecquet, probably through Pascal, gained admittance to Mersenne's circle, and by the evidence of their correspondence, they became quite close. Pecquet also made the acquaintance of Mersenne's nephew Pierre, at that time a student at the Paris Faculty. Through Mersenne, Pecquet frequented Bourdelot's academy, which was later described as being "une véritable eschole" of medicine, and according to some sources, he gave anatomical demonstrations there by the mid-1640s.[141] When his patron died in 1645, he performed the postmortem with the marquise's physician and the Saint-Côme surgeon Paul Emmerez.[142]

Pecquet's dissecting skills were well advanced when he entered the Paris Medical Faculty in 1647. His talents, however, did not pay the faculty's high fees. For this he needed a new patron. Mersenne introduced Pecquet in 1648 to François Fouquet (1611–73), physician of the Paris Faculty, bishop of Agde, and elder brother of Nicolas Fouquet (1615–80), an administrator who was rapidly rising in Mazarin's bureaucracy. Both brothers had attended the Collège de Clermont. Pecquet wrote to Mersenne from Agde in August 1648 thanking him profusely for the introduction. François Fouquet in turn introduced the young medical student to a circle of nobles and collectors that intersected with the intellectual circles he already knew. Fouquet's father, also named François, was a famous collector of books and objects, and Peiresc had admired his collection on a visit in 1618.[143]

Patron and client each had something to offer. François Fouquet was interested in the new natural philosophy and Pecquet kept him informed about recent experiments on the vacuum. Pecquet asked Mersenne where Fouquet could get lenses like the ones Mersenne had obtained in Florence.[144] Supported by Fouquet, Pecquet spent much of his time between 1647 and 1650 dissecting over one hundred different animals, many of them alive, most likely at Fouquet's residence in Paris. When Pecquet published his account of the thoracic duct as *Experimenta nova anatomica* in January 1651, he dedicated the book to François Fouquet. It was, he said, "born in your house."[145] Cureau de la Chambre had managed to find himself a sinecure at the Jardin du roi by writing books for Pierre Séguier. François Fouquet (and later his brother Nicolas) supported Pecquet to dissect. Apparently anatomy carried enough cultural capital to make it worthwhile to sponsor what was, after all, a pretty messy business in one's own house.

Marin Mersenne died in September 1648, shortly after he introduced Pecquet to Fouquet. After a flurry of activity surrounding the vacuum from the summer of 1647 to the spring of 1648, when Roberval, Auzout, and Pascal experimented with tubes and troughs of mercury, Bourdelot's academy abruptly ended as the unrest of the Fronde spread through Paris in the second half of 1648. Beginning as a financial crisis, the Fronde soon developed into a revolt against the government of the regent and her chief adviser Mazarin. The Fouquets strongly supported Mazarin, but the prince de Condé became a leader of the opposition.[146] The political upheaval of the Fronde upended patronage relationships. Following the prince's arrest in January 1650, Bourdelot, displaying an unusual skill in self-preservation, moved to Sweden, and may have taken with him a copy of Pecquet's book, which Queen Christina's physicians knew about by 1652. Unlike Descartes, he was a great success with Queen Christina, whom he cured of a mysterious ailment, and he presided over an academy much as he had in Paris until his return in 1653.[147]

Mersenne's circle—including Gassendi, Roberval, and the Pascals—continued to meet, led by his friend François Le Pailleur (whom Pierre Gauja described as a "dilettante fort savant"). Blaise Pascal most likely described the Puy-de-Dôme experiment to this group in the autumn of 1648.[148] Pecquet too may have attended these meetings, as well as Montmor, the master of requests for the king and, like Le Pailleur, an amateur of science. When Le Pailleur died in 1654, Montmor took over the reins of the circle.

With the publication of *Experimenta nova anatomica* in 1651, Pecquet's position at the Paris Faculty was no longer tenable. He does not seem to have received permission from the faculty to publish, and since he was a student, this constituted a breach of etiquette.[149] In the midst of the deanship of archconservative Gui Patin as well as the anticirculation polemics of Riolan, Pecquet must have recognized that the faculty was not about to give him such permission, or, after he published, a degree. There is no evidence that he performed any disputations. Although Claude Perrault was a professor of physiology in the faculty during 1651–52, we know nothing of his thoughts or actions in this period. Perrault's only publications from this era are some verses written with his brothers in the burlesque style, and a 1653 work called *The Walls of Troy or the Origins of Burlesque*, dedicated with what his biographer calls "heavy irony" to the wooden leg of the poet Scarron. However, these verses are significant, as we shall see in chapter 5, as an early example of Perrault's skepticism toward the ancients.[150]

By the summer of 1651, Pecquet had decamped to Montpellier. Montpellier's medical faculty had gained the disdain of the Paris Faculty for its support of chemical remedies, and although it was by no means a radical institution, it was far less strict in its admission policies than Paris. Moreover, as Pierre Gilis noted in his account of Pecquet, "It was always ready to support those whom the Paris Faculty attacked," and Riolan had recently attacked the rival school's "Hermétique et Émétique" therapeutic philosophy in his *Curieuses recherches sur les escholes en médecine, de Paris, et de Montpelier*

(1651).[151] By the following spring, Pecquet had a medical degree. Although the Danish physician and anatomist Thomas Bartholin had described Montpellier in the early 1640s as "dormant" in the study of anatomy, things had changed in the ensuing decade.[152]

Pecquet was a celebrity in Montpellier, where he gave public anatomical demonstrations, and he remained there, with occasional trips to Paris, until 1654. He continued dissecting and experimenting in Montpellier, witnessed by the eminent professor Lazare Rivière (1580–1655) and the surgeon Jean Martet.[153] Martet published a short book in 1652 detailing Pecquet's conclusions and his own experiments on live animals, probably undertaken with Pecquet, to demonstrate the lacteals, the thoracic duct, and the circulation.[154] Martet laid out precisely the tools and animals needed to duplicate Pecquet's experiments.[155] Pecquet mentioned his work with Rivière and Martet in the second edition of *Experimenta nova anatomica*, published in 1654.

By the 1650s, the four anatomists who would enter the Paris Academy of Sciences a decade hence followed widely differing career paths. Cureau de la Chambre, as we shall further see in chapter 3, devoted his time to writing large tomes on human and animal cognition. Louis Gayant continued his career as a surgeon of Saint Côme and assisted Mentel and Pecquet. Claude Perrault led a career as a faculty physician, practicing medicine in Paris and writing poems with his brothers. Of the four, only Pecquet had made a name for himself among the practitioners of the new science. But each of them intersected with the others at the faculty, the academies of Mersenne and Bourdelot, and as members of the Republic of Letters, and the republic is where they found patrons. As we shall see in chapter 2, they would also be united by the most important discovery in seventeenth-century science, the circulation of the blood.

TWO

The Anatomical Origins of the Paris Academy of Sciences

By the time the Geneva physicians Daniel Le Clerc and Jacques Manget published their enormous 1685 compendium of anatomical literature, *Bibliotheca anatomica*, the discoveries of the 1620s of the circulation of the blood by William Harvey and the lacteal veins by Gaspare Aselli had become "normal science" (in Thomas Kuhn's phrase), inspiring several decades of investigation. "Within the space of forty or fifty Years last past," they wrote, "there have been more Discoveries made in Anatomy, than all the preceding Ages could pretend to." Niels Stensen had already asserted twenty years earlier, "We owe to the dissection of animals nearly all of the new discoveries of this century."[1] Dissection of living and dead animals had become a widespread technique, a standard method of obtaining new knowledge, and an important focus of the new experimental science.

Anatomy occupied a peculiar place in seventeenth-century culture and natural philosophy. In the broadest sense of the term, "anatomy," as noun and verb, was a metaphor for all of early modern natural philosophy. As sixteenth-century naturalists penned "histories," so seventeenth-century naturalists wrote "anatomies" of multiple topics; in the *New Organon*, Francis Bacon wrote that the new knowledge he sought could not be found "without a very diligent dissection and anatomy of the world."[2] Not only could the act of dissection itself be experimental, but anatomy was, as a methodological principle, equated with the analysis or taking-apart that formed one aspect of experimental practice.[3] While the descriptive natural history of animals had always included dissection, during the seventeenth century a new enterprise of comparative anatomy emerged that combined natural history and dissection into a powerful method of discovery.[4] At its foundation in 1666, the Paris Academy of Sciences declared anatomy, along with astronomy, mechanics, and chemistry, to be one of its signature fields of

Figure 2.1 Title page, Le Clerc and Manget, *Bibliotheca anatomica*, 1685.
(Wellcome Library, London.)

inquiry. This research, documented in various publications, particularly the
Mémoires pour servir à l'histoire naturelle des animaux (1671 and 1676), con-
firmed an emerging discipline of the comparative natural history of animals.
    The discoveries of Harvey and Aselli unleashed an accelerating outpour-
ing of activity that culminated in the work of Jean Pecquet, first published in
Paris in 1651, which in turn led in the next few years to the work of Thomas
Bartholin and Olof Rudbeck on the lymphatic system and Francis Glisson

on the function of the liver, among others. As we have seen, from the 1630s onward, many people dissected, and many more witnessed dissection. Indeed, a majority of the practitioners of the new science across Europe engaged in this activity at some point in their careers. By the mid-seventeenth century, comparative anatomy emerged as an experimental discipline as well as a descriptive one. Pecquet's work combined dissection and the mechanical philosophy and set the stage for the mechanical theories of bodily function that dominated the second half of the century. While Descartes's *Discours de la méthode* also contributed to the mechanization of the body, his major work on this topic, *Traité de l'homme*, remained unpublished until the 1660s, and emphasized theory more than anatomical practice.

At the same time, new works of descriptive natural history, including the 1651 publication of the work of the Spanish naturalist Francisco Hernández as edited by members of the Accademia dei Lincei, contributed to a broader knowledge of animals. The 1651 *Rerum medicarum novae Hispaniae thesaurus* included not only the descriptions of Hernández but a number of dissections by the Lincean Johann Faber. Led by Claude Perrault, the anatomical projects of the Paris Academy rejoined natural history to the new experimental anatomy. Although Le Clerc and Manget explicitly excluded discussions of animal anatomy, claiming that their work was devoted to the human body and to medical applications of modern anatomical discoveries, they could not avoid talking about animals: not only were most of the discoveries they lauded dependent on experiments on animals, but they in fact included many works devoted solely to animals. As the main subjects of anatomical investigation, animals played an essential role in the development of experimental methods.

1. *Historia* and *Scientia*

What did it mean to experiment with animals in the seventeenth century? There is much ambiguity surrounding the terms "demonstration," "experience," and "experiment" in this period, further complicated by linguistic ambiguity: "*expérience*" in French and "*experientia*" in Latin could mean what we know in modern English as either experience or experiment. The medieval term "*experimentum*" referred to the empirical knowledge of a single thing, usually a medical cure.[5] Early modern definitions of experiment and the experimental method are many and various. Observation and direct empirical experience of nature, rather than specific techniques or an overall method, dominate early modern accounts of experiments and experimenting.[6] Such observation could be random or controlled. Francis Bacon viewed

Figure 2.2 Title page, Hernández et al., *Rerum medicarum novae Hispaniae thesaurus*, 1651. (Wellcome Library, London.)

experimentation as a method of forcing nature to reveal her secrets, one aspect of the collection of facts—the natural history—that would then fulfill his inductive method of reaching knowledge. Galileo famously referred to an experiment as a question put before nature, using the phrase *"facere periculum,"* "to put to the test."[7] In his published works he emphasized measurement and precision, an exact correspondence between mathematical theory and experimental result. But the precise agreement between theory and demonstration that Galileo wished to claim did not correspond to his actual experimental results, which were much more ambiguous. Indeed, one can argue that in the *Discorsi* in particular, Galileo attempted to figure out just what an experiment was.[8] Domenico Bertoloni Meli points out, "Without a canon or even shared views on experimental error . . . different practitioners could have seen, and indeed did see, the same experimental data either confirming or refuting a theory."[9] These difficulties were compounded in dissection, when it was difficult even to confirm what one could see.

Thus, the very definition of experimenting in the seventeenth century was in flux, and practitioners employed what they termed experiments in differing ways. Descartes, for example, did not clearly distinguish putting to the test from other varieties of experience. While the role of *"expérience"* in Descartes's natural philosophy has been much debated, he recorded a great number of observations and interventions, including many dissections.[10] Robert Boyle emphasized the predictive quality of experiments as their main purpose in his definition of a hypothesis:

> That it enable a skilful Naturalist to foretell future Phenomena, by their Congruity or Incongruity to it; and especially the Events of such Experiments as are aptly devised to Examine it; as Things that ought or ought not to be Consequent to it.[11]

Experience, to these men, included human manipulation of physical objects toward a particular goal (which could be demonstrative, predictive, or rhetorical), but also included simple observation of the ordinary course of nature without manipulation, such as commonly occurred in the natural history of plants and animals. In contrast, most modern experiments involve foresight, planning, and the prediction of an outcome (the hypothesis) and entail active intervention in natural processes, not simply passive observation, although there are experiments that are primarily exploratory. By this definition, anatomy and natural history are mainly descriptive rather than experimental practices. However, seventeenth-century men did not clearly distinguish either experiment from experience, or the dissection of dead bodies from that of living animals, further complicating explanations of what constituted experimentation in this period.

The natural history of animals cannot be separated from anatomical practice. Naturalists routinely dissected and at times vivisected the animals they observed; and anatomists routinely employed the descriptive tropes of natural history. The intertwined practices of anatomy and natural history hinged on the distinction between *historia* and *scientia*. In Scholastic philosophy, *historia* referred to a description that did not offer causes, a *demonstratio quia* ("demonstration of which") as opposed to a *demonstratio propter quid* ("demonstration on account of which"). True knowledge, or *scientia*, included knowledge of final causes, the *propter quid*. In anatomical studies, the facts of *historia* derived from *autopsia*, meaning "to see with one's own eyes." *Autopsia*, according to the French anatomist André du Laurens, included the dissection of dead humans and "of animals both dead and alive, to observe the motion of the internal parts."[12] Contemporaries often defined dissection as experimental: an active intervention to reveal the body. Francis Bacon's *New Atlantis* included animals reserved for experimental purposes, for "dissections and trials" as well as for experiments in generation and regeneration.[13] In 1681, almost a century after du Laurens wrote, the Cartesian Bernard Lamy equated dissection with experimentation: "To look for the facts of nature, that is to make experiments; for example, the dissections of animals, plants, fish, to open the nature that has been closed to us up to now."[14]

Among modern historians, Peter Dear has offered the most detailed definition of early modern experimentation. Dear defines an experiment in terms of singularity and contrivance, and linguistically, as a historical account of a particular event. He distinguishes experimentation as an event outside the ordinary course of nature from more generalized experience or demonstration, which displays nature without contrivance.[15] This distinction of experience from experiment might be framed as a distinction between natural history and experimentation as different ways of gaining knowledge of nature. But if activity that could be called experimental was rare among naturalists who observed plants, it was commonplace among those who observed animals.[16] Viewing natural history and dissection as distinct activities fails to capture anatomical practices that were at once experimental and descriptive, and fails to capture the practice of the natural history of animals, which most often included dissection. Following Dear's definition, although dissection involves planning and contrivance, and intervenes in the body to display a nature that is ordinarily not visible, it is not necessarily experimental as opposed to merely demonstrative or didactic, which conforms to a definition of early modern natural history as a descriptive and didactic activity offered by Brian Ogilvie. Simply revealing to sight what had been hidden should not be a singular event. But often it was. Variations abounded among animals and between animals and humans, not to mention among degrees of human skill in performing the dissection.

William Harvey commented that he never dissected an animal without finding something new and unexpected.[17] Dissection was complex and difficult, and dependent on many contingencies of time, place, subject, light, instruments, and skill, to name only the most obvious. In addition, it could only reveal information by destroying its subject; like the chymists' use of fire, the dissector's knife permanently transformed the body it touched. The violence inherent in dissection, especially of live animals, was, as we shall see, a factor in the acceptance of the knowledge it produced.

Without question, the dissection of live animals was indeed experimental in Dear's sense. This kind of dissection could reveal functions as well as forms, and always acted outside the ordinary course of nature. The Roman medical writer Celsus had described the resulting condition as "preternatural," a term much employed in the seventeenth century, meaning exactly this: not supernatural, which only God could cause, but outside the everyday actions of the natural world.[18] In addition, practitioners found the results of the dissection of live animals to be even less predictable than those of dissection of dead animals, varying from species to species and even from individual to individual. The joining of anatomical demonstration with the dissection of live animals, and the validation of *historia* as a method that created knowledge and did not simply describe experience, made anatomical knowledge into science in the seventeenth century.

This new practice of anatomy depended on animals for both practical and symbolic reasons. Anatomists could easily acquire the domestic animals such as dogs, cats, and pigs that they commonly used. In many respects, they resembled humans enough to demonstrate human function, and when they differed, the differences were also instructive. When the teaching of anatomy revived in the West in the twelfth century, the first demonstration subjects were animals.[19] The humanist physicians of the sixteenth century knew that both Aristotle and Galen had relied upon the structure of animals in order to talk about the human body, and the humanist conception of knowledge as an interconnected whole supported this idea.[20]

Caspar Bauhin asserted in the dedication to his *Theatrum anatomicum* (1605) that man was "animal admirandum" and the proper subject for natural philosophy, much as Shakespeare at about the same time referred to man as "the paragon of Animals." The language of the *Theatrum anatomicum* displays frequent slippage from "human" to "animal," repeating the phrase "as it is in other animals" (quam fit in aliis animalibus), where "animal" encompasses both animal and human, though Bauhin referred to animals alone as "brutes."[21] Seventy-five years later, the anatomist Guillaume Lamy likewise referred to "l'homme ou les autres animaux" as interchangeable, and it was this assumption that provided the basis of much of the interrogation into the nature of life in the seventeenth century.[22] When Galen's *On Anatomical Procedures* became part of the humanist harvest of lost classical texts, it further encouraged the ongoing use of animals. Galen's dissections of living and dead animals provided the basis for his discussion of the human body. Although Vesalius criticized some of his conclusions and their basis in animal anatomy, anatomists from the sixteenth century onward followed Galen in using animals as proxies for humans in anatomical teaching and public demonstration, as well as in private dissections.

The similarity of animal (especially mammal) and human form confirmed an overall divine plan of nature that was for human benefit. That animals did not have souls and therefore could not participate in the afterlife had been established by Thomas Aquinas in the thirteenth century. Accepting Aristotle's hierarchical concept of nature as a chain of being, Thomas explained in the *Summa theologica* that human superiority was based on the possession of reason, which implied the existence of an immortal soul. Since animals clearly lacked reason, they also lacked souls. Therefore, humans had no moral obligations to animals, although a good Christian, being compassionate, would not be unnecessarily cruel to animals. Descartes's "animal machine" brought this notion to a logical conclusion. But questions nonetheless remained about animal consciousness and capacity for suffering, and, as we shall see, these arose with particular intensity in the seventeenth century as animals became common experimental subjects.[23]

2. From Natural History to Comparative Anatomy

At the same time, the humanist reevaluation of Aristotle's *Historia animalium*, a work of much description but little theory, led many naturalists to turn toward a more Aristotelian evaluation of the animal as valuable in itself and not simply as an inferior copy of the human. According to these naturalists, all of creation exhibited God's goodness and creative energy, and one could aspire to attain knowledge of its entirety.[24] Animal dissection, therefore, need not be confined to those animals deemed most like humans. Indeed, Harvey later noted that dissection of a variety of animals best displayed the purposes of nature.[25] Aristotle and Galen did not always agree with each other, but their methods, which focused on dissection and detailed description—that is, *historia*—provided a common thread.

Like human anatomy, the natural history of animals (and plants) began as a bookish discipline. If we think of natural history as a register of facts, an inventory, the catalog seems its natural expression.[26] In the same way that the rediscovery and translation into Latin of Galen's *On Anatomical Procedures* helped to revitalize practices of human and animal dissection, Pierre Gilles's translation from Greek into Latin of *De natura animalium* of the Roman Aelian (Claudius Aelianus, ca. 175–235), along with printed editions of Aristotle's *Historia animalium* and *De partibus animalium* in the fifteenth-century Latin translations of Theodore Gaza and the *Naturalis historiae* of Pliny the Elder (ca. 23–79) led to a flourishing of the natural history of animals, with numerous books appearing from the 1550s onward.[27] Books and manuscripts, texts and illustrations, continued to play critical roles alongside observation and experimentation for the next two centuries. Humanist practices of reading and note taking constituted tools of cognition as much as did direct observation.[28]

New books of natural history overlapped significantly in content and authorship with books ostensibly devoted to human anatomy. The physicians who wrote many of the works on the natural history of animals also actively dissected animals and humans. By the time Marco Aurelio Severino published his *Zootomia democritea* in 1646, a new discipline of comparative anatomy had begun to take shape. In addition, unlike most works on human anatomy, many works on the natural history of animals relied heavily on illustration, beginning with the Zürich physician and humanist Conrad Gessner's encyclopedic *Historiae animalium* (1551–58), which illustrated each animal he discussed.[29]

Gessner (1516–65), like most Renaissance naturalists, worked on botany and philology as well as animals. He intended his work to be a new and comprehensive replacement for ancient and medieval accounts, and he included much material from these earlier works; he combed the ancients, going so far as to issue a new edition of Aelian's works in 1556.[30] But he also looked to more recent accounts, and his descriptions and illustrations came from a variety of sources, including live and preserved specimens and contemporary drawings, as well as older representations. We may think of his and similar natural history texts as a commonplace book of collected observations from a variety of sources. He cited the evidence of his own and others' dissections, for example, in his account of the dolphin, which relies heavily on the work of Guillaume Rondelet. This heterogeneous quality of Gessner's work, including its close attention to names and naming, reflects its humanist origins, and his emphasis on use—literary, symbolic, medical, or culinary—placed it firmly within the Plineian tradition. One consequence of Gessner's illustrations was to constrain the heterogeneous verbal accounts that constituted the work.[31] Indeed, Asúa and French claim that "the pictorial dimension of the *Historiae animalium* has a life of its own," and variations on Gessner's images appear again and again in the seventeenth century. Only with John Johnstone's natural histories of the 1650s do we see a decisive break with Gessner's style of illustration of a single image attached to a text, and this break was only partial.[32]

At the same time, Europeans gradually recognized over the course of the sixteenth century that many animals and plants from the New World and other far outposts had no analogues in the Old World. In the long run, this recognition upended the established classical notion of a chain of being and of a logical and familiar order in nature, though this understanding came slowly.[33] In the middle of the century, the first volume of Gessner's *Historiae animalium* (1551) included only one New World animal, the opossum, although subsequent volumes added a few more.[34] In 1605, the Flemish naturalist Charles de l'Écluse (Clusius, 1526–1609) brought together material from a number of sixteenth-century treatises on exotic (particularly New World) plants and animals in his *Exoticorum libri decem*. By 1651 William Harvey could claim that, unlike the ancients, "to Us the Theatre of the World is now open, and by the Sedulity of Travellers, we wel know, not only the Place, Habits and Manners of its Inhabitants; but also, what Animals, what Vegetables, what Minerals every Region is furnished withal."[35]

Apart from Gessner, sixteenth-century physician-naturalists—including the Frenchmen Rondelet and Pierre Belon (1517–64), the Dutchman Volcher Coiter (1534–76), and the Italian Ulisse Aldrovandi (1522–1605)—routinely dissected animals in work that commingled natural history with human and animal anatomy. Rondelet, professor of anatomy at Montpellier, compared marine animals to one another and compared the dolphin, which he recognized as what we would call a mammal, to humans and pigs, the latter being a common anatomical surrogate for the human body. Belon's *Histoire naturelle des estranges poissons marins* (1551) likewise described dissections and comparisons of *cetacea* to *pisces*.[36] In his work on birds, *L'histoire de la nature des oyseaux* (1555), he famously compared the skeletons of a human and a bird. While Gessner's multivolume work covered the entire animal kingdom, Belon and Rondelet talked about a single class of animals, focusing on anatomy as the mechanism for "scientific" (Belon's word) description. They omitted much, but not all, of the mythological and symbolic meanings that Gessner's "philological" work included, and minimized reference to human uses. Belon asserted that the ancients did not initiate the practice of dissection in order to find uses (particularly medical uses) for animals, but simply to learn more about them. He followed the Aristotelian notion of the essential unity of "animal" as a broad concept that encompassed many kinds.[37]

Coiter, on the other hand, did not confine his work to a single kind of mores improbi fictiᵍ,& populari notitia,quondam notati fuerint. Vnde Ariftophanes in Vefpis rifum Megaricum,de quo dictum eft alibi,taxat.Suidas refert & hunc uerficulum, fed tacito (ut fo= let)authoris nomine,ᾀν᾽ἰϛυ ἠμῖμ Μεγαρικὸ τις μηχανή. Rifum feu iocum Megaricum intelligit molliti= em Megarenfium.Refertur à Diogeniano (& Suida,)Erafmus. Megaricæ fphinges , meretrices: unde forfitan molles etiam σφίγκτια nominati funt,Suidas & Varinus.Vocabulũ quidem σφίγκτια magis probârim cum gamma fcribi quàm fine eo:utroᵍ enim modo reperitur,ut iam citaui.

¶Icon. Sphingem Aegyptij qui fcalpturam exercent,& Thebanæ fabulæ,biformem nobis re= præfentant,ex corpore uirginis & leonis cum grauitate & compofitione ipfam architectantes,Aeli= anus. Scyles Scytharum rex in Boryfthenitarum urbe ædes magnificas habebat,circum quas è la= pide candido fphinges & grypes ftabant, Herodotus libro 4. Athenis in arce,in templo quod Par thenôna uocant,tum aliæ quædam de generatione Mineruæ effigies habentur , tum Neptuni cum Mineruæ contentio de terra , Simulachrum ipfum (Mineruæ) ex ebore & auro factum eft,& in medio galeæ Sphingis imago expreffa, Paufanias. Pyramides tres funt inter Memphim & Del ta oppida , uíce appofito quem uocant Bufirin, Ante has eft fphinx, uel magis miranda , qua fyl= ueftria funt accolentium. Amafin regem putant in ea conditum,& uolunt in uectam uideri. Eft au= tem faxo naturali elaborata , & lubrica, Capitis monftri ambitus per frontem centum duos pedes colligit,longitudo pedum CXLIII.eft, altitudo à uentre ad fummum apicem in capite LXII.Plinius 36.12. Per totam longitudinem(templi Heliopolitani in Aegypto) protinus ex utraque latitudinis parte funt pofitæ lapideæ fphinges,uigenis cubitis,uel paulo pluribus inter fe diftâtes:& alter fphin gum ordo eft à dextra,alter à finiftra. Poft fphinges eft ueftibulum ingens , procedenti ulterius ali= ud ueftibulum,poftea aliud:& neᵍ ueftibulorum,neᵍ fphingum diftinitus eft numerus,Strabo li= bro 17. Aegyptij ante templa (in propylæis)Sphingem aftituebant, quo argumento indicarent, theologicam ipforum fapientiam obfcuriorem,fabulisᵍ ita fæpe conuelatã,ut ueritatis ueftigia uix interlucerent,Cælius Rhodig.& Calcagninus ex Plutarcho de Ifide. Amafis rex Aegypti ueftibu= lum Mineruæ fecit in Sai, opus admirandum,ubi etiam ingentes coloffos & immanes androfphin= gas pofuit,Herodotus lib.2. Signis quæ uocantur Corinthia, plerique in tantum capiuntur, ut fe= cum circumferant, ficut Hortenfius orator fphingem Verri reo ablatam. Propter quam Cicero illo iudicio in altercatione neganti ei fe ænigmata intelligere, refpondit debere,quoniam fphingem do= mi haberet,Plinius. Octauius Auguftus in diplomatibus libellisᵍ & epiftolis fignandis,initio fphinge ufus eft,mox imagine Magni Alexandri,nouiffimè fua, Suetonius.

¶Alciati emblema in fubmouendam ignorantiam,per dialogifmum.
Quod monftrũ id:Sphinx eft,Cur cãdida uirginis ora,   Et uolucrum pennas,crura leonis habet?
Hanc faciẽ affumpfit rerũ ignorantia:tanti   Sellicet eft triplex caufa & origo mali.
Sunt quos ingeniẽ leue,funt quos blãda uoluptas,   Sunt & quos faciunt corda fuperba rudes,
At quibus eft notũ,quid Delphica littera poffit,   Præcipitis monftri guttura dira fecant.
Nanᵍ uir ipfe, bipesᵍ tripesᵍ & quadrupes idẽ eft, Primaᵍ prudentis laurea, noffe uirum,
Hæc ille per uirgineam partem uoluptatem intelligens:per crura leonis,fuperbiam:per alas , leutta= tem. ¶Eft & Sphinx fabula ueteris cuiufdam poëtæ,Aefchyli ni fallor.

DE SIMIVVLPA, SIC ENIM FINGO

NOMEN, NE SIT ANONYMOS HAEC BESTIA:
cuius imaginem addidi,qualis in tabulis Geographi=
cis depingi folet.

It qui noftra memoria Payram regionem luftrarunt,beftiam dicunt fe uidiffe quadrupe= dem,ex anteriore parte uulpem,ex pofteriore fimiam: præterquàm quòd humanis pedi= bus fit,& noctuæ auribus:& fubter communem uentrem,inftar marfupij alium uentrem

Figure 2.3 *Simivulpa* (opossum). Gessner, *Historia animalium*, 1551.
(Linda Hall Library of Science, Engineering, and Technology.)

animal. To his mentor Gabriele Falloppio's work on the human skeleton, Coiter added comparisons to a number of other animals, including birds and reptiles. In eschewing the encyclopedic goals of Gessner and l'Écluse, Belon, Rondelet, and Coiter focused more closely on their own observations and dissections. Belon in particular traveled widely. But Gessner's attitude was shared by du Laurens and other anatomists, who argued that while direct observation was best, there were many other ways to gain knowledge, including from books.[38]

At the University of Bologna, Aldrovandi had been among the first in Europe to teach natural history as a part of the medical curriculum. His approach to *materia medica* included minerals and animals as well as plants; he founded Bologna's botanic garden and established a famous cabinet of animal, plant, and mineral specimens. This cabinet, as well as Aldrovandi's own observations and dissections, formed the basis for his multivolume work on the natural history of animals. The first volumes, on birds and insects, appeared during his lifetime. His students drew from the prodigious manuscripts he left at his death to produce further volumes on mollusks and arthropods, quadrupeds, fish, and reptiles, the latter of which appeared in 1640. Additional volumes on other topics, including monsters, appeared until 1668.

Like Gessner, Aldrovandi's sensibility was encyclopedic, and he too tracked myths, uses, and symbols in his descriptions. His index, like that of du Laurens, was of "things and words": observation and texts held equal ground. Aldrovandi included substantially more anatomical description and detail than had Gessner, and even depicted the skeletons of some of the animals he described. Nonetheless, his work definitely remained within the encyclopedic tradition rather than the narrower focus of Belon or Coiter.

A strong French thread may be traced through the natural history of animals in the sixteenth century and into the seventeenth. Belon, who wrote in French rather than Latin, received an MD from the Paris Faculty in 1540, where Gessner had briefly studied a few years earlier. Rondelet received his degree from Montpellier and taught anatomy there for several years. Coiter was among his students. Jacques Daléchamps (1513–88), another Montpellier MD (1547) and student of Rondelet, translated Galen's *Anatomical Procedures* into French in 1569, and six years earlier had issued an edition of Pliny. Pliny had already been translated into French by the grammarian Louis Meigret (1510–58). This peculiarly French preference for the national language, manifested most obviously in the Académie française, became even more marked in the later seventeenth century.

Philippe Glardon refers to a "no-man's-land" of seventeenth-century natural history that followed the plethora of works in the sixteenth.[39] The works of sixteenth-century naturalists continued to be reprinted—or in the case of most of Aldrovandi's animal works, printed for the first time—into the seventeenth century.[40] Gessner's works in particular appeared in numerous manifestations, from Thomas Topsell's English edition of 1607 to a variety of epitomes that bore only passing resemblance to the original. All of these works exhibited to varying degrees a comparative approach in which the human body was not a necessary referent. But tensions between particulars and universals and between human and comparative anatomy remained. After Aldrovandi, few new works of encyclopedic natural history appeared, with the important exception of John Johnstone's works in the 1650s, until Claude Perrault began to publish the Paris Academy's observations in the late 1660s, drawing on and citing the work of all of these authors.

Although new encyclopedic natural histories on the model of Gessner and Aldrovandi were few, other formats of animal description took their place. One new model came from the Bolognese lawyer Carlo Ruini (1530–98), whose magisterial 1598 treatise on the horse was perhaps the first to look at a single species with the level of attention paid to human anatomy. Aldrovandi cited it in his account of the horse.[41] The *Anatomia del cavallo*—written in the vernacular rather than in Latin—looked like an atlas of human anatomy, with full-page illustrations of dissected bodies and body parts, and Ruini constantly compared the horse's anatomy to that of the human body. His narrative followed the course of a human dissection from head to toes (or hooves) in exacting detail. Continuing the medical model, the treatise also discussed diseases and illnesses of horses. Ruini included none of the information on habitat, behavior, or indeed etymology or mythology found in Gessner or Aldrovandi. Within a year the French royal physician Jean Héroard (1551–1628) published his *Hippostologie*. Much shorter, with few but beautiful illustrations reminiscent of Leonardo, it treated only the bones of the horse. Like Ruini, Héroard wrote in the vernacular, in this case French.[42]

The Paduan anatomy professor Fabrici d'Acquapendente placed the comparison of human and animal anatomy at the center of his work. He proposed a "*totius animalis fabricae theatrum*," a "theater of the whole animal structure," which would describe not only humans but all animals and would encompass both form and function. He carried out this project in the context of his lectures on anatomy, which therefore necessarily included the dissection of live animals to demonstrate function.[43] Fabrici, a devoted

Aristotelian, conceived of the *historia* as the first step in a process that would lead to causal knowledge. The *historia* to Fabrici was the investigation of a particular organ or function in as many different animals as possible. He then went on, following Aristotle, to find the action of the part, and its use or cause—the *propter quid*, which would constitute a demonstration of the truth of his description, a general truth that applied to all animals.[44]

Fabrici viewed animals as a category rather than as individuals, with humans always the basis of comparison. For example, in his 1600 treatise on vision, voice, and hearing, the focus is on human organs and human function. Fabrici used "animals"—most often "quadrupeds"—for purposes of comparison only; the placement of the ear differed in nonhuman animals and quadrupeds, he noted, because the heads of quadrupeds inclined toward the ground while human heads were upright. His illustrations depicted human structures. Although his examination of the organs of speech included a section on animals, this was a philosophical disquisition on whether animals have speech rather than an examination of the speech organs of animals.[45]

In contrast, Fabrici's student and deputy Giulio Casserio (1552?–1616) compared the vocal and auditory organs of a number of different animals in his *De vocis auditusque organis historia anatomica* (1600–1601). Like Coiter and Belon, Casserio examined many species of animals, humans among them. Unlike them, he focused on a single set of structures rather than a single class of animals, and his striking illustrations display the strong similarities across species: as the anatomist penetrated more and more deeply into the organism, specific differences fell away and the observer was left with a collection of bones and tissues that looked remarkably alike; even the mouse and the frog held similarities. Only the insects were utterly different.[46]

3. Harvey and Aselli

Casserio's works merged natural history and anatomical texts. William Harvey employed the historical methods of Fabrici and the comparative anatomy of Casserio as he developed a powerful experimental method from the dissection techniques that had evolved over the sixteenth century to a peak of skill.[47] Like Fabrici, Harvey began with the premise that the observation of many particular animals would ultimately lead to conclusions applicable to all animals, but he did not aim for a demonstrative "*theatrum animalis.*" Rather, in his hands, this natural history, as he called it, became a method of discovery and justification of new knowledge about all animals.

Animals became for Harvey the center of an extensive research project whose ultimate outcome was the discovery of the circulation of the blood, a phenomenon that, in Jerome Bylebyl's words, "was quite novel in more than a strictly factual sense: first, the circulation could not have been predicted from, nor was it reducible to, the externally apparent activities of whole organisms; and second, the idea was antithetical to, rather than supportive of, the prevailing medical theories and practices."[48] Harvey discovered this new knowledge by means of visual demonstration and experiments.

Harvey's methods included dissection of a wide variety of living and dead animals, not only mammals but also reptiles, amphibians, birds, and fish. Following Aristotle's premise of the essential unity of living form, lower animals could have as much explanatory value as higher ones. The lectures he delivered in 1616 to accompany the dissection of a human cadaver at the Royal College of Physicians of London referred to a variety of observations and experiments on animals; the historian F. J. Cole counted 128 different kinds of animals and almost 500 individuals. Harvey's 1628 *On the Motion of the Heart and Blood in Animals* (*De motu cordis*) included 49 varieties and over 120 individuals.[49] Harvey's historical method and dissection techniques became the model for the investigation of life.

Harvey's work definitively established the usefulness of animals as experimental subjects and not simply as anatomical models. Anatomists such as du Laurens, Bauhin, and Riolan the younger demonstrated on live animals to display known facts that could not easily be seen in a human corpse, but the human body retained its primacy and dissection of dead animals retained its priority over that of living ones.[50] The discovery of the circulation of the blood established experimentation on live animals and not merely demonstration on living or dead animals as the preferred method of discovery for establishing new knowledge about the human and animal body up to the present day. By the end of the seventeenth century, as Le Clerc and Manget confirmed, experiments on and observations of living organisms had become routine. Harvey's methods of systematic investigation set him apart from those who revered the ancients, including his mentor Fabrici. Fabrici had argued that *historia* in itself could not lead to causal knowledge—to natural philosophy or *scientia*—but Harvey asserted that the description of particulars could lead to true knowledge without the discovery of causes.[51] Indeed, Harvey claimed that the discovery of causes was secondary in this investigative process.

In *De motu cordis*, Harvey narrated his discovery of the circulation as a discovery of a new way of seeing, an accumulation of descriptions. He established the new knowledge of the circulation by means of repeated, systematic dissection, which gave visual proof, confirmed by public witnesses, of the phenomenon of the circulation. This method was not a Baconian induction, where a long list of instances at some point coalesced into a causal explanation.[52] Harvey consciously did not provide a causal explanation. Rather, observation, the method of natural history, was the critical element in his demonstration: in his dedicatory letter to the London College of Physicians, Harvey noted that over a period of years he had "confirmed [the circulation] in your presence by numerous ocular demonstrations [*ocularibus demonstrationibus*]." He employed this phrase several times. These demonstrations, repeatedly witnessed over time, could not but lead to a "reasonable acceptance" of this new theory. Throughout the book, he kept to his methodological theme: "anatomical dissection, manifold experiments, careful observation." These tools and the historical method revealed the "previously unrecognized circular pathway" of the blood.[53] Thus, he said, "the exact opposite to the commonly accepted view is *seen*." Previous investigators saw wrongly by looking primarily at the lungs rather than the heart and did not accumulate Harvey's mass of empirical evidence.[54]

Harvey did not clearly distinguish between demonstrations on dead animals and humans and experiments on live animals, although the majority of what he labels experiments was indeed performed on live animals. These included tying off vessels to observe the unidirectional flow of blood and observing the motion of the heart under various circumstances. Harvey also cited demonstrations on live animals that other anatomists had performed. Bauhin and Riolan, for example, had each described the complex motion of the heart in a living animal.[55] The accumulation of these instances, argued Harvey, constituted new knowledge. Yet the citation of Bauhin and Riolan (and in other passages, du Laurens, Colombo, Fabrici, and of course Aristotle and Galen) also links Harvey to the humanist textual tradition of natural history.

Even before Harvey published his work on the circulation of the blood, the Milanese physician Gaspare Aselli had viewed what he called the lacteal veins in a live dog in 1622. Aselli, professor of anatomy at the University of Pavia, demonstrated the mesenteric nerves and their function in a dog that had been recently fed, before an audience of four physicians at the anatomical theater of the university. He noticed something new: a number of white filaments or vessels, which, when pricked, released a milky fluid. He called these vessels "milky or white veins" and proceeded to explore them and their function in other animals, including cats, lambs, cows, pigs, and a horse, employing the term "*factum periculum*."[56] In each case he found that the veins became apparent shortly after the animal had been fed.

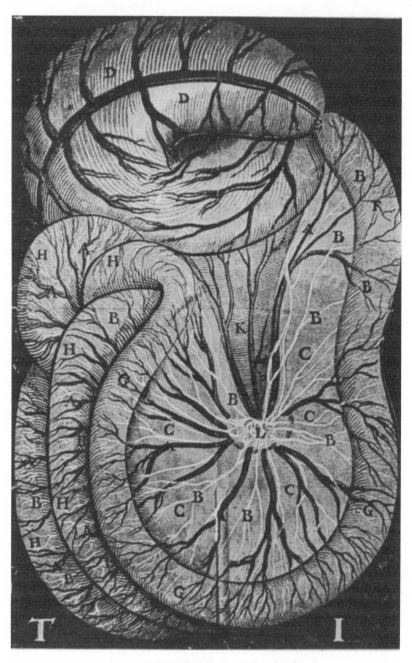

Figure 2.4 Lacteal veins. Aselli, *De lactibus sive lacteis venis*, 1627.
(Wellcome Library, London.)

In *De lactibus sive lacteis venis*, published posthumously in 1627, Aselli concluded that these vessels carried nourishment or chyle, defined as digested food, and that therefore they must terminate at the liver, where, according to Galenic theory, the blood was formed.[57] Like Harvey, Aselli built his argument on a careful accumulation of a series of dissections of living and dead animals performed before reliable witnesses, emphasizing the importance of *historia* in the discovery of new knowledge while acknowledging—unlike Harvey—that this knowledge was provisional.[58] In addition, he stated unequivocally that the results of such animal dissection applied to humans, even though he was, of course, not able to experiment on live humans: since he had established the existence of the lacteal veins in so many animals, that man should alone not have such structures would be an unacceptable shortcoming.[59]

Many commented on the discovery of lacteals, and some duplicated Aselli's experiments. A few believed that the lacteals would also eventually serve as a test case for the theory of circulation, and by the early 1640s, Aselli's and Harvey's works often appeared in print together. The roles in the production of blood of the lacteals, the mesenteric and portal veins and arteries, and the liver continued to be debated though the 1640s. Harvey expressed skepticism about the lacteals, which he believed contained not chyle but milk. He continued to believe that chyle mixed with the blood in the mesenteric veins, which in turn led to the liver.[60] Pecquet's discovery of the thoracic duct in 1650 definitively connected the circulation and the lacteals, and shortly thereafter Thomas Bartholin concluded that the lacteals were in fact part of a larger vascular system he called the lymphatic.

4. Debates about the Circulation, Especially in France

During the 1630s and 1640s, the curious across Europe used yet more animals to explore the circulation and the lacteal vessels. As we saw in chapter 1, Jacques Mentel, a physician in the Paris Faculty of Medicine, dissected live dogs before his students to investigate the lacteals beginning in 1629. In addition, savants not previously noted for their anatomical interests began to dissect. Pierre Gassendi had witnessed dissections that purported to demonstrate the pores in the septum in the early 1620s, which he described in his 1630 critique of the English Paracelsian Robert Fludd, *Epistolica exercitatio, in qua principia philosophiae Roberti Fluddi medici reteguntur*. Fludd was an early supporter of Harvey's theory, and Marin Mersenne had sent Gassendi a copy of *De motu cordis* shortly after its publication.[61] A few months later, he arranged to have a copy sent to Peiresc, commenting, "His opinion of the continual circulation of the blood by the arteries and veins is quite likely and established." But he did not agree with other aspects of Harvey's theory.[62] In 1634, Peiresc and Gassendi witnessed a dissection at the medical faculty in Aix-en-Provence, which Peiresc recounted in a letter to Jacques Dupuy in Paris. He carefully instructed the jailer to feed the prisoner well before execution; as a member of the Parlement of Aix, Peiresc also had a hand in getting the corpse released to the faculty. He further instructed the surgeon where to cut so as to observe the lacteals while they were still full, "as I had proved previously with a sheep."[63] A month later, Gassendi described this dissection to his friend Elia Diodati, adding that he and Peiresc had since viewed the lacteals in a cat "that had died, or at least was no longer moving, before we opened it."[64]

Descartes also learned of Harvey's work from Mersenne and read *De motu cordis* in 1632.[65] He presented his mechanistic interpretation of the circulation in his *Discours de la méthode*, published in Leiden in 1637. Descartes could not accept Harvey's notion of a self-moving heart whose contraction, the systole, was the dominant motion that caused the heartbeat. Such a lack of causation—here mechanical causation—smacked to Descartes of ancient notions of "faculty" or "vital soul," and he proposed instead that the heart acted simply to provide heat, a product of fermentation, which was a mechanical process. This heat vaporized the incoming blood, causing it to expand and propel the blood onward. Thus, the essential motion was expansion or diastole rather than systole. Descartes advised his readers to witness the dissection of an animal's heart and lungs as a reference for his description. This advice, coupled with his detailed description of the heart and its motion, gave the impression that Descartes based his explanation, as had Harvey, on dissection and experimentation. But his concept of the circulation, as he explained, owed more to logic than to dissection:

> Now those who are ignorant of the force of mathematical demonstrations and unaccustomed to distinguishing true reasons from probable may be tempted to reject this explanation without examining it. To prevent this, I would advise them that the movement I have just explained follows from the mere arrangement of the parts of the heart (which can be seen with the naked eye), from the heat in the heart (which can be felt with the fingers), and from the nature of the blood (which can be known through observation). This movement follows just as necessarily as the movement of a clock follows from the force, position, and shape of its counterweights and wheels.[66]

Looking at the shape of the heart and feeling its heat were observations, not experiments. To Descartes, these observed facts served to illustrate a causal explanation; knowledge did not arise from the facts alone. Only later, in the *Description du corps humain* (1648), did Descartes attempt to find experimental proof for the circulation in animals.[67]

Few accepted the Cartesian variation on Harvey's theme. One who did was the Utrecht professor Hendrik de Roy, known as Regius, who led disputations on the topic of the circulation between 1639 and 1641. Regius's lectures ended in at least one instance in a riot. Descartes's friend and correspondent, the Dutch physician Vopiscus Fortunatus Plemp (1601–71), on the other hand, rejected Descartes's version in the late 1630s, but accepted Harvey's in the 1640s, based on his own experiments on dogs. Plemp instructed Descartes in dissection in the mid-1630s, and the latter's anatomical notes detail numerous dissections through the 1640s.[68] Descartes, who suppressed the publication of *Le monde*—his treatise on the mechanization of nature and the human body—because he feared the fate of Galileo, chose the theory of the circulation as his sole published example of the operation of the mechanical philosophy in the body. The Catholic Church, and other Christian churches, apparently found neither the circulation, nor the methods employed to reach this discovery, to be heretical. In the sixteenth century, the works of Realdo Colombo, which included his description of the pulmonary circulation, had received the Catholic Church's imprimatur, and, as far as I have been able to discern, no anatomical works ever appeared on the *Index of Prohibited Books*.[69]

Descartes had witnessed the lectures in Leiden of the independent anatomist Frans de le Boë, usually known as Franciscus Sylvius (1614–72), who demonstrated the theory of the circulation in the late 1630s by means of dissections of live dogs in the university's botanic garden. These lectures coincided with the reprinting of Harvey's work in Leiden in 1639 and convinced the anatomy professor Jan de Wale (Johannes Walaeus, 1604–49) of the theory's validity. He chose an English student, Roger Drake, to present the first medical dissertation in Europe in support of the circulation early in 1640. In the same year a new edition of Aselli's treatise on the lacteals appeared.[70]

Walaeus went on to defend the circulation in two letters addressed to Thomas Bartholin, who had been in Leiden between 1637 and 1640. He described new experiments on dead and living dogs (over one hundred, he claimed) to demonstrate that blood flowed from the arteries to the veins and from smaller veins to larger ones. He also argued that the liver did not have the capacity to produce the quantity of blood that Harvey had estimated to flow through the heart. These letters appeared as a pamphlet in 1640 and then in Thomas Bartholin's new edition of his father Caspar's *Institutiones anatomicae*, published in Leiden in 1641.[71] However, Bartholin did not revise the text itself to reflect the theory of the circulation until several years later. Walaeus continued to experiment and revised his letters, appending a new version to a 1643 edition of Harvey's works. Harvey, Aselli, and a third version of the letters appeared together in an edition of the works of the Flemish anatomist Adriaan van den Spiegel (1578–1625) in 1645. The latest version described new experiments on digestion that showed that the lacteal and blood vessels did not connect in the mesentery.[72]

The great anatomist Riolan the younger had left Paris in 1633 to serve as the physician to the queen mother, Marie de'Medici. When he returned to the Paris Faculty nearly a decade later, he discovered a profoundly changed world in which the faculty appeared to follow rather than lead medical theory. Riolan recognized that Harvey's theory of the circulation—coupled with a non-Galenic interpretation of the lacteals—threatened the faculty on several levels. Not only did it challenge the bases of Galenic therapeutics, particularly diet and bloodletting, but more seriously, it challenged the foundations of knowledge. The faculty had to take a stand, and the 1642 defense of the first faculty thesis on the topic of the circulation concluded with a resounding negative.[73]

In 1645, three years after his return to the Paris Faculty, Riolan presided over the second thesis to discuss the theory of the circulation. The title of Jean Maurin's thesis was "An propter motum sanguinis in corde circulatorium, mutanda Galeni methodus?" ("Whether because of the circular motion of the blood in the heart, the Galenic method is altered?"), "*methodus*" referring to Galenic therapeutic methods.[74] Again, the conclusion was negative. Maurin's thesis presented for the first time Riolan's compromise theory of the circulation, which was not unlike Tycho Brahe's compromise with Copernicanism of a half century earlier. Where Tycho had kept the earth in the center of the universe but made the other planets rotate around the sun, Riolan kept the central role of the liver as generator of blood and admitted only a circulation in the aorta and the vena cava; as Galenic theory demanded, the tissues absorbed the rest of the blood. While Riolan acknowledged that the blood must circulate, he argued that this circulation was slow—all of the blood in the body only passed through the heart two or three times a day, whereas Harvey argued this passage took place every two hours—and did not include the lungs. The pores in the septum remained. With this compromise, Riolan preserved Galen's therapeutics "in good repair and

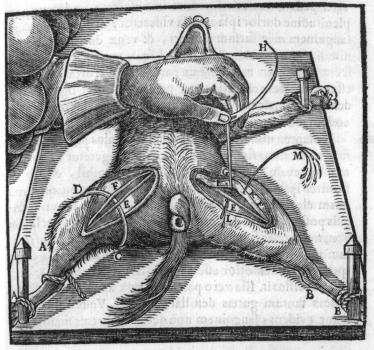

FIGURÆ EXPLICATIO.

A. *Crus canis dextrum.*   B. *Crus canis sinistrum.*

C. D. *Ligatura subiecta arteriæ & venæ, qua femur firmiter constringitur, expressa in dextro crure, ne literarum linearumque confusio in sinistro crure spectatorem posset turbare.*

E. *Arteria cruralis.* F. *Vena cruralis.*

G. *Filum quo constricta est vena & est elevata.*

H. *Acus, cui filum est traiectum.*

I. *Venæ pars superior & detumescens.*

K. *Venæ pars inferior à ligatura intumescens.*

L. *Guttæ sanguinis, quæ, é superiori parte venæ vulnerata, sensim distillant.*

M. *Rivulus sanguinis, qui, inferiori venæ parte vulnerata, continuo exilit.*

F 2   vero

Figure 2.5 Experiment on a live dog showing the unidirectional flow of blood.
Walaeus, *Epistola prima*, 1647 edition.
(Wellcome Library, London.)

inviolate" (*sarta tecta & inviolata*), "as much in the making of blood, as in the revealing of disease, by reason of causes and places."[75] Where Harvey had left the issue of causation open, Riolan backed anatomical facts with Galenic therapeutics, offering both a cause and a purpose for the circulation. His opposition to Harvey's theory of circulation centered on this lack of causal proofs, underwritten by therapeutics (especially bloodletting) and the role of the liver as generator of blood.[76]

Riolan defended Maurin's thesis on the basis of his deep knowledge of human and animal anatomy. In the course of the disputation, Riolan defended his theory against members of the faculty who opposed any sort of circulatory theory as well as some who supported Harvey.[77] Thirty-two *objectiones* to the thesis are recorded. It must have been a lively day of discussion and argument. In answer to more than one *objectio*, Riolan repeated his long-held belief that new knowledge was possible. He defended the practice of dissection of live animals and the use of animals, even lower animals, to make statements about human function, using examples from his own work. The expulsion of the blood from the arteries, he said, "is observed in the dissection of living animals."[78]

Riolan continued to debate Harvey in the late 1640s and into the 1650s; indeed, his critiques were the only ones to which Harvey replied.[79] A contemporary of Harvey and Riolan, the Neapolitan physician and anatomist Marco Aurelio Severino (1580–1656), published his major anatomical work, *Zootomia democritea*, in 1645. Thomas Bartholin spent the years from 1643 to 1645 in Naples studying with Severino, and their correspondence makes it plain that Severino (who had met Harvey around 1637) supported the theory of the circulation. However, it did not appear in *Zootomia democritea*.[80] Nonetheless, Severino's work was the most significant effort in comparative anatomy since Casserio nearly a half century earlier, and it was much broader, aiming toward an "*anatome generalis totius animantium opificii,*" a general anatomy of all the workings of animals, implying function as well as form while echoing Fabrici's claim of a "*Theatrum totius animalis fabricae.*"

Severino's evocation of Democritus in his title revealed his atomistic philosophy. The term "anatomy," he said, derived from the Greek άν' άτομα, which he translated as "*resolutio in indivisibilia,*" to resolve into indivisible (atoms).[81] He vigorously defended the usefulness to medicine and natural philosophy of "*zootomia,*" the dissection of animals, but he devoted only about a third of the book to actual dissection. Severino described his dissections of dozens of familiar animals, ranging from snails and crickets through

in duos majores ramos finditur, fed venæ arteriofæ bre-
viores funt.

Uterus cum vefica connexus infernè figuram hanc
habet; pudendum rectum auc æ-
quale non eft, fed medio fui tube-
rofum, ut olivæ nucleus; corpus
uteri, ubi os incipit anguftius, mox
obliquè ampliatur, & in duo cornua finditur fimùl arctè conjun-
cta, quæ parte pofteriore curvan-
tur & revolvuntur, ut arietis cornua, Externè, tum parte dextrâ,
tum finiftra transparent ofcula_
cotyledonum, rectâ quâdam lineâ
per longitudinem difpofita; Internè verò variè eft con-
camerata, & multis quafi cellulis diftincta. Laryngis car-
tilagines tres, quales à Cafferio defcribuntur.

Capræ pridem ablactata & vacca non
gravidæ uterus,

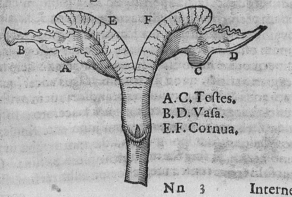

A. C. Teftes.
B. D. Vafa.
E. F. Cornua,

Internę

Figure 2.6 Uterus of goat. Severino, *Zootomia democritaea*, 1646.
(Bibliothèque interuniversitaire de santé, Paris.)

the chain of being up to horses, dogs, and monkeys. Although he claimed to be anti-Aristotelian on the basis of his atomistic, Democritean philosophy, he agreed with Fabrici on the "uniformity of the animal body," and agreed with his fellow anatomists that the true subject of anatomy was the human body.[82] Severino's many illustrations are of organs, not of animals. The parts were more important than the whole; a reader could not tell (other than through the text) from what animal a particular organ came, and one illustration served for several animals; an image of a goat's uterus also applied to a cow. While his observations were not unique, Severino graphically demonstrated the essential unity of animals and a truly comparative outlook. But his example gained few followers until Claude Perrault began his work at the Paris Academy two decades later.

By the mid-1640s, debates about the circulation extended across Paris. In 1647, the physician Abraham Du Prat, a close associate of Gassendi, published his French translation of Bartholin's revised anatomy text, using the 1645 edition, which included the letters of Walaeus. Bartholin still had not updated the chapter on the heart, but he had inserted an appendix that described the theories of Descartes and Harvey, endorsing the latter.[83]

Walaeus had criticized Gassendi's account of his observations of the pores in the septum, first published in 1630 and republished in 1635, 1639, and 1641. It was obvious, said Walaeus, that the so-called pores were in fact made by the dissector's probe as he sought to find them.[84] Gassendi dissected in Paris over the winter of 1640–41 but found no evidence to alter his opposition to the circulation. Du Prat wrote Gassendi in the fall of 1644 as he began to translate Bartholin's *Institutiones anatomicae*, asking if he might append Gassendi's account of the septum to Walaeus's letters. In his reply to Du Prat, Gassendi stood by his story, but admitted it may have been a singular event, although he believed it was not. His account did not appear in Du Prat's translation.[85]

Debates in Paris between 1645 and 1648 at the academies of Mersenne and Bourdelot also failed to change his mind. In the fall of 1647, Samuel Sorbière, another member of Mersenne's circle, addressed a book to Du Prat that detailed "the reasons that our common friend brings against the passage of the chyle by the lacteal veins, and against the circulation of the blood by the arteries." The common friend was Gassendi, and the book responded to Du Prat's translation.[86] Sorbière presented this work as the transcript of an interview with Gassendi; according to Sorbière, as much as Gassendi wished to be convinced by Harvey, he was still not quite able to accept the circulation. Gassendi repeated once more his story of the pores in the septum, remaining convinced that what he saw was not an artifact of the dissection itself, and described many other dissections he performed or witnessed on humans and animals, including the dissection with Peiresc of the hanged man in Aix. These experiments showed him, he said, that diastole and not systole was the active phase of the heartbeat and that only a small amount of blood entered the heart at each beat.[87]

Gassendi objected less to Harvey's experimental evidence than to what that evidence could demonstrate. It could not, he thought, demonstrate causes. When Sorbière expressed his admiration for Walaeus to Gassendi, "He [Gassendi] answered me smiling, 'is what Harvey said on that matter a demonstration?'" Sorbière, flustered, said it may not have been a demonstration, but because of the "clearness" ("netteté") of Harvey's thought and "the beautiful economy that he gives to the body," few, he thought, would require "the rigor of the term demonstration." Harvey cut up an eel's heart and showed that each portion still beat; but, Gassendi objected, this demonstration, however pleasing, did not show what caused the heart to beat, and Harvey offered no theory about this. Gassendi proposed an atomist solution, in which the motion of the minute parts of the body caused heat, which then propelled the heart, somewhat resembling the ideas of Descartes. But he did not outline a full theory.[88]

Gassendi likewise expressed skepticism about the contents of the lacteals. Although, with Peiresc, he had observed them to be full after a meal, he was not sure they were full of chyle. The milky substance did not resemble the contents of the stomach, and he thought it might be produced by a gland, perhaps the pancreas, and therefore was a product of the blood, not the precursor of the blood. Or, he speculated, it might be fat from the mesentery that served as a natural lubricant to keep the system from drying out during the extreme heat of digestion. Moreover, in his own experience of the dissection (or vivisection) of a horse, he did not witness the chyle to move through the lacteals. On the other hand, he disagreed with Aselli that the valves in the lacteals pointed toward the liver.[89]

Gassendi continued to view digestion as a Galenic coction produced by heat, and one of the difficulties he found in linking the lacteals to the circulation was that they concerned different faculties of the body. In Galenic physiology, digestion belonged to the category of the natural faculty, seated in the liver, which encompassed nutrition, growth, and reproduction. The action of the heart belonged to the vital faculty of which it was the seat. The idea that the chyle could bypass the liver and go directly to the heart made no sense in this system. Nonetheless, Gassendi expressed his willingness to be persuaded of the validity of the new theories. "Natural knowledge," he told Sorbière, "which had been neglected for centuries, begins in our time to be better cultivated," and he was optimistic that better proofs would be found.[90]

5. Pecquet, the Vacuum, and the Thoracic Duct

When Pecquet entered the circles around Mersenne and Bourdelot in the mid-1640s, the existence of the vacuum dominated their discussions. Galileo's disciple Evangelista Torricelli (1608–48) proposed an experiment in 1643 that investigated a phenomenon that Galileo had noted in his *Discorsi*: that a suction pump could raise water only to a certain height. Galileo argued that an internal force of the vacuum allowed the column of water to rise until, stretched to its limit, it collapsed of its own weight. Torricelli filled tubes of various lengths with mercury and inverted each of them over a bowl filled with the same fluid. In each instance the mercury fell in the tube to the same height, leaving an empty space at the top of the tube, which, Torricelli argued, had to be a vacuum, despite long-standing arguments that a vacuum could not exist in nature—arguments that Descartes had repeated. In a letter to·his friend Michelangelo Ricci in June 1644, Torricelli explained the behavior of the mercury as a result of the pressure of the air on the mercury in the bowl: the mercury rose in the tube only to the extent that it maintained equilibrium with the external air. Mersenne saw the tube, but not this experiment, during his trip to Italy in 1644–45.[91]

Mersenne was unable to find tubes strong enough to repeat the experiment, but Pecquet's old friends Blaise Pascal and Adrien Auzout and his new acquaintance Gilles-Personne de Roberval all worked on this problem. Pascal heard about Torricelli's experiments in the fall of 1646 from the royal engineer Pierre Petit (1598–1677), a member of Mersenne's circle. Petit, working with Blaise Pascal and his father, Étienne, succeeded in replicating Torricelli's experiment.[92] Blaise Pascal's *Expériences nouvelles touchant le vuide*, published the next year, described these and other experiments. In that year, Auzout offered further experiments, placing one Torricellian barometer within another, showing that in the absence of air pressure, the mercury did not ascend into the tube at all. In 1648, the famous Puy-de-Dôme experiment, performed by Pascal's brother-in-law Florin Périer, demonstrated that air pressure decreased with altitude. Périer found that the level of mercury in the tube was lower at the top of the mountain than at the bottom. At around the same time, Roberval tied the end of a carp's bladder and placed it inside the top of a Torricellian tube. When the mercury descended in the tube, the bladder inflated. Pecquet witnessed the latter experiment at one of Bourdelot's gatherings, where, as we saw in chapter 1, the vacuum dominated discussion in 1647–48.[93]

These experiments had a profound impact on Pecquet and his ideas about the relationship between the circulation of the blood and the lacteals. Upon entering the Paris Faculty as a medical student in 1647, Pecquet immediately embarked upon an ambitious research program of comparative anatomy outside the classroom. By his own account, he set to work over the next three years dissecting and vivisecting over one hundred animals of a variety of species—dogs, cattle, pigs, deer, geese, and horses—employing the techniques of Harvey and Walaeus. Completely reversing the medical goals of knowledge, Pecquet later complained that the knowledge he had gained from dissecting human cadavers was "mute and cold," and that he only gained "true knowledge" (*veram scientiam*) by cutting apart live animals.[94] Unlike the previous generation, Pecquet took it for granted that the knowledge he gained from dissection of living and dead animals was indeed *scientia*. Witnessed and assisted at various times by Adrien Auzout and Louis Gayant as well as recent Paris Faculty graduates Pierre Mercenne and Jacques Duval and faculty professor Jacques Mentel, Pecquet sought the secrets of digestion, the role of the lacteals, and further proof of the circulation of the blood.

Pecquet published his own "*anatomia nova*" in 1651, the *Experimenta nova anatomica*, which established the existence of a *cisterna* (or *receptaculum*) *chyli* near the kidneys, which was the true destination of the lacteal veins, not, as Aselli had assumed, the liver. He further traced a structure that looked something like a ladder that paralleled the spine, reaching from the *cisterna* to the thoracic duct, which, he argued, directed the chyle to the subclavian vein and thence to the heart, definitively linking the lacteals to the circulation. Pecquet praised Riolan, who had described how he traced the course of the lacteals in a living animal. Riolan had concluded that the milky veins led not only to the liver but also to the pancreas and even to the vena cava.[95] In a Grand Guignol of gore, Pecquet gleefully dismantled the ideas of his predecessors as he vividly described the dissections that led to his discovery, which he then illustrated in highly sanitized, even schematic, format. Riolan, like Fabrici, had instructed his students to dissect slowly and painstakingly; even when opening a live animal, the goal was to minimize blood loss. If the anatomist inadvertently touched a blood vessel with an instrument, Riolan advised cutting the vessel in two so that it retracted under the skin and did not bleed on the surface.[96] Pecquet, in contrast, began his researches:

> Having cloven asunder the Thorax of a Great Hound, I begun [*sic*] my view of the contained parts without delay; I pluckt out the Heart, having cut asunder those Vessels wherewith it was tied to the rest of the body; The abundance of blood, which immediately flow'd, did at present stop my prying sight; that being spent, I did wonder to see flowing in the Pipe of the *Vena cava* at its connection to the right Ventricle a milkie liquor, casting it self out by inter-mission.[97]

But in addition to the dramatic anatomical *historia*, Pecquet went on to find the mechanical causes of these actions. The second part of Pecquet's book, devoted to his proofs of the circulation of the blood, featured yet more experiments involving several living animals and much spouting blood with the aim of convincing even the most doubtful. Riolan, said Pecquet, had claimed that the blood of the vena porta could not and did not enter the vena cava. Such was his authority, Pecquet wrote, that he himself had hesitated to investigate further, "if my eyes, which I appointed strict Judges of the game, had not withstood it, reproaching my laziness that fail'd in the midst of the course." Pecquet disproved Riolan's notion by means of several ligation experiments, carefully noting that his evidence was only that provided by αυτοψία, *autopsia*.[98]

Pecquet asserted that the second half of his treatise would not only demonstrate the circulation of the blood but show its cause, proving that the purpose of the liver was simply to act as a strainer.[99] The cause of the circulation was not a simple attraction or suction, as some had claimed, and which he refuted with additional experimental evidence. Rather, the overarching cause was the mechanical force of air pressure, as recently demonstrated by experiments on the Torricellian vacuum. The blanket of air that surrounded the earth was, said Pecquet, composed of particles with a power of "Spontaneous dilatation [which I call *Elater*]."[100] In a lengthy digression, he then described several experiments that displayed this power of air. Roberval's experiment of the carp's bladder within the Torricellian tube played a central role here. The bladder's expansion in the vacuum of the tube proved Pecquet's contention of the air's infinite elasticity and capacity for expansion. Likewise, Pascal's 1648 Puy-de-Dôme experiment demonstrated that the spongy (or "woolly") blanket of air that surrounds the earth is less compressed and therefore exerts less pressure at higher elevations. Pecquet translated these results into an argument that air pressure and the natural elasticity of the vessels could themselves account for the motion of the heart and blood, and by extension that of the chyle, without reference to attractive faculties. While respiration and the diaphragm also contributed to

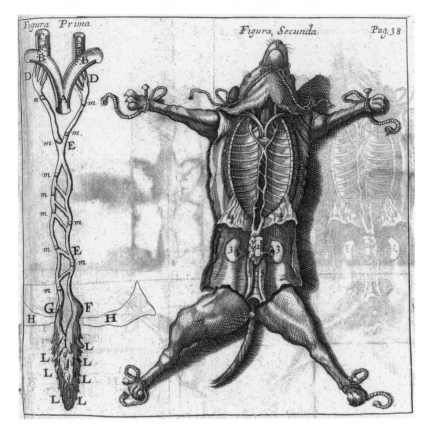

Figure 2.7 Demonstration of the thoracic duct. Pecquet, *Experimenta nova anatomica*, 1651. (Wellcome Library, London.)

these motions, they were not the sole causes. Pecquet concluded this second section—which was three times the length of the first, which had announced his discovery—with an account of the function of the liver, since he had "usurped [its] Glory of Blood-making which undeservedly it retain'd for so many ages."[101] The liver, he said, helped to heat the stomach and therefore aid the digestive process by the heat of the blood that flowed through it, but its main function was to strain the bile from the blood.

Pecquet's discovery, if not his causal explanations, electrified the younger generation of physician-anatomists, and a few of the older generation such as Mentel as well, though researchers for the next several years focused on the implications of the thoracic duct and not on the role of the vacuum in circulation. At least one of the many reprints of the *Experimenta nova anatomica* simply omitted its second part.[102] Among the flurry of works that followed the discovery of the thoracic duct was Thomas Bartholin's *Vasa lymphatica* in 1653. In his *De lacteis thoracicis* of 1652, he had expressed skepticism about Pecquet's discovery, but a year later he not only accepted it but extended it. Bartholin demonstrated that the lacteals were in fact part of a larger vascular system that he labeled the lymphatic system, which led to the thoracic duct. He dedicated this work to Riolan, but undercut the respect of the dedication by closing his work with a mock funeral ode to the dethroned liver.[103] Within a few years of Riolan's debates with Harvey in the late 1640s, the picture had utterly changed. Pecquet and Bartholin (and others across Europe) completely reinterpreted the role of the liver from blood maker to a much less exalted digestive role and discovered an entirely new vascular system. Once more, *autopsia*—detailed and repeated dissection of living and dead animals and of dead humans—had created new knowledge.

6. Salons, Academies, and Dissection in Paris, 1650–66

According to Samuel Sorbière, Gassendi became convinced of the truth of the circulation with Pecquet's demonstration of the thoracic duct. Sorbière recounted this change of heart in the posthumous edition of Gassendi's works he jointly edited with Montmor. Pecquet's use of living animals convinced Gassendi that the thoracic duct was real and acted as Pecquet described, which, following his arguments, meant that the circulation must operate as Harvey had surmised and that the lacteal veins indeed carried chyle, and carried it to the heart. In addition, Pecquet's mechanical explanation of the motion of the blood gave Gassendi the cause he required; in this case at least, the arguments about the vacuum had found their mark. With Sorbière, Gassendi witnessed Pecquet's experiments, most likely at Montmor's house.[104]

The playwright Molière's (1622–73) *Précieuses ridicules* premiered in Paris in the fall of 1659, poking fun at the superficially learned men and women who frequented the salons and academies of the city. Paris itself, said Molière, was a "cabinet of marvels," and each salon, each academy, offered something different.[105] Molière had attended Montmor's scientific salon, which met at his house on the rue Sainte-Avoye (now the rue du Temple). It began by 1654, tracing a line of descent from Mersenne via Le Pailleur, but the "Académie Montmor" only gained a formal organization three years later. Montmor was a longtime correspondent of Mersenne, who had dedicated his *Harmonicorum libri* to him, as well as a charter member of the Académie française, and an extremely wealthy man of the court and of

Paris. A fervent Cartesian, he had transcribed Descartes's *Principles of Philosophy* into Latin verse, a perfect example of the intersection of the new science and the Republic of Letters. He offered his country house to Descartes, who declined; but Gassendi accepted Montmor's offer and lived at the rue Sainte-Avoye from May 1653 until his death in October 1655. Like Bourdelot and Peiresc, Montmor moved easily among courtly and literary salons, antiquarians and book collectors, physicians and natural philosophers. His book collection was renowned. He had delivered a treatise (now lost), "De l'utilité des conférences," to the Académie française in 1635, and now in the 1650s he decided to follow his own prescription.[106] The traveler Balthazar de Monconys (1611–65) described meetings *chez* Montmor in 1656 that included Cureau de la Chambre, Pecquet, Jean Chapelain, Samuel Sorbière, and Pierre Petit. The vacuum was among the topics discussed.[107] Pascal, Roberval, Gui Patin, Claude Perrault, Carcavi, and others frequently dropped by as well. Also in attendance was the Cartesian mathematician Jacques Rohault (1620–75), who started his own *conférences* in 1657, focusing on experimental demonstrations.[108]

Gui Patin reported to Bartholin in July 1654 that Pecquet was back in Paris working on his revisions to *Experimenta nova anatomica* and still under the patronage of Bishop Fouquet.[109] But Pecquet moved into the household of the bishop's brother Nicolas as his personal physician by the next year, when Nicolas Fouquet purchased an estate at Saint-Mandé, near the royal castle of Vincennes. Fouquet soon rebuilt it into a showcase of splendid buildings and gardens, including an *orangerie* and a library of twenty-seven thousand volumes, second only to that of Mazarin.[110] Pecquet became his resident savant, directing Fouquet's salons and concocting remedies for Mme. Fouquet in his "*apothicairerie*," all the while still dissecting and engaging in controversy with natural philosophers across Europe. Probably for Pecquet's use, Fouquet purchased the library of faculty physician René Moreau. As Fouquet positioned himself to succeed Cardinal Mazarin as chief minister, he developed a network of clients in the arts and sciences who owed him allegiance. Apart from Pecquet, these included Carcavi, whom he appointed his librarian; Cureau de la Chambre; and Charles Perrault, Claude's younger brother.[111] Following the example set by Mazarin in the 1640s, Fouquet proposed to open his library to the public. But already he was working on an even more magnificent residence at Vaux-le-Vicomte, in the countryside southeast of Paris, where he moved in 1658.[112]

In 1657, Montmor charged Sorbière and Du Prat to draw up a set of rules for a more formal academy, seeking standards for civility and upholding the ideals of *honnêtêté*. The group, they realized, needed to learn how to listen, how to disagree without acrimony, how to separate intellectual debate from political divisions. Many of the nine rules dealt with procedural issues in an effort to ensure that all opinions were heard in a dignified and organized manner. But a few dealt with the more general aims of the academy, which differentiated it from a salon: not the "vain exercise of the mind on useless subtleties, but . . . clearer knowledge of the works of God, and the improvement of the conveniences of life, in the Arts and Sciences which seek to establish them." Correspondence with other scholars was encouraged, but unlike Rohault, whose wildly popular lectures attracted hundreds, Montmor's academy was to be a closed body, with new members admitted only on a vote of two-thirds of the current members.[113]

The membership solidified to include many of Mersenne's circle and others from Bourdelot's. Sorbière, Du Prat, and Auzout were members, and Pecquet frequently gave demonstrations. Bourdelot and the courtier Melchisédech Thévenot (1620?–92) also attended. Despite the rules, civility was not easily imposed, and the fierce arguments that had characterized the earlier meetings of the group persisted. Although science was in theory politically neutral, divisions between Cartesians and Gassendists and theorists versus experimentalists divided the company. The mathematician Roberval was ejected after insulting Montmor in 1658.[114]

No minutes exist, so evidence of what occurred at these gatherings is spotty. Although Sorbière claimed that he was charged to "dresser les Mémoires" of the academy, he published only a few of his own "Discours prononcé dans une Assemblée chez monsieur de Montmor" in a volume of collected letters and essays in 1660.[115] Henry Oldenburg (1619–77), future secretary of the Royal Society of London, resided in Paris from early 1659 until mid-1660 and regularly attended Montmor's academy. But his correspondence mentions it infrequently, far less than his encounters with chymists and alchemists. In June 1659 he noted that the topic of discussion *chez* Montmor was spontaneous generation, and Du Prat told Oldenburg about a theory that the nerves functioned for nutrition as well as for sensation and movement. A few months later, the topic was winds, and Pecquet recounted a person who, it seems, belched on command when touched.[116] In November, Oldenburg told Samuel Hartlib that he was going to translate into French the Dutch anatomist Lodewijk de Bils's work on preserving cadavers for presentation at Montmor's.[117]

The journal of the young Dutch natural philosopher Christiaan Huygens (1629–95) from his visit to Paris in the winter of 1660–61 provides a few more glimpses of the academy's anatomical activities. Already well known as a savant, Huygens could, like the *précieuses*, spend his time in Paris dining with various other savants (with experimenting sometimes part of the after-dinner activity), viewing cabinets, libraries, and art collections, and attending salons, including Montmor's and at least one of the Saturday gatherings of the novelist Madeleine de Scudéry.[118] On 7 December 1660, he saw at Montmor's a skeleton in which one could see "all the nerves, veins, arteries, the heart, the eyes, made of wire covered in silk." At the same meeting, Pecquet described his theory "that the food of the body is distributed by the nerves," which appears to be a version of the theory Du Prat had described to Oldenburg a year earlier. On 4 January 1661, Bourdelot spoke about the gout, his views opposed by Pecquet.[119] The following week, "Bourdelot harangua encore de la goutte," and he spoke yet again on the topic—"and very well," said Huygens—a month later.[120] On 8 March, Pecquet spoke on the generation of the chick in the egg and was booed ("fut sifflé"), reflecting the strong opinions that persisted even after the Roberval fiasco.[121]

A series of letters Sorbière wrote to Cardinal Mazarin in February 1659 provides by far the most detailed account of anatomical activities at Montmor's gatherings. The letters describe a session led by Pecquet on the circulation and the thoracic duct. That Sorbière felt this demonstration was worth no fewer than seven letters to Mazarin indicates the position Pecquet held in the world of Parisian science, as well as Sorbière's own ambitions for the Montmor academy. He wrote in fact eight letters, the fifth being a general disquisition on the public value of medical knowledge, a transparent attempt to gain Mazarin's patronage for the academy.[122] Pecquet demonstrated before the academy the passage of the chyle, the circulation of the blood, and the lymphatic system by means of dissecting three live animals, although Sorbière does not tell us what kind of animals. Sorbière detailed how Pecquet came to his great discovery, and then described his specific performance before the academy. In the final letter in the series, Sorbière praised Pecquet, who "only a student, saw better than the wise old men his masters," and noted that the eminence of Pecquet's supporters spoke well for the importance of his ideas.[123]

But the vagaries of patronage became startlingly clear to Pecquet when Mazarin died in March 1661 and Nicolas Fouquet lost the struggle to succeed him to Jean-Baptiste Colbert. Colbert had been plotting against Fouquet for some time, but Fouquet's conspicuous consumption and patronage of the arts, which rivaled the king's, did not help his case. He was arrested and charged with embezzlement, and his estates, including the dazzling Vaux-le-Vicomte, confiscated.[124] Pecquet's loyalty to Fouquet was so profound that he, along with Fouquet's valet, stayed in the Bastille with their employer for over three years during Fouquet's trial, from September 1661 to February

1665. Ultimately, Fouquet was condemned to life in prison and exiled to the fortress of Pignerol in the Piedmont, where he remained until his death in 1680. Pecquet was not allowed to accompany him, and after a year of exile in Dieppe, he was allowed to return to Paris in 1666.[125] The scramble for patronage that followed the fall of Fouquet led several of Montmor's circle, including Charles and Claude Perrault, to shift their loyalties to Colbert.

Among the five Perrault brothers, the youngest, Charles—fifteen years younger than Claude—was the most ambitious. Like Cureau de la Chambre, Charles Perrault wrote poetry, though mainly in French, and his writing attracted the attention first of Nicolas Fouquet and then, after Fouquet's fall in 1661, of Jean-Baptiste Colbert. By 1664 Charles Perrault was Colbert's right-hand man—he eventually seconded him in the powerful position of *contrôleur des bâtiments*—and he played a central role in the establishment of what became the Academy of Sciences. Claude's biographer Antoine Picon refers to Charles as an "arriviste sans scrupule," and I would not call him far off the mark.

Oded Rabinovitch has recently shown the close relationships and interactions among the Perrault brothers, all of whom achieved considerable success. The advancement of the family and not merely of an individual was key to social mobility in this era, and success could be measured in generations rather than years, although this process greatly accelerated in the seventeenth century. Each member of the family played a role: the Perraults' father, Pierre, born in Tours, was an *avocat* for the Parlement of Paris, as was the eldest son, Jean. Claude was a physician, Nicolas a priest and theologian at the Sorbonne, Pierre a financier. Charles also trained as an *avocat* but then worked as Pierre's assistant. Pierre fell from his position in 1664 following the purge of Fouquet's followers. But Charles had already caught the eye of Colbert with his poetry, and soon attached himself to the rapidly rising Colbert. Not personal generosity but the ethos of the bourgeois family led Charles to take his brothers along when he rose in status. Historians have looked commonly at inheritance from fathers to sons; but what demographers call the horizontal relationships among siblings (as well as among in-laws and cousins) was equally important.[126]

In his massive 1770 *Histoire de l'anatomie*, Antoine Portal claimed that Paris was "devoid of anatomists" ("dépourvu d'Anatomistes") after the death of Riolan in 1657 until Duverney emerged on the scene in the 1670s.[127] Although dissection continued to flourish in Pecquet's absence, Frenchmen did little of it. In December 1663, Auzout asked Huygens to bring the surgeon Bruynsteen and his instruments to Montmor's to remove the spleens "from a dog or two" and show the salivary glands in the head of a cow or calf.[128] By the summer of 1664 the academy had degenerated into acrimony once more with frequent disputes between those who wished to experiment and those who preferred debate, and in June 1664 Huygens reported that "the Montmor Academy ha[s] ended forever."[129] But, he added, there are so many well-intentioned men, another academy will be born from this one.

As the Montmor group began to fall apart, one of its members, the courtier and scholar Thévenot, began to invite natural philosophers to his home at Issy, south of Paris, to look through telescopes and perform experiments. Thévenot had been one of Montmor's circle who wanted more experimenting and less talk. Like Montmor a book collector, he eventually replaced Pierre de Carcavi, another member of the circle, as the royal librarian. Carcavi had been Fouquet's librarian and managed to enter Colbert's service after Fouquet's fall. When the young Danish physician Niels Stensen (1638–86) came to Paris in the fall of 1664, Thévenot supported him in return for experimental demonstrations to his academy, which included many of Montmor's regulars, such as Auzout, Petit, and Frenicle de Bessy. For the better part of a year, Stensen lived in Thévenot's house in Paris, with occasional excursions to Issy, and participated in the fortnightly meetings of his circle. Already living at Thévenot's was a young Dutch medical student, Jan Swammerdam (1637–80), who had met Stensen when they were students in Leiden. Another Leiden-educated anatomist, Regnier de Graaf (1641–73), joined Stensen and Swammerdam in Paris in the summer of 1665.[130]

Stensen, or Steno, as he became better known from the Latinized version of his name, had studied with Thomas Bartholin in Copenhagen and was therefore well informed about current research on the circulation, the lymphatic system, and the thoracic duct. In 1659 Stensen left Copenhagen for the Netherlands. Under the guidance of Gerard Blaes in Amsterdam, he discovered the duct to the parotid glands in a dissection of a sheep's head in April 1660. By the time he reached Paris in November 1664, he had received a medical degree from the University of Leiden and had published two books, *Observationes anatomicae* (1662) on his work on the glands of the head, and *De musculis et glandulis observationum specimen* (1664) on muscles, in which he asserted that the heart was indeed no more than a muscle and not the source of animal heat or vital spirits, or the site of blood making.[131] The latter work was reviewed in the new *Journal des sçavans* in March 1665. The reviewer did not agree with Stensen's conclusions: they "overturn all that is most certain [*constant*] in Medicine," he wrote, adding that the heart was not composed of muscle fibers like those in the rest of the body but of a "particular substance." This idea of particularity—whether of body parts, blood, diseases, or individuals—derived from Galen, as we shall see further below. Nonetheless, the reviewer wrote, Stensen's great skill in dissection revealed so many new and unexpected things that "one must suspend one's judgment," even if the anatomist did not entirely prove what he claimed. The reviewer noted that this "sçavant Danois" was presently in Paris, where he dissected daily in the presence of various learned individuals, as well as at the medical faculty.[132]

Swammerdam, an equally talented dissector, had arrived in Paris in September 1664. For his Leiden thesis on respiration, he had dissected a number of dead and living dogs. In Paris he repeated these dissections before Thévenot's academy and in other venues. Swammerdam also did various injection experiments, much like those concurrently performed at the Royal Society.[133]

Early in 1665, Stensen dissected the human brain, a project fraught with technical, philosophical, and theological difficulties. Du Laurens in the 1590s had called the brain "the true seat of the soul," and Riolan took this as a matter of fact. Stensen agreed: "It is very certain that [the brain] is the principal organ of our soul." His detailed account of this dissection, published in Paris four years later, critiqued his contemporaries while admitting that even after numerous dissections he had not entirely penetrated the secrets of this organ. Like other anatomists, he too emphasized "les desseins de son grand Architecte" as a reason to study the body, but the soul and its location earned no further mention.[134]

The printer, Robert de Ninville, dedicated *Discours de Monsieur Stenon, sur L'anatomie du cerveau* to Cureau de la Chambre, but Stensen's work bore little relation to Cureau de la Chambre's speculations on the mind. Taking Descartes in the recently published *Traité de l'homme* at his word—that he described only a mechanical model and not an actual human—Stensen found much to praise. "No one else has explained mechanically all the actions of man and particularly those of the brain. . . . Monsieur des Cartes speaks to us only of a machine, [but] nonetheless makes us see the insufficiency of what others teach us." But the value of the *Traité de l'homme* remained on the level of inspiration, for anatomists such as Sylvius had already recognized its many errors. Chief among these was the location and role of the pineal gland. By means of painstaking anatomical detail, Stensen demolished Cartesian claims and noted that the gland occurred in all animals, not only in humans. He had found it in a horse in 1663.[135]

It is not clear where Thévenot obtained the human body Stensen opened, nor do we know who did the several drawings of the brain that accompanied his text. The review in the single issue of the *Journal des sçavans* published in 1670 took an ironic tone: "The Author of the Discourse avows ingeniously that after all the dissections that he has done, he still knows nothing about the structure of the Brain."[136]

Although Thévenot continued to support Stensen and Swammerdam after they left Paris, he could not afford to establish a fully fledged experimental enterprise, and his academy was in some ways subsumed by the Paris Academy of Sciences. Other sites for anatomy persisted: for example, the physician Jean-Baptiste Denis (1643–1704), a graduate of Montpellier, held regular public *conférences* at his house from about 1664 onward.[137] But when Regnier de Graaf came to Paris in the summer of 1665, he headed for the house of the indefatigable Abbé Bourdelot.

Bourdelot had returned to France in 1653, and through the influence of Queen Christina, he was granted the living of the abbey of Massay near Berry in central France, thereby gaining the sobriquet of *"Abbé."* In the 1650s he practiced medicine, attended Montmor's gatherings, and gained some well-known patients, including Marie du Rabutin-Chantal, Marquise de Sevigné (1626–96), who was also a close friend of Pecquet and frequented Fouquet's salon. When the Grand Condé returned from exile in 1659, Bourdelot resumed his role as the prince's personal physician and the prince resumed his role as patron. By 1664, Bourdelot had regenerated his academy, this time meeting once a week (Monday or Tuesday; the sources differ) in his own rooms in the rue de Tournon on the Left Bank near Saint-Sulpice.

Many of the same savants who frequented Montmor and Thévenot visited Bourdelot's academy, including Auzout, Petit, Roberval, Claude Perrault, and Huygens. Bourdelot's interests remained broad, and his meetings, like those of Molière's *femmes savantes*, tended to "mêler le beau langage et les hautes sciences." As he had done in the 1640s, he began each gathering with music, and enjoined discussion around the dinner table. Given the arguments that had disrupted Montmor's meetings, this was a wise policy.[138] At about the same time, Henri Justel (1620–93), a Protestant royal administrator and book collector, began his weekly *conférences* at his home on rue des Fossés Monsieur le Prince—quite close to Bourdelot. Many of the same people visited Justel, but his discussions included politics and literature. Justel was especially known for his correspondence network, and foreigners, especially the English, often were present. Bourdelot himself was frequently seen there. Justel held his meetings until 1680, when increasing pressure on Protestants in France compelled him to emigrate from France to England, among the first of the Huguenot migration. There he later became royal librarian.[139]

In April 1664, Huygens brought the surgeon Bruynsteen to Bourdelot's to dissect one of three dogs whose spleens had been excised a few months earlier, probably at Montmor's the previous winter.[140] The dogs had survived perfectly well without their spleens, and the dissection showed no evident abnormality. When Regnier de Graaf arrived the following year, he had already published his thesis on pancreatic juice, notable for the chemical theories of his mentor, the Leiden physician Sylvius, but also for the extraordinary anatomical skill required to extract the pancreatic juice. In Paris, de Graaf extended his research to demonstrate that the pancreas was itself a gland and not, as Thomas Bartholin had argued, an excretory organ for the spleen. He removed the spleen from yet another dog and, two months later, was able to collect pancreatic juice from the animal. De Graaf published these discoveries in a revised French edition of his thesis, published in 1666 and dedicated to Jean Chapelain, another of his patrons, a poet, courtier, and confidant of Colbert. Back in Paris a year later, he performed additional experiments at Bourdelot's, including excising the pancreas as well as the spleen from yet another dog. Despite the best efforts of Bourdelot's servants, the dog soon died.[141] De Graaf also dissected at the homes of Montmor and Denis, and he dedicated his 1668 collected works to Montmor.[142]

7. The Anatomical Origins of the Paris Academy

Following the fall of Fouquet, a patronage vacuum existed in French society, which Louis and Colbert endeavored to fill.[143] Among Colbert's projects was the Paris Academy of Sciences, founded in 1666.[144] Colbert, following the example of Richelieu, wished to exert royal control over intellectual life and to glorify the crown. Jean Chapelain, a founding member of the Académie française and head of the "Petite Académie" of inscriptions and belles-lettres, had introduced Charles Perrault to Colbert by early in 1663. The Petite Académie, founded in that year, concerned itself with every aspect of the commemoration of the king and the celebration of his *gloire*, particularly medals and inscriptions. Colbert soon named Charles secretary of the Petite Académie, and when in the following year he became *contrôleur des bâtiments*, or supervisor of the king's buildings, he named Charles as his assistant.[145] Supported by Charles, Colbert proposed the development of a general academy of arts and sciences, a humanist project that looked back to the Dupuys rather than to the emergent Royal Society for inspiration. The royal engraver Sébastien Leclerc (1637–1714) later presented the king with an engraving representing this ideal "Académie des sciences et des beaux-arts." The engraving was modeled on that monument of humanist iconography Raphael's 1510 *School of Athens*. But Raphael's philosophers talked while Leclerc's acted. His academy was filled with globes, skeletons, maps,

Figure 2.8 Sébastien Leclerc, *L'Académie des sciences et des beaux-arts*, 1698.
(© The Metropolitan Museum of Art / Art Resource, New York.)

instruments, and experiments. It looked forward to an era of telescopes and dissections rather than back to Aristotle and Plato. Where Raphael had depicted a few books whose texts were being hotly disputed, Leclerc showed an entire library, although it was on one side and not at the center.[146]

While this plan for a general academy fell through, the Academy of Sciences that came into being by the end of 1666 retained elements of that original plan, not least in its membership. It was both a body of technical advisers to the crown and a much broader intellectual assembly of savants. The medal that celebrated the foundation of the academy displayed "Minerva surrounded by symbols of astronomy, anatomy, and chemistry," and these three sciences dominated the early academy.[147]

Early in 1663, Samuel Sorbière addressed the Montmor Academy on the topic of the support of experimental science; "Only Kings and wealthy sovereigns," he said, had the resources to support "une Académie physique." He sent a copy of his address to Colbert.[148] Around the same time, Christiaan Huygens publicized a plan for an academy of arts and sciences with a utilitarian and empirical emphasis. Its members would experiment, observe the heavens and the earth with telescopes and microscopes, "learn more particularly the construction and the movements of the human body by means of chemistry, anatomy, and medicine," invent new machines, and learn the secrets of trades, arts, and artisans. Reflecting his distaste for the verbal battles that plagued the Montmor academy, Huygens's document forbade discussion of religion or politics and limited mention of metaphysics or moral philosophy.[149] Three years later, in a *mémoire* to Colbert, Huygens described the "principal occupation" of an academy as the creation of a natural history according to the designs of Francis Bacon: "Such a history consists of experiments and observations and it is the unique means to come to the knowledge of causes": only by means of the historical method could one reach philosophical knowledge. The dissection of animals, he added, "[is] certainly necessary to this design."[150] He was among the first to be named to the new Paris Academy of Sciences in May 1666; in the next month, Colbert chose Auzout, Roberval, Carcavi, the *abbé* Jean Picard (1620–82, astronomer and follower of Gassendi), Frenicle de Bessy, and the mathematician Jacques Buot (1623?–78). These men constituted the "mathématique" section (according to Charles Perrault's plan).[151] Charles Perrault certainly played a role in these appointments, as did Chapelain, who advised Colbert on who should receive royal pensions. Like Chapelain, these men took part in the salon culture of the *honnêtes hommes*.

The four anatomists whom Colbert appointed to the Paris Academy of Sciences between October and December 1666 had made few appearances among anatomical activities of the mid-1660s. Claude Perrault occasionally crops up, but not as an anatomist; Cureau de la Chambre, still a client of Séguier, was the subject of dedications but little else. Gayant did not appear at all, and Pecquet was in the Bastille for most of this period. Thévenot listed the four candidates in "physique" in October 1666 as Perrault, Cureau de la Chambre, Gayant, and the botanist/chemist Samuel Duclos (1598–1685), and Henri Justel named the same four in a letter to Henry Oldenburg dated 5 October, although he singled out "Gayen pour l'Anatomie." Neither of them mentioned Pecquet, who was named later along with the botanist Nicolas Marchant (d. 1678) and the apothecary and chemist Claude Bourdelin (1621–99). Chapelain claimed that Pecquet was only chosen after Stensen declined.[152] Of these seven, five were physicians: Cureau de la Chambre, Perrault, Duclos, Pecquet, and Marchant, but at this time only Pecquet and possibly Perrault had recently practiced medicine. Sometime in 1667 or 1668 the *abbé* Edmé Mariotte joined the "physiciens."[153]

Of the four anatomists, only Pecquet and Gayant were practicing anatomists, and only Pecquet was known as an avatar of the new philosophy. In fact, at the time of their appointments, Pecquet was the only practicing natural philosopher of the four, and the only one whose fame reached outside France. Claude Perrault fulfilled his duties as a member of the Paris faculty and attended the academies of Montmor and Bourdelot. But his appointment as a charter member of the Paris Academy owed more to the assiduous career-building efforts of his brother than to his own merits. Cureau de la Chambre's talents were always more literary than medical. Most if not all of these men, like the *mathématiques*, had participated in the academies and salons of the 1640s and 1650s, and, as we shall see, the academy followed the relatively egalitarian and discursive style of the salons. But at the same time, it was a particularly extravagant product of the increasingly elaborate and top-down system of patronage that dominated intellectual life in Louis XIV's Paris.

Although Pecquet, aged forty-four, was hardly "elderly," as Jacques Roger suggests, his appointment nonetheless demonstrates that in this case scientific fame trumped political liability (and possibly social liability: Pecquet was an alcoholic, an addiction that killed him not so many years later). Gayant, as a supporter of Pecquet and a highly skilled and well-known anatomist, presumably also gained admission for his skill. As Roger Hahn points out, Cureau de la Chambre would have fit equally well in the original plan of a "General Academy."[154] But also notable were those who were omitted: Paul Emmerez was as well known a surgeon as Gayant; Jacques Mentel had performed far more dissections than Perrault; another faculty physician, Claude Tardy (1601–71), was a Thévenot regular and had published several procirculation works in the 1660s; Guy-Crescent Fagon (1638–1718), a rising star of the Paris Faculty, had defended the first procirculation thesis there just a few years earlier; Jean-Baptiste Denis was a royal physician and well known for his anatomical interests. Talent was not the only criterion for admission; Chapelain and Charles Perrault chose men they knew. That the academy met in the royal library enforced its decidedly humanist cast. Yet within a decade, the Paris Academy would become a leader in the new science, not least in its applications to the human and animal body, and the leader of this effort, somewhat improbably, was Claude Perrault. The next chapter looks at the experimental program he established.

THREE

The Animal Projects of
the Paris Academy of Sciences

1. The Academy Opens; Perrault's Plans for the Anatomy Project

The foundation of the Paris Academy of Sciences in 1666 enthroned a
particularly Parisian brand of anatomical research. At its foundation, the
academy divided itself into two sections: "mathématique," which included
astronomy, algebra, and geometry, met on Wednesdays; "physique," which
included mechanics, chemistry, botany, and anatomy, met on Saturdays.[1]
But the entire membership—only fifteen at the outset, soon enlarged to
twenty-one—was encouraged to attend both sections. Not only did physicians, botanists, and anatomists participate in astronomical observations
and discussions, but astronomers and mathematicians participated in dissections and botanical and chemical analysis. The disciplinary spectrum of
the early academy was a continuum, and members freely crossed disciplinary lines. Moreover, most of its members came from independent salons and
academies rather than the university faculties, and the eclecticism of these
groups carried over to the academy.

The academy occupied several rooms in a large house on the rue Vivienne on the Right Bank, not far from Colbert's own *hôtel*. (See fig. 1.1.)
The earliest organizational meetings in 1666 had in fact taken place at Colbert's home.[2] The travel writer Germain Brice commented in 1685 that "the
house . . . hath but an ordinary appearance; and one would hardly believe
from the outside, that it contains so many curious things."[3] At the same
time the academy moved into number 8 rue Vivienne, the Bibliothèque
du roi, the Royal Library, also moved there from its earlier quarters at the
convent of the Cordeliers, its juxtaposition a seeming affirmation of Colbert's and Charles Perrault's earlier vision of a humanist academy of arts and
sciences. Indeed, this juxtaposition was not accidental; a *mémoire* from 1667
noted that the king provided the academy "all the assistance they could

wish, beginning with establishing them at the King's Library, where nothing is wanting for them with regard to books and manuscripts." Although the *mémoire* went on to describe the building of a laboratory in the same place, Colbert obviously believed that a good library was essential to the academy's operations.[4] Colbert's librarian, Pierre de Carcavi, one of the stalwarts of Mersenne's and then Montmor's circle and previously librarian to Nicolas Fouquet, became the king's librarian in 1666.[5] He now established his residence at the rue Vivienne, overseeing a library that contained not only books and manuscripts but engravings, medals, and antiquities. Brice referred to it as the "Cabinet du Roi," and by the time he wrote in the 1680s, it included many natural history specimens.[6] The relationship between the library and the academy remained close: the librarians shared living space in the rue Vivienne with *pensionnaires* such as Huygens, and the king's librarian in the 1690s, the *abbé* Jean-Paul Bignon, was at the same time president of the academy.

The academy appears to have occupied five small rooms on the ground floor of number 8; its meetings, according to Fontenelle, were held in "a little packed room" in the house.[7] From August 1666, Huygens, the first foreign *pensionnaire* of the academy, lived in the same building, in four rooms upstairs. Colbert purchased number 10 next door in 1669, which allowed more space to both the library and the academy, and the Imprimerie du roi, the royal printing press, moved to a building between numbers 10 and 12 by 1667 along with the royal engravers.[8] As we shall see, publication and the paper technologies of printing and engraving were central to the academy's anatomy project.

Although it was a royal institution, the academy set its agenda with minimal oversight from Colbert, who exercised little influence on its activities.[9] But he had already set parameters that in many ways determined the kinds of work the academy could undertake. Perhaps the most important of these was the collective ideal: from the outset, the academy portrayed itself as a collective body ("la compagnie," which suggests its corporate status, although it was not to attain this status legally until 1713), which arrived at decisions by consensus and collaborated on all its projects. Any publication would be in the name of the academy rather than of any individual. The collective ideal was intended to minimize conflict and the individual pursuit of fame in favor of the greater benefit of the kingdom and, of course, the greater prestige of the king. The notion of a company implied sociability and trust among its members. Establishing constraints on both individualism and disputation, the collective ideal promoted the civility that had been so sorely lacking in Montmor's academy. But not all members were equal: scientific stars such as Huygens and later the astronomer Giovanni Domenico Cassini (1625–1712) who had been recruited to the academy to bring it prestige published in their own names, even in works that the academy sponsored. While the collective ideal recognized the differing knowledge and expertise of the academy's members, it glossed over their differing stations and levels of compensation. Compensation varied widely: Huygens was granted 6,000 livres per year, Cassini 9,000, while Pecquet and Gayant each received 1,200.[10]

The ideal of the "compagnie" also fostered an in-group mentality that Colbert encouraged, which limited the sharing of ideas with outside bodies. In exchange for secure pensions and institutional support, the academicians gave up some of the free commerce in ideas that characterized the new science of the seventeenth century and that had been a feature of the private academies, at times, as in Montmor's academy, to their detriment.[11] The academy had no written rules or charter, its meetings and its minutes were secret, and publication was at first limited. When publication did occur, it was supposed to be in the name of the academy and not of any individual. The collective ideal led to tension from the outset and was never entirely followed; by the 1680s, it had collapsed, and the "renouvellement" of the academy in 1699 established statutes that guided the academy's operation as well as rules for the preservation of its "registres," "traités," and other papers. It became more of a corporation and less either a salon or an extension of the court.[12]

Claude Perrault announced a project of "anatomical observation" at an early meeting of the academy in 1667.[13] He took charge immediately of the section on "physique," a topic deemed to be easier and less abstract than the mathematical sciences. Perrault's sudden assumption of center stage at the age of fifty-three, after over twenty-five years of self-effacement as an inoffensive member of the Paris Faculty, presents an interesting conundrum. Although all of the Perrault brothers had done well, Charles, the youngest, had done the best. By the 1660s, as we have seen, he was in the inner circles of government, and following the ethics of the bourgeois family, he dragged his brothers along with him, particularly Claude. Although Claude could not have been completely idle during the twenty-five years since he received his MD, he had not burst upon the scientific scene like Pecquet.[14]

Did Charles push Claude to take a leading role, or had he been waiting for this chance? He was already becoming known as an architect: while his role in the colonnade of the Louvre continues to be debated, it is certain that he designed the Paris Observatory. Like the colonnade, its construction began in 1667. Charles played a major role in bringing Claude's architectural

Figure 3.1 Claude Perrault, pointing to the colonnade of the Louvre, ca. 1670.
(Bibliothèque nationale de France. Photo credit: Snark / Art Resource, New York.)

plans to the attention of Colbert, and given his role in the "Petite Académie" and the abortive plans for a general academy, it is likely that Charles accelerated the advance of Claude's career.[15]

Claude Perrault eagerly seized the opportunity. On 15 January 1667, he outlined to the academy his plan of work for *Physique*, which would consist of two main topics, "the most useful and the most curious in natural philosophy": anatomy and botany.[16] Several different versions of this address exist. The sources for the history of the early academy vary considerably, with multiple paper trails that reflect a self-conscious recognition of the value of keeping records as well as the multiple constituencies that wished to appropriate the academy's message. The academy's minutes, the *procès-verbaux*, although the closest to the events in time, were redacted and recopied by the academy's secretaries (Jean-Baptiste Du Hamel and Jean Gallois) from eyewitness accounts. Annual *pochettes de séance*, presumably also compiled by the secretary, included copies of papers delivered at the academy's meetings. But the *pochettes* were often incomplete (and entirely missing for some years) and did not coincide exactly with the *procès-verbaux*.[17] Two histories of the academy, by its first secretary Du Hamel in 1698 and later secretary Fontenelle in 1733, drew on the *procès-verbaux* but differ in their emphases and attention to detail, although Fontenelle claimed that he drew his account of the academy's early years from Du Hamel.[18] In the case of Perrault's address, the *procès-verbaux* and the much longer manuscript in the *pochette* for 1667 coincide on the main points, although the language is not identical. Du Hamel's history offered only a brief summary, apparently drawn from the *procès-verbaux*. But Fontenelle's account in the 1733 *Histoire de l'Académie royale des sciences* differs considerably from these, making Perrault appear to be much more of an empiricist.[19]

According to the *procès-verbaux*, Perrault admitted that the truths unveiled by anatomy may not be the most "éclatante" but they are the most solid, and the foundation of all the well-being of humans.[20] The dissection of the human body and those of other animals, he continued, ought to be considered the basis of all knowledge of this subject. In the manuscript in the *pochette* he added, "One cannot doubt that Anatomy is one of the principal parts of medicine and the most necessary for the cure of indisposed bodies. Just as one cannot repair a building if one does not know its construction, so one cannot remedy the ills of the body without knowing how it is made." The architectural metaphor, which Perrault often employed, implied a static, morphological anatomy, but his anatomical program was not merely structural. Perrault criticized the ancients for looking only at the most unusual parts rather than the body as a whole: "We have the good fortune in this century to have discovered many important things which were unknown to the ancients," and there were many things yet to be found.[21]

The circulation of the blood, the principle that gives and sustains life, must form the basis of research on the nature of life. Under the rubric of "anatomy," Perrault outlined a research agenda for the academy consisting of twelve topics, beginning with the circulation and the structure of the heart, followed by a list of the most significant issues it had given rise to in the previous twenty years. These were both structural and functional, and included sanguification, or how blood was made. Although Pecquet had shown the passage of the chyle through the thoracic duct, he did not show how blood came to be made from that chyle, nor the origins of various other fluids and secretions of the body, including milk, urine, and glandular secretions. The functioning of the glandular and lymphatic system, and the use of the spleen and of bile, also fell into this general category. Other topics included the head and the brain, respiration, and generation. Perrault distinguished two varieties of anatomy, one dealing with form, "things of fact," and the other with function, or "things of law."[22]

According to Fontenelle, Perrault valued observation above reasoning, arguing that observation must always be prior to reason. He did not mention Descartes, whose *Le monde* had finally appeared only a few years earlier, and indeed he outlined an anti-Cartesian project and methodology. Dissection ("of human bodies and those of other animals") and contemplation together formed the basis for truth. Fontenelle's Perrault characterizes the anatomist as one who employs "equally his eyes and his reason, but nonetheless keeping some advantage for the eyes over reason, for it is not worthwhile either to torment himself too much in looking for mechanical parts and dispositions whose uselessness can be shown by reason . . . or to neglect to assure oneself of things, as much as possible, by all the experiments that art can imagine."[23] Reason could disprove, but not prove. Experimental evidence, not reason alone, provided the key to truth.

But these words do not appear in either the *procès-verbaux* or the *pochette de séance*, which portray a much more cautious and conservative approach. In these documents Perrault says almost exactly the opposite: after discussing the importance of dissection, he added,

> It is nonetheless true to say that the eyes are not the only guides in this research, and Reason also furnishes light to lead one which does not merely serve to clarify the use of the parts that one has found, but even the necessity or the probability of those that one hopes to discover and that must happen often enough that outside its [i.e., reason's] counsel one works uselessly for organs and passages that Reason has judged to be not at all necessary.[24]

Up to this time, Perrault continued, there was no definitive anatomical evidence that there are vessels that conduct anything but blood to the breasts. But reason tells us that there is most likely a canal that conducts chyle, which is probably what composes milk, to the breasts. Reason therefore tells us what to look for so we do not search fruitlessly like Democritus (and Severino?), who, said Perrault, lost a lot of time. "Il est necessaire de joindre toujours les observations avec le Raisonnement" ("It is always necessary to join observation with reasoning"), said Perrault, and then showed himself a true son of the Paris Faculty of Medicine: "Thus will one exercise as much prudence and modesty so as not to run carelessly after these new observations that have had the strength [force] and candor [sincerité] to undo the ancient warnings [préventions]." Pecquet was presumably in the room when Perrault declared that the chyle may indeed flow to the liver, and that Bartholin's discovery of the lymphatic system certainly complicated the picture but did not convince him that all chyle goes through the thoracic duct.[25] But further experiments would reveal the truth.

Nonetheless, as we shall see in chapter 5, Perrault followed a mechanical program that relied on the Galilean mechanics Pecquet had employed. Like Pecquet, he sought the physical principles behind the actions of the body. Cartesian vortical systems had no place in this program. Even though Perrault sounded rather Cartesian in his emphasis on reason, he in fact followed a resolutely inductive and particularist epistemology rather than Cartesian deduction.[26] The most important activity was the collection of facts; as Claire Salomon-Bayet has commented, Perrault and the academy "feared the escape of any curious or useful truth." The academy functioned above all as a cabinet of facts.[27]

Rather than seeking causes, Perrault viewed the animal project as a natural history based on observation and experiments and focused on dissection and vivisection. Although Perrault also noted the importance of chemical analysis of substances such as the blood, this in fact played a very small role in the anatomy program. Duclos and later Claude Bourdelin, who led the chemical operations, focused their analysis on plants and mineral waters. The anatomical program aimed to reproduce, and ideally extend, the new physiological experiments that anatomists across Europe pursued in the wake of Harvey's discovery of the circulation, Aselli's discovery of the lacteal veins, and Pecquet's joining of the two in his discovery of the thoracic duct. With this program, the academy inserted itself into the trajectory of modern science: led by Harvey, experimentation on animals became the marker of modernity in the examination of life. Such projects, said Perrault, required proper instruments, a suitable setting, and both human cadavers and live animals. Particularly to be desired would be corpses of women who had recently given birth. Among the animals, dogs, pigs, sheep, calves, donkeys, cows, and horses could be used. Required instruments included the usual scalpels, scissors, and sponges, as well as chemicals for experiments in effervescence and coagulation of blood and other body fluids. The omission of cats from Perrault's list of animals is surprising, but may be explained by the fact that the academy's patron Colbert was a great lover of cats, as had been his predecessor Richelieu; according to the eighteenth-century chronicler of cats Paradis de Moncrif, Colbert always had cats wandering around his office.[28]

2. Transfusion

The population of stray dogs in Paris, already under siege for anatomical experiments across the city, was about to be diminished even further as the Paris Academy of Sciences began its anatomical program. On 22 January 1667, a "grand chien" was opened, probably by Pecquet, to show the lacteal veins. On the same day commenced the first of many attempts to transfuse blood between two animals, in this case two dogs.[29] We do not know who proposed the transfusion experiments, but the academicians planned both of these demonstrations for maximum impact, to demonstrate the central role Perrault felt anatomy must play in the new academy, as he had stated a few days earlier. Although four of the original twenty-one appointees to the academy entered as anatomists, the balance among the membership was definitely weighted toward the physical sciences.[30] The high-profile foreigners who gained pensions and apartments at the rue Vivienne were Christiaan Huygens and, a few years later, Giovanni Domenico Cassini, both best known for astronomy and mechanics, rather than Stensen or de Graaf. Moreover, a recent comet and an eclipse had allowed the astronomers to display their expertise. Perrault's observatory, which was originally intended to house the academy itself, figured prominently in the 1671 image of a mythical visit of Louis XIV and Colbert to the academy and in a cartoon for a tapestry from the same period that shows Colbert presenting the academy to Louis.[31]

The demonstration of the lacteal veins fit Perrault's proposed agenda and showcased Pecquet, still a star although he had published nothing in over a decade. Blood transfusion, however, did not exactly correspond to anything on Perrault's list except perhaps the circulation of the blood. But it was more dramatic than the lacteals, and it highlighted multiple methodological, philosophical, and political objectives. Blood transfusion was empirical and experimental, but, like seeing the lacteal veins, it required great skill and precision; it only made sense in the context of the theory of the circulation; and it placed the French in direct competition with the En-

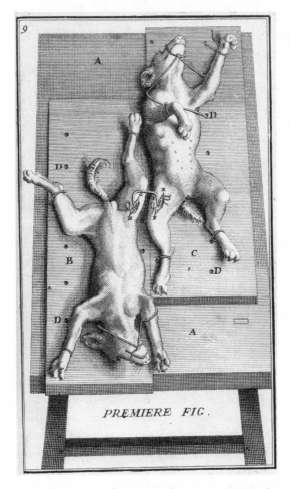

Figure 3.2 Two dogs undergoing transfusion. Perrault, *Essais*, 1688 (original in PV 1 [1667]). (Bibliothèque interuniversitaire de santé, Paris.)

glish. King Charles II had granted the Royal Society of London a charter in 1662, and its journal, the *Philosophical Transactions*, was already, after a little over a year of publication, a must-read among natural philosophers across Europe. The *Journal des sçavans*, founded in 1665 as a general learned publication, regularly quoted the "journal anglais," and the academy maintained a file of *Philosophical Transactions* articles translated into French. But the Paris Academy's excursions into blood transfusions in the first half of 1667 highlighted several differences between the 21 appointed Parisians and the 150 elected Londoners, particularly the Parisians' insistence on collaboration and anonymity. It also revealed that the academy was not the dominant force in Parisian science it aspired to be.

By the mid-1660s, blood transfusion had been a topic of discussion for nearly a decade. From the early 1640s onward, physicians suggested that one could inject substances into the bloodstream and that these substances would then circulate throughout the body. The therapeutic possibilities seemed endless. Around 1656, the English natural philosopher Christopher Wren (1632–1723) began to experiment in Oxford with injecting various substances into dogs.[32] These experiments continued at the early Royal Society, and as we have seen, Swammerdam did injection experiments during his sojourn in Paris in 1665. Many of these experiments ranged from silly to cruel, but the therapeutic value of injecting opiates, for example, seemed manifest. The progression from injection to transfusion was a small step: at one of Montmor's assemblies in July 1658, the Cartesian monk Robert Desgabets proposed transfusing blood from one animal to another and detailed how to carry this out.[33]

France and England vied for priority in these experiments, although it was clear that the English were the first to transfuse blood. The English physician Richard Lower (1631–90) transfused the blood of one dog directly to another in February 1666.[34] A few months later, his colleagues at the Royal Society collected blood from one dog in a bladder and then injected it into another dog. Descriptions of these experiments appeared in the *Philosophical Transactions* at the end of 1666, and then in the *Journal des sçavans* at the end of January 1667. By then Perrault and his colleagues at the Paris Academy had themselves begun transfusions.[35] However, these experiments were only published long after the fact, and many contemporaries, both in and outside France, were quite unaware of them.

While the manuscript minutes of the academy described the experiments as they happened, the first printed account was Perrault's own, which drew on the minutes, but only appeared in 1688 in the fourth volume of his *Essais de physique*. Du Hamel's 1698 history of the academy added some details, and Fontenelle's *Histoire* drew its account mainly from Perrault's. All of these accounts were somehow mediated: the academy's minutes, as we have seen, were redacted copies, and Perrault's, Du Hamel's, and Fontenelle's works appeared decades after the work had been done. They also differ in style. The minutes and Perrault's account share a plain, straightforward descriptive style, while Du Hamel is more discursive and theoretical; as with his account of Perrault's address, Fontenelle selected his facts to give a particular character to the academy.[36]

Even if they were written with a large measure of hindsight, these accounts reveal a different style of doing natural philosophy than practiced in England. The English may have had "all the glory" of the new discovery of transfusion, said Fontenelle, but the French perfected the work day by day, and even Henry Oldenburg, the Royal Society's secretary, acknowledged that the French excelled in precision and introduced quantification into the procedure.[37] The French, moreover, fully collaborated. A number of individuals performed the Royal Society's transfusions, and were so credited in the *Philosophical Transactions*. In Paris, the *procès-verbaux* identified the surgeon Gayant as doing most of the actual work of cutting open the dogs, finding the proper veins or arteries, and attaching the tubes, assisted by Perrault, Pecquet, and Auzout. But apart from Perrault's, no names appeared in any of the published accounts. Perrault's 1688 account removed names that had appeared in the *procès-verbaux*.[38]

Perrault recounted seven attempts at animal-to-animal transfusion between January and March 1667. He detailed seven failures: the recipient dog died in the first attempt, and in subsequent attempts, in contrast to the Royal Society's accounts, the recipient was visibly "enfeebled" rather than revitalized by the operation. The last attempt, on 21 March, included "une précaution infaillible": the quantity of blood that the animals gave and received would be measured. The dogs were weighed before and after the transfusion, and a measured amount of blood was removed from the recipient dog before he was transfused. Perrault concluded that this dog had had six ounces of venous blood removed and five and a half ounces of arterial blood added. The dog nonetheless died the next day.[39]

In his 1688 preface, Perrault credited the academy's experiments with "exactitude and with the means capable of giving certain knowledge of what we have done: which is not at all practiced in the experiments made by foreigners or elsewhere in Paris." The proper instruments were as important as measurement; the first experiment failed in part because the tubes had not been made properly. Perrault designed special curved tubes to convey the blood more efficiently and with less clotting. He also instructed that the tubes be heated to reduce clotting further. Lower in London had employed quills and later straight metal tubes. Perrault also proposed to use two tubes at once, so the transfusion process was mutual, but this does not seem to have been tried.[40]

Perrault omitted not only names but places from his account. Although the first transfusion took place at the Bibliothèque du roi, only one more was performed there, at the end of January 1667. The rest were performed "chez M. Gayant." While it is not surprising that these noisy and bloody experiments were transferred to a less public (and presumably less physically confining) location, this change of venue also highlighted the academy's tenuous position in Parisian science as one academy among many. Although Fouquet was gone, the king was not the only patron of natural philosophy. Gayant continued to perform transfusions at his house—successfully—after the academy abandoned the project. And he was not the only one to do so. Encouraged by Montmor, the physician Jean-Baptiste Denis began transfusion experiments shortly before the academy ceased its project. He was assisted by Paul Emmerez, the surgeon who had assisted Pecquet in the dissection of his patron the marquise twenty years earlier.[41] Beginning in June 1667 and into 1668, Denis attempted animal to human transfusions. His theory, which the English shared, was that the blood of young animals, particularly lambs, was purer and more wholesome than human blood and could be more easily assimilated than other sorts of blood. The obvious Christian symbolism of the blood of the lamb went unmentioned. Following the example of Denis, members of the Royal Society attempted the transfusion of a human in November 1667. The ensuing scandals following the death of one of Denis's experimental subjects in January 1668 led to the involvement of the Parlement of Paris, which shut down human transfusion in Paris in April 1668. The English also abandoned it, although it continued elsewhere.[42]

Although Perrault supervised the academy's work, he was skeptical about the outcome, and his skepticism reflects both his background as a physician of the Paris Faculty and mistrust (not unlike that of Descartes) of complicated instruments and procedures.[43] The doubts he expressed in 1688 may have included some hindsight after the spectacular failure of Denis's human experiments, but skepticism about everything was a founding creed of the Paris Academy. "The reign of words and ends is past, one wants things," wrote Fontenelle of the academy's goals.[44] Descartes had based his philosophy on an initial skepticism about the world, but the academy's approach was hardly Cartesian, and Fontenelle's official history barely mentioned Descartes. In his preface to the academy's 1671 *Memoires pour servir à l'histoire naturelle des animaux*, Perrault criticized those who made general propositions without full recourse to observed facts; he referred to Aristotle in this context, but he could as easily have meant Descartes. One should establish a fact by means of many particular examples, said Perrault, before proceeding to the next step or making large claims of universal applications.[45] By this standard, the blood transfusion experiments never reached the first step of establishing a fact. A Cartesian might argue that the conformation of the parts should allow for successful transfusion, but the experiments did not work. Perrault described nearly insurmountable technical barriers: blood clotted in the tubes and in the recipient's veins, and the siphons and tubes were so inefficient that very little blood was transferred in most cases. In the last academy experiment in April, when the dogs were weighed and the utmost care was taken to ensure that blood was in fact transferred, the recipient dog nonetheless died the next day.

Joseph Schiller explains transfusion as an example of an anatomical technique used to solve a physiological problem, noting that the physiological tools did not yet exist to analyze the blood itself. Perrault therefore resorted to an essentially Galenic idea of the blood and its formation. Although Pecquet had shifted the location of sanguification from the liver to the heart (a theory Perrault did not at this point accept), he had not changed the overall theory of blood formation, which remained rooted in the idea that the blood of each animal was specific to itself and its individual temperament, and could not be interchanged with that of others, even of the same species.[46] This notion of specificity was a particular point of debate in the many pamphlets on transfusion that appeared in 1667 and 1668. Once again employing an architectural metaphor, Perrault explained that a stone shaped for a particular function in a wall could not be placed anywhere else without severe consequences. Therefore, the analogy that Denis proposed between transfusion and the circulation of maternal blood to the fetus was, said Perrault, invalid: although maternal blood certainly resembled that of the fetus, it was not identical, and had to undergo certain refinements in the vessels of the placenta to "prepare and rectify" it before it could nourish the fetus.[47] The radical individualism that Perrault later expressed with regard to species had its origin in this essentially Galenic and medical recognition of the specificity of the individual.

As a physician of the Paris Faculty, Perrault also had good therapeutic arguments against blood transfusion. The popularity of bloodletting—one of the faculty's standard therapies—showed no signs of abating. Even though the concept of the circulation of the blood would have, it seems, discredited the practice, Harvey himself continued to proclaim its efficacy. Declaring that the vital principle resided in the blood, he nonetheless maintained that "dayly experience shewes, that *Letting blood* is a safe cure for several *Diseases*, and the chiefest of Universal *Remedies*: because the default, or superfluity of the *blood* is the seminary of most *distempers*."[48] Galen had staunchly promoted venesection, and it is not surprising that the Galenic Paris Faculty upheld his opinion. The Hippocratic revival of the second half of the sixteenth century had little effect on the faculty's support of bloodletting, and Hippocrates was co-opted as a proponent even though the Hippocratic treatises in fact recommended it only rarely.[49] The self-proclaimed Hippocratic physician Claude Tardy enthusiastically promoted both the circulation of the blood and transfusion.[50]

Perrault frequently bled his patients. When his brother Jean came down with a fever during a trip they took to Bordeaux in 1669, Claude, together with a number of Bordeaux physicians, undertook the usual remedies. Over the month of his illness, Jean was bled at least ten times. At the beginning of Jean's illness, Claude had sustained an injury to his temple in a coach accident, and he demanded a bleeding for himself as well.[51] The fact that Jean died at the end of the month had no impact on Claude's views. With regard to transfusion, Claude Perrault maintained that the animals from whom blood had been taken were healthier than those who had received blood.

There were also good political reasons to cease blood transfusion experiments. At least one pamphleteer claimed that Colbert had little enthusiasm for it.[52] Even though Colbert's hand over the academy was hardly dictatorial, it would not have been a good idea to oppose his wishes directly. An explosion of pamphlets and treatises on transfusion in Paris from the late spring of 1667 into 1668 also accorded ill with the academy's preference for anonymity and avoidance of conflict. Blood transfusion therefore fit neither Perrault's epistemology nor his goals for the academy. As a technique, transfusion afforded little new knowledge. The circulation did not need more proof, and transfusion added little to the store of facts about the human and animal body. However dramatic and compelling as a spectacle, transfusion did not yield either a therapeutic or an epistemological payoff. Pecquet inserted a plea for further work on transfusion in his yearly report to the academy in November 1669.[53] But it was ignored.

3. Publication and the Production of Credit

None of the credit that accrued from the transfusion experiments went to the academy. Instead, Denis and his patrons Bourdelot and Montmor (none of whom were academicians) amassed credit, as did Gayant.[54] Increasingly—although not exclusively—credit came with publication, and the academy's efforts went unpublished. By the time the academy published its new regulations in 1699, publication, along with invention and discovery, had become a prerequisite for nomination. It was proof of scientific credibility.[55] This was not yet the case in 1667: Perrault himself had neither publications nor inventions to his name, unless architecture counted as a science. In addition, the collaborative ideal of the academy sought to give credit to the patron—the king—rather than to the individual scientist. Transfusion immediately strained this ideal. Although the academy agreed that any discovery made there could be published in the *Journal des sçavans,* it had to be presented to the membership first.[56] Collaboration meant that all members of the academy had to agree to the publication of their work, and all did not agree on the outcome of the transfusion experiments. The main dissension was between Perrault, who thought the experiments failed, and Gayant, who went on to transfuse successfully on his own.[57] No mention of the academy's transfusion experiments appeared in the *Journal des sçavans.* Henry Oldenburg grudgingly gave credit to the French in the *Philosophical Transactions* but did not mention the academy, naming only Gayant and Denis as practitioners.[58] By the time Oldenburg wrote in the autumn of 1667, Denis was famous, or even infamous, for his attempts to transfuse animal blood into humans.

Although the *Philosophical Transactions* had no official standing with the Royal Society, the journal was edited by the society's secretary, and many of its articles reported society activities. The society's decision to pay Oldenburg a salary in 1669 acknowledged the importance of the *Transactions.*[59] The Paris Academy had no such publication outlet and was further hampered in its attempts at publication by its policies of secrecy and collaboration. As we shall see in chapter 4, the academy's proximity to the king's printers and engravers led to a number of publications in the 1670s, but rapid publication of research results was limited. Yet even before the academy opened its doors on the rue Vivienne, there was a Parisian periodical that published scientific articles. The first issue of the *Journal des sçavans* appeared in January 1665, before the *Philosophical Transactions,* whose first issue was dated 6 March. The lawyer Denis (or Denys) De Sallo (1626–69) founded the *Journal des sçavans*—with Colbert's blessing—as a general-interest periodical, a kind of public commonplace book that reviewed works on theology, philology, and history as well as natural philosophy. De Sallo had indeed for some time composed such commonplace books. The *Journal* differed both from Théophraste Renaudot's frankly propagandistic *Gazette de France* of the 1630s and from the gossipy *Mercure de France,* which began to appear in the 1670s.[60]

De Sallo, under the pseudonym "le Sieur de Hedouville," issued the journal once a week. "The design of this journal," said the preface to the first issue, "is to make known the news of the Republic of Letters," a far broader remit than that of the *Philosophical Transactions.* De Sallo enumerated what he would cover: reviews of books; obituaries of noted *savants;* descriptions of all kinds of scientific work, including experiments, inventions, and observations; principal decisions of the secular and ecclesiastic courts; and in addition, "anything worthy of the curiosity of *gens de lettres.*"[61] De Sallo was assisted in his editorial work by Académie française stalwart Jean Chapelain and the *abbé* Jean Gallois (1632–1707), who had taught Colbert's children.[62]

The first number included reviews of works on history, philology, and theology, as well as a review of Descartes's *Traité de l'homme*, which declared "admirable" his explanation of the formation of the fetus. In March appeared a review of Stensen's work on muscles and glands, preceded by a lengthy review of a controversial work on the Council of Trent, and succeeded by a review of a new production of a play. De Sallo had strong opinions about the Jesuits and the church in Rome, praising theological works that had been placed on the *Index of Prohibited Books* and generally upholding a Gallican position that privileged the French state over the pope.[63]

Publication of the *Journal des sçavans* ceased abruptly at the end of March 1665 (its final article was, ironically, a review of the new *Philosophical Transactions*). It reappeared in January 1666 with a new editor, Jean Gallois, and a new preface addressed to the king as its protector. A brief note to readers regretted the hiatus, which it attributed vaguely to "some people" who objected to the "too great liberty" of some reviews.[64] The journal kept its eclectic content but the reviews of theological works were fewer and strictly descriptive, while the scientific content increased. Most of the final issue for 1666 consisted of a lengthy illustrated review of Robert Hooke's *Micrographia*. In all this Colbert's hand may be seen, and Gallois's appointment to the Academy of Sciences in April 1668 as assistant to its secretary Jean-Baptiste Du Hamel indicated Colbert wished the *Journal* to assume the quasi-official status for the academy that the *Transactions* held for the Royal Society.[65] Although Gallois remained editor until 1675, he was no Henry Oldenburg, and by 1669 the yearly page count had dwindled from over five hundred to under fifty. It continued at this low level until 1673, the year of Gallois's election to the Académie française, when it did not appear at all.

In March 1667, the academy had declared that

> if anyone makes a new discovery in *physique* or mathematics, after they have communicated it to the *Compagnie* one is able to print it or put [it] in the Journal, to name the person who has made the discovery and to note that he has communicated it to the people who have assembled expressly for research on natural things or to express that in another way which the Assembly will agree on.[66]

Although the *Journal des sçavans* therefore possessed official status, plans to publish the academy's work within its pages did not long succeed. The tension between collaboration and individual credit as well as the multiple sites of anatomy inside and outside the academy were evident in an article that appeared in the *Journal des sçavans* in the following month describing experiments performed at the end of March by Pecquet, Gayant, and Perrault on the connection between the thoracic duct and the emulgent or renal vein. The essay was in the form of a letter from Pecquet to Pierre de Carcavi. According to Fontenelle, these experiments, on the corpse of a woman who had recently given birth, occurred at the academy, but from the context this may not have been the case, since Pecquet asked Carcavi to report them to the academy if he saw fit.[67] The original purpose of the experiments was to find the vessels that "transport the chyle to the breasts." But "the subject not being well-disposed," the research turned to a more general exploration of the lacteal vessels, which resulted in a new discovery, the connection between the "thoracic canal" and the emulgent vein, subsequently proved by several more experiments conducted by Gayant.[68]

The article did not state where the experiments took place; Pecquet cited in passing a dissection that had taken place in February "en l'Assemblée," which might imply that the present dissections took place elsewhere—most likely *chez* Gayant. According to Pecquet, these experiments continued work he had begun in the 1650s and published in the 1654 edition of *Experimenta nova anatomica*. Pecquet here seemed to test the rules of anonymity: did these experiments indeed take place at the academy? There is no evidence in the disorganized 1667 minutes of these experiments, or that the *compagnie* agreed on publication. Yet surely there was a fine line between the activities of academicians inside and outside the walls of the Bibliothèque du roi. Pecquet's trial run of the publication policy, if that is what this essay was, was not repeated, and none of the subsequent reports on "la Compagnie" mentioned any members by name other than Huygens and later, Cassini. Yet Gayant appears to have taken matters into his own hands, for within a year this account, credited to him, appeared in Latin in a collection of works by "celebrated anatomists." Also in this collection was a defense of French priority in human transfusion and *Clysmatica nova*, an account of blood transfusion by Johann Sigismund Elzholtz, physician to the elector of Brandenburg.[69] The policy of anonymity would be difficult if not impossible to enforce.

Appearances by the *compagnie* on anatomical matters in the *Journal des sçavans* were very occasional, with only half a dozen items between 1667 and 1669, most of them reviews of works published elsewhere. On the other hand, the first of several articles describing the transfusion experiments of Denis had appeared in the issue prior to Pecquet's account. By the early 1670s, Denis had been offering his own *conférences* for some time, and he published accounts of it in his own journal, which ran monthly, under various titles, between 1672 and 1674.[70] Unlike the *Journal des sçavans*, which was printed by Jean Cusson, a well-known printer in the Latin Quarter, Denis employed the services of a royal printer, Frédéric Léonard, who also printed the academy's first publications.[71]

4. Anatomical Projects *dans tout Paris*

The cramped rooms at the rue Vivienne were one of many places where dissection took place in Paris. Apart from Gayant's residence in the Latin Quarter, other sites included the more luxurious town house of Montmor in the Marais. Bourdelot's academy, in his house near Saint-Sulpice, continued to flourish, and at the same time that Denis gained attention for his blood transfusion experiments, Bourdelot hosted a new and very young anatomist from Provence who repeated the experiments that de Graaf, Stensen, and Swammerdam had offered a few years previously. Only nineteen, Joseph-Guichard Duverney, with a fresh medical degree from Avignon, even duplicated Stensen's dissection of the brain, which had not yet been published.[72] Duverney very likely dissected the ostrich that Henri Justel saw at Bourdelot's and described to Oldenburg in the autumn of 1667. Several years later, he dissected many more ostriches for the academy.[73]

Duverney also dissected for the circle of Denis.[74] Denis began his *conférences* around 1664, and took advantage of a hiatus in the publication of the *Journal des sçavans* to issue his own *Recueil des mémoires et conférences*, which had been "presentées à Monseigneur le Dauphin."[75] The *mémoires* consisted of reviews and reports of various scientific activities. The emphasis was on medical topics and particularly the circulation of the blood, but other topics included a French translation of Newton's account of his reflecting telescope.[76] The *conférences* of Denis took place at his lodgings on the quai des Grands Augustins on the Left Bank at two o'clock on Saturday afternoons (thus overlapping with the academy's meetings) and followed a format dictated by the host, who introduced a topic and then opened the floor to discussion. These topics included observations from his research and medical practice, as well as reports from the work of others, often in the form of letters. Dissections of living and dead animals took place regularly. One of Denis's major concerns was to disprove Descartes's theory of the circulation in favor of Harvey's, which he accomplished with experiments on "plusieurs animaux." Denis did not give any names in his *Recueil*, but we know that academicians as well as others attended.[77]

The academy could, and did, pursue many projects at once. Blood transfusion was only a sideshow to the anatomy project as a whole. Over the course of 1667 and into 1668, the anatomy project devolved into two, and then three, different but related pursuits. Fundamental research on human anatomy employed human cadavers and also used living and dead domestic animals, particularly dogs. Beginning in June 1667, the company also began to dissect various exotic animals provided to the academy by the royal menageries at Versailles and Vincennes. Alongside these projects was the dissection and occasionally vivisection of a large number of common wild animals such as hares, badgers, and weasels; for much of 1668, these latter dissections took place every week. All of these projects were connected, since the dissection of all sorts of animals served as evidence in the ongoing debates about human and animal anatomy and physiology that revolved around the circulation of the blood. Beginning later in 1667, work on the nature of vision and the structure of the eye combined optics and dissection. Huygens designed a new and improved air pump, which he introduced to the academy in April 1668, and the first experiment involved inserting a live mouse and pumping out the air; after the third attempt, the mouse was declared dead and dissected. A pig's bladder, tied at each end, was inserted for the second experiment, reminiscent of Roberval's experiment of a carp's bladder in a tube of mercury that Pecquet had recounted in the *Experimenta nova anatomica*.[78]

Research on human anatomy began early in February 1667, when Gayant dissected the body of a young woman who had been hanged a few hours after having a meal. The body, obtained "by order of the King" according to Perrault, was still warm. However, the lacteal veins were difficult to see, and the thoracic canal equally obscure. The plan to trace the formation and passage of milk to the breasts proved impossible. Gayant therefore conducted many other experiments, including pumping air into the lungs and colored water into the veins, and thoroughly dissected all the parts, including the eyes.[79] This may have been the case reported by Pecquet in April. This dissection took place over a number of days at the rue Vivienne; writing to his brother during a break in the activity, Huygens said the dissection occurred "in the house."[80] Witnesses to this and subsequent dissections included not only Pecquet and Perrault, but also Auzout and several other members, including Duclos, Du Hamel, Carcavi, Huygens, Frenicle, and Picard. In the next few weeks Gayant dissected a human fetus, two dogs, and a human head, which Duclos had sent to Carcavi. The dissection of the head appears to have taken place elsewhere, and Pecquet reported on it to the assembly. One of the two dogs revived when air was pumped into its lungs, and it was sewn up and released and "felt quite well."[81] In addition, the transfusion

Figure 3.3 Dissection scene at the Paris Academy. Gayant dissects a fox. Claude Perrault, seated at the right, holds a book (*Histoire des animaux*). Claude Perrault, ed., *Mémoires pour servir à l'histoire naturelle des animaux*, 1676. (Linda Hall Library of Science, Engineering, and Technology.)

experiments continued, at least for a time. This furious rate of dissection continued into the 1670s. Perrault's energetic programming was amply rewarded: his "gratification" in 1666 and 1667 had been 1,500 livres, but in 1668 he was granted 2,000 livres, putting him on the same plane as Cureau de la Chambre and Duclos, the latter of whom had shown himself to be equally energetic. As we have seen, Pecquet and Gayant, in contrast, each received 1,200 livres.[82]

The supply of bodies for dissection continued to be a source of contention between the Paris Faculty of Medicine (which by law received the bodies of executed criminals), the surgeons, and now the Paris Academy. It is not clear how the academy obtained human corpses in these early years other than by an exceptional royal order. Sometime in 1667, Auzout suggested getting the dissectors or physicians of the Hôtel-Dieu to "open the bodies of several who will die there, or to have several eyes when we need them."[83] The air pump experiments in 1668 were interrupted when the body of a recently hanged woman was transported to the academy, requiring immediate dissection.[84] This attention to the human body distinguished the Paris Academy from the Royal Society, and made plain the advantages of royal sponsorship.

5. The Royal Menageries and the Supply of Animals for Dissection

Human bodies remained rare while animals were easily obtained, particularly the domestic animals that served as subjects for vivisection and other kinds of experiments. Stray dogs abounded on the streets of Paris. The king's menageries at Vincennes and Versailles had multiple functions, among which was the provision of food animals for the royal household, and the academy likely received such animals from them. It would be interesting to know if the academy's animals were somehow unsuitable to be eaten: too old or sick, for example. But I have found no comments on this. Animal parts, such as the innumerable eyes that the academicians dissected, could easily be obtained from butchers. As we have seen, the rue de la Boucherie, along the Seine next to the Châtelet, still housed most of the city's butchers, and at least one of the academy's anatomists purchased animal parts from them.[85]

The menagerie at Vincennes, a medieval fortress southeast of Paris, was established by the royal minister Mazarin in 1654 primarily to stock the king's table; indeed, the definition of "*mesnagerie*" in the first edition of the *Dictionnaire de l'Académie française* (1694) referred solely to a place in the country where one fattened domestic animals. Mazarin's protégé Colbert supervised the construction, which was designed by Louis Le Vau (1612–70). The same architect and even the same crew constructed Fouquet's estate at Vaux-le-Vicomte a few years later.[86] In 1658 a "sérail pour nos bestes" was built at Vincennes to house wild animals, including birds of prey, for the staging of animal combats.[87] Some of these animals had been housed at the Tuileries. Germain Brice, whose *Description nouvelle de Paris* first appeared in 1684, described it thus:

> At the entrance into the Park is the *Menagerie* or place where they keep several sorts of Wild Beasts, which oftentimes they cause to fight together in a Court in which are Galleries that serve the Spectators to stand in and see without danger.

Brice noted that Fouquet's former residence at Saint-Mandé was just behind the menagerie.[88]

Louis XIV's son the Dauphin, born in 1661, was particularly fond of viewing these combats as he grew up. Some of these animals, such as the female tiger presented by the Moroccan ambassador in 1682, had come to Paris as gifts from foreign monarchs, while others were purchased or otherwise acquired by foreign travelers or on collecting expeditions.[89] Although the economic value of plants made them more common objects of early modern naturalists' regard, animals provided more visible evidence of royal and colonial power when they were brought back to Europe. Exotic animals as a form of diplomatic exchange dated back to antiquity. Aristocrats and monarchs had long collected rare animals for display and for combat, and early modern monarchs believed they recreated imperial Roman practices.[90] By the late 1660s, the Vincennes menagerie contained a number of large, fierce mammals including lions and tigers and various smaller types of wild cats such as the serval and the lynx, as well as elephants, bears, and other animals. Battles pitted these animals against each other as well as against many of the domestic animals still in residence at Vincennes. While fights between dogs and bears were common street entertainment, fights between cows and tigers were not. The *Mercure galant* in August 1682 described a day in the life of the Dauphin, which began with a hunt in the Parc de Saint-Mandé and ended at the menagerie at Vincennes, where he and his party witnessed several animal combats, featuring a cow that overpowered a tigress and went on to fight a lion, a tiger, and a wolf.[91]

When Louis XIV decided in the early 1660s to renovate his father's hunting lodge at Versailles, southwest of Paris, he specified that a menagerie be included in the plans. Designed by Le Vau, it was among the first buildings at Versailles to be completed. By the summer of 1664, according to the *Gazette de France*, animals were already on exhibit to admiring visitors, and the menagerie's buildings were mostly complete by late in 1668.[92] Both the novelist Madeleine de Scudéry (1601–1701) and the fabulist Jean de la Fontaine (1621–95) wrote about the menagerie a year later. Scudéry described a stroll through the grounds, with ample opportunity for praise of the wisdom of the king, including a comparison to Alexander the Great and a mention of the Academy of Sciences. Scudéry and La Fontaine each particularly noted the "Demoiselles de Numidie," Egyptian crowned cranes.[93] The inhabitants of the Versailles menagerie, viewed from above in the octagonal tower, included domestic food animals as at Vincennes, as well as many other animals: birds, smaller mammals, and animals such as deer that were hunted, rather than hunters, dominated.[94] The dividing line between these birds and those destined for the table was quite fine; birds we would now consider inedible, such as peacocks and herons, were at this time still commonly eaten, particularly by the nobility.[95] Other animals at Versailles included an elephant and several monkeys. The animals that came to these menageries were difficult and expensive to obtain, transport, and maintain, making them even more desirable. Political, imperial, and aesthetic considerations jostled for attention.[96]

Colbert paid close attention to the provision of animals for the menageries and particularly for Versailles. He ordered colonial governors to send specific animals, particularly exotic birds, and trading vessels and scientific expeditions were expected to seek out and bring back exotic animals. One captain brought back in a single voyage civet cats, a crocodile, and several birds; when Jean Richer (1630–96) traveled to Cayenne in 1672 to make astronomical observations, he attempted to bring back a crocodile with him, but it died on the way.[97] Colbert also commissioned an animal buyer named Monier who traveled annually to the Levant to purchase animals. During twenty-three years between 1671 and 1694, Monier made forty-one voyages and spent two hundred thousand livres on the purchase and transport of animals, mainly birds and other decorative rather than fierce animals. Care of the animals once they reached Versailles was mainly the responsibility of Alexandre Bontemps, the *intendant* of the *château*, although Colbert also charged Claude Perrault on at least one occasion and the royal physician Gui-Crescent Fagon on others to ensure the safe passage of animals to Versailles.[98]

In the spring of 1667, Adrien Auzout suggested a policy that would provide even more, and different, animals to the academy: when animals died

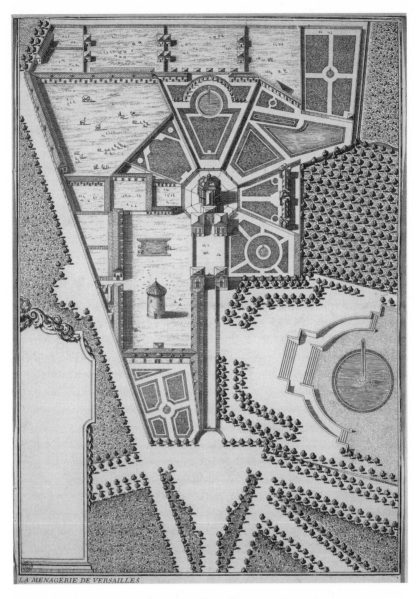

LA MENAGERIE DE VERSAILLES

Figure 3.4 Bird's-eye view of Versailles menagerie, 1663–64.
(Getty Images.)

at the king's menageries at Vincennes or Versailles they would be brought to the academy for dissection as soon as possible. Animals died regularly, particularly over the winter; the "grand hyver" of 1670 led to many deaths in the menageries.[99] When one of the Vincennes lions died in June of 1667, he, and a thresher shark caught in Brittany at the same time, became the first two of about one hundred different exotic species (well over four hundred individuals) to be dissected by the academy over the next thirty years.[100] These animals were conveyed at first to the academy's rooms in the Bibliothèque du roi, and later to the Jardin du roi, the King's Garden, where Joseph-Guichard Duverney, who became the academy's anatomist in 1674, presided as professor of anatomy after 1680. At times the anatomist came to Versailles: when the elephant, which had been a gift from the king of Portugal in 1668, died thirteen years later, Duverney came out to Versailles to dissect it before the king and the Dauphin. When a tiger died after a combat with two elephants the next year at Vincennes, its corpse was thrown into a nearby field, and members of the academy came out to inspect it.[101]

Provision of both animal and human bodies was irregular, and depending on when corpses became available, members were sometimes called on nonmeeting days, even on Sundays. When the king obtained a chameleon from Egypt as a gift in September 1668, he ordered the company to an extra meeting on a Thursday to examine it, and when a female tiger died at Vincennes a year later, members were summoned on a Sunday for her dissection.[102]

The distinction between wild and domestic, exotic and familiar, remained hazy; the academy did not use the term "exotic," but "foreign," "*étrange.*" The use of domesticated animals for vivisection, such as dogs, cats, sheep, and pigs, had been commonplace at least since the time of Vesalius a century and a half earlier. But the use of wild but native animals such as foxes and badgers was new. In March 1669, for example, the company witnessed the dissection of a live young fox; although it was not identified as alive, the peristaltic motion of its intestines was noted.[103] Where did these animals come from? Paris by 1670 was a large urban area of over half a million people. Its population had broken through two sets of city walls since the Middle Ages and threatened to overrun a third. But it was concentrated in a relatively small area, mostly on the Right Bank of the Seine. Vincennes is now just outside the latest Paris wall, the freeway known as the *périphérique,* but in the seventeenth century it was in the country. Versailles, about ten kilometers from central Paris, was even more rural, two hours away by coach. Wild animals, then, were not so far away.[104]

Figure 3.5 Nicolas Langlois, View of the Versailles menagerie.
(© RMN-Grand Palais / Art Resource, New York.)

A central figure in all of this, and nearly as invisible to history as the animals themselves, is Claude-Antoine Couplet (1642–1722). Variously described as "caretaker," "purchaser," "huissier," and "treasurer," Couplet was one of five "élèves" of the academy, who served as what Steven Shapin has called the "invisible technicians" who assisted with instruments and experiments.[105] With Carcavi and the chemist Bourdelin, he had the power to purchase items for the academy and also to reimburse the expenditures of its members.[106] He resided with his family at the observatory, and cared for its instruments. Among Couplet's duties was to "find" (*chercher* or *re-couvrer*) animals. Stray dogs and cats were easily found, and presumably he obtained pigs, sheep, cows, and horses at the menageries or other farms. But how did he obtain badgers, weasels, polecats, and foxes? It is unlikely that he hunted for them himself, so he must have contracted with hunters to find these animals to order. On 2 June 1668, for example, the minutes of the academy recorded a resolution "to make at the next Assembly the dissection of a badger, and gave the order to Monsieur Couplet to get one." He also supplied domestic animals for the royal menageries, and reported back to the academy about the exotic animals at Versailles.[107]

6. Defining the Animal

Academicians dissected many live and dead animals, and described them in great detail. These descriptions themselves constituted a statement of method. But they did not explicitly address a philosophical stance on the cognitive or moral status of animals. Nor did they address the religious significance of the presence or absence of an animal soul. Nonetheless, these questions were very much on the minds of the early academicians, particularly Marin Cureau de la Chambre and Claude Perrault.[108]

It was not coincidental that the earliest pamphlets of the academy took the form of letters to Cureau de la Chambre, the senior member among the four anatomists originally appointed to the academy, and a link between the original idea of a "general academy" and the Academy of Sciences. As we saw in chapter 1, he enjoyed patronage in very high circles—he had long been physician to the Chancellor Pierre Séguier, and he was in addition one of the many physicians-in-ordinary to the king. But perhaps most important, Cureau de la Chambre had staked out a distinctively anti-Cartesian view of animal soul and cognition beginning in the 1640s. His views in turn influenced Perrault's own mechanistic but not Cartesian theories, which he detailed in his *Essais de physique* beginning in 1680.

Although it is unlikely that Cureau de la Chambre ever performed a dissection, he was very interested in animals, and in what demarcated animals from humans. Pierre Bayle referred to him as "one of the most illustrious Peripatetics of this century," while admitting that his ideas were not conventionally Aristotelian.[109] His magnum opus was his treatise on the passions, published in five volumes between 1640 and 1662. The first volume of this work outlined an even larger project that he referred to as "l'art de connoistre les hommes," the art of knowing men.[110] This knowledge would penetrate into every aspect of humanness: the passions, the soul, vices and virtues, customs and morals, the idea of beauty, the influence of climate, and not least, the relationship between humans and animals. All of this moreover related to his ideas about physiognomy and chiromancy. Out of this mixture of topics came his very influential comments on animals.

His approach to the passions was metaphysical rather than anatomical. The passions, in Galenic medicine, were one of the six "nonnaturals," external as opposed to innate influences on health, and were sometimes defined vaguely as "the affections of the mind." Both Cureau de la Chambre and Descartes, in his 1649 *Passions de l'âme*, defined them more specifically as particular emotions such as love, hate, desire, and sadness. They caused the humors, or the animal spirits (a natural thing), to move. Cureau de la

Chambre referred to them as "the actions of the virtues and vices," but also as "the movements of the Appetite," specifically the sensitive as opposed to the intellectual appetite.[111] The passions, then, were functions of the senses rather than the mind.

Although it seems that Cureau de la Chambre aimed toward a senses-versus-mind dualism, that is not the case. His discussion of the soul, both in this book and in Le système de l'âme of 1664, made it clear that he viewed both thinking and feeling, as well as life itself, as *conscious* acts: "Everything which is living knows, and all that knows is living."[112] His discussion of instinct and animal intelligence in volume 2, first published in 1645, shed further light on this notion. Beginning with the chain of being and the idea of a hierarchy of intelligence, he argued that instincts were a form of innate knowledge.

Cureau de la Chambre included a lengthy discourse at the end of this volume on "the courageous passions," passions that animals clearly exhibited, which addressed the question of animal intelligence. He distinguished the sensitive from the rational soul; while the sensitive soul ranked below the rational, the function of each was knowledge or perhaps what we would call consciousness (*"connoissance"* is a difficult word to translate, defined by the *Dictionnaire de l'Académie française* [1694] variously as an idea of something, as the power of judgment, and as the usage or function of the soul [l'âme]). Cureau de la Chambre declared, "Because to feel, to conceive, to judge, to reason, all that is nothing but to know."[113]

Because perception required an action of the soul, one could not receive sense impressions passively. Sensation, therefore, was a form of knowledge or consciousness. And this in turn required the involvement of imagination and memory. Cureau de la Chambre concluded that animals therefore possessed a degree of reason because they possessed a "sensitive soul" ("âme sensitive") that included a degree of memory. If not the equal of humans, neither were they entirely unconscious. The fact that they could learn to do certain actions showed that. Cureau de la Chambre argued that passions therefore resulted from consciousness; if animals exhibited passions such as anger or fear, then they must possess a kind of reason and even the ability to choose between good and evil, at least to some extent. Christian theology since Thomas Aquinas had considered this ability to choose as a hallmark of humanness.[114]

Although Cureau de la Chambre never mentioned Descartes, it is hard to believe that he was not aware of Descartes's arguments about animal consciousness in the *Discourse on Method*, published eight years earlier. But unlike the famous "theriophiles" of a half century earlier, Pierre Charron and Michel de Montaigne, Cureau de la Chambre did not draw moral lessons from animals. He readily acknowledged that even if they possessed a kind of reason, animals did not possess the "perfect and universal reason" of humans that assured their immortality.[115] In this he agreed with Descartes.

Pierre Chanet, a physician of La Rochelle, responded to Cureau de la Chambre in 1646 in *De l'instinct et de la connoissance des animaux*. Chanet had penned a critique of Montaigne and Charron a few years earlier (*Considerations de la sagesse de Charon [sic]*, 1644) that argued that both humans and animals possessed instinct, which led us to eat when we were hungry or follow a path previously followed. But only humans had reason, which he defined as "a faculty of proving a thing by another different thing, and which proceeds by deliberation to a goal": in other words, the capacity to make logical inferences.[116] He argued similarly in response to Cureau de la Chambre, adding that if humans and animals were essentially equal, it would overturn the order of nature, and destroy that "belle gradation" that nature observes. He hastened to add that he did not attribute all animal actions to instinct; "they have as many external senses as we do; they also have memory, imagination, and a motive faculty." But they did not have reason, which is "proper and specific to man." One proof that beasts possessed reason would be if they could talk. But they could not.[117]

The question of animal cognition also appeared in a work by the bibliophile and scholar Gabriel Naudé (1600–1653), Mazarin's librarian. In 1647 he published a book entitled *Quod animalia bruta ratione utantur melius homine* (That brute animals make better use of reason than man), by the Renaissance humanist Girolamo Rorario (1485–1556). Rorario gave many examples of animal sagacity, beginning with a performing dog he had witnessed, and concluded not only that animals appeared to have intelligence and foresight but that they could learn more quickly than humans.[118] At precisely the time when animals were being used as never before in experiments, some people began to question if animals were truly as "brutish" as they had been portrayed. The slippery distinction between human and animal that characterized early seventeenth-century anatomical texts—as we have seen, Bauhin and du Laurens both encompassed humans when they used the term "animal"—continued later into the century. In the 1670s, the Gassendist physician Guillaume Lamy in referring to "l'homme ou les autres animaux," noted that man "is not as elevated above animals as he imagines."[119]

In response to Chanet, Cureau de la Chambre in 1648 published a more extended account of his views on animal intelligence, *Traité de la connoissance des animaux, où tout ce que a esté dict Pour, & Contre Le Raisonnement des Bestes, est examiné.* Like his contemporary Pierre Gassendi, Cureau de la Chambre argued that the idea that humans needed to be distinguished from animals was itself specious. He continued to distinguish two levels of reason, however—a reason based on observation and particulars that animals held and a higher reasoning based on the notion of universals, which is uniquely human. Cureau de la Chambre argued that attributing animal actions to instinct rather than to reason (and free will) led to a theological as well as a philosophical problem, implying that God had somehow left animals incomplete, but also requiring a kind of occasionalism—a direct intervention of God—to enable instinct. Creating animals with a degree of reason eliminated these problems.[120]

The third of Cureau de la Chambre's refutations of Chanet in this lengthy book concerned the vexed question of animal speech. If beasts have reason, he argued, they must be able to communicate—a neat reversal of Descartes's argument in the *Discourse on Method*. "One would have to be extremely stupid," said Cureau, "not to notice that animals have voices," and that those voices differed according to the emotions they felt and expressed. It was obvious that animals communicated with each other, and domesticated animals also communicated with humans, not only by voice but by gesture and facial expression, reminding the reader of the centrality of physiognomy to Cureau de la Chambre's thought. He dismissed as meaningless arguments against animal speech and cognition based on physiological differences between humans and animals, but he did not draw any ethical conclusions about how humans should therefore behave toward animals. He also continued to ignore the moral lessons Montaigne and Charron had drawn from animal behavior, although he expressed his admiration for those authors. And he never mentioned Descartes, addressing his remarks entirely to Chanet, even though he agreed with Descartes that animals did not possess an immortal soul, which God reserved to humans.[121]

A year later, Descartes's *Passions de l'âme* offered a completely different—and mechanical—explanation of the passions, the soul, and the mind. Descartes intended to eliminate the Aristotelian vegetative soul that governed bodily function in favor of mechanisms that performed these functions without conscious intervention. The passions are "motions of the soul, caused by the movement of the spirits." Therefore, they are mechanically induced, but then, in humans, acted upon by the rational soul, which is the seat of reason. Descartes admitted that some emotions, such as sadness, could be induced entirely by the rational soul without any external stimulus—you can think yourself into being sad, in other words. By this argument, then, animals were perfectly capable of experiencing the passions, but they did not have the rational capacity of humans to then act upon them by moving the body in some way. Joy in an animal was pure emotion, without any rational goal or meaning.[122] Descartes believed that his mechanistic characterization of animal soul confirmed orthodox religious views of human exclusiveness and the immateriality of the soul and avoided the pitfalls of occasionalism.[123] Indeed, in his dedication of *Principia philosophiae* (1644) to the theology faculty of the Sorbonne, he referred to himself as a "Christian philosopher." But in avoiding a God who intervened too much, did he substitute a God who intervened too little? By invoking mechanism as a cause of human and animal emotion, Descartes relegated God and salvation to the sidelines.

When Claude Perrault set up a program of animal anatomy at the Paris Academy, therefore, he had a substantial legacy of discussion on animal cognition to consider, including both its scientific and theological implications. His colleague Cureau de la Chambre's work on the soul had appeared only a few years earlier, and his works on the passions had been widely praised and often reprinted. Descartes's *Le monde* and *Traité de l'homme* likewise had appeared quite recently. Through the 1660s and into the 1670s, the Cartesian beast-machine was fiercely debated. The Cartesian Géraud de Cordemoy (1626–84) argued against animal speech in his 1668 *Discours physique de la parole*. Two years previously, his *Discernement du corps et de l'âme en six discours (The distinction of the body and the soul in six discourses)* had argued that occasionalism could account for the appearance of animal cognition. Any evidence of animal intelligence as expressed in speech, he explained, was simply the direct action of God.[124] The Cartesian cleric Nicolas Malebranche (1638–1715) offered an Augustinian interpretation of Descartes in his 1674 *Recherche de la vérité*, which included his famous defense of the "beast-machine" theory:

> Thus, in animals, there is neither intelligence nor souls as ordinarily meant. They eat without pleasure, cry without pain, grow without knowing it; they desire nothing, fear nothing, know nothing; and if they act in a manner that demonstrates intelligence, it is because God, having made them in order to preserve them, made their bodies in such a way that they mechanically avoid what is capable of destroying them.[125]

The "beast-machine" theory, defended by Malebranche and by Antoine Arnauld (1612–94) and Pierre Nicole (1625–95) of the Jansenist convent at Port-Royal, has been much discussed by historians, and the public and lengthy debates in the 1670s and 1680s between Arnauld and Malebranche on the theological implications of Cartesianism were widely known among intellectuals.[126] Jansenism, a movement within Catholicism based on the work of the Flemish theologian Cornelis Jansen (1585–1638), took the position of Saint Augustine that emphasized an omnipotent God. By this reading, free will was largely illusory and salvation was predetermined by God alone. Grace could not be earned, and salvation was not available to all. At the center of Jansenism was the convent of Port-Royal, and Arnauld became its spokesman in 1643. The Jesuits who dominated the theological faculty at the Sorbonne fiercely opposed Jansenism, and Arnauld engaged in continuous controversy until a papal bull condemned certain of its doctrines as heretical in 1655. Nicolas Perrault, brother of Claude and Charles and a member of the theology faculty, was expelled in 1656 for his Jansenist views, as was Arnauld. In that year appeared Pascal's defense of Jansenism in the *Lettres provinciales*.[127]

Although Jansenism remained an amorphous set of ideas, it was nonetheless viewed as a political as well as a religious threat; Jansenists had tended to side with the anti-Mazarin forces during the Fronde, and the increasing censorship of them (and of Descartes) in the 1670s can be traced to the hardening position of the king on religious orthodoxy that culminated in the revocation of the Edict of Nantes in 1685. The perceived link between Cartesianism and Jansenism was ambiguous but lasting. In 1676, Mme. de Sévigné referred to a scholar she had met as "Jansenist, that is to say, a perfect Cartesian."[128] The link hinged on free will and the mind-body relationship, which in turn had implications for such theological issues as transubstantiation. Was the body of Christ truly present in the Eucharist? One implication of Descartes's thought, which he expressed in private correspondence, was that there could only be a spiritual presence. This heretical notion was used as a reason to ban the teaching of Descartes's works starting in the 1670s. Urged on by the crown, the Paris theology faculty formally condemned Cartesianism in 1671; the medical faculty followed two years later.[129]

Perrault and the academy completely ignored the beast-machine and Cartesianism in general. In part this owed to the academy's characteristic aversion to controversy, an aversion fervently shared by Perrault. But political and religious considerations also played a role. Many of Descartes's works had been placed on the *Index of Forbidden Books* in 1663, including *Passions de l'âme*, and Perrault may have sought a Christian and non-Jansenist alternative to Cartesian mechanism.[130] But the *Index of Prohibited Books* held no sway in Gallican France, which operated its own censorship. *Traité de l'homme* was published in Paris in 1664, followed by *Le monde* and editions of Descartes's letters and other works through the 1660s and into the

1670s: a second edition of *Traité de l'homme* appeared in Paris in 1677. In addition, the Cartesian Jacques Rohault continued to give *conférences*, and Descartes himself was reburied at the church of Ste. Geneviève du Mont in 1667, although the chancellor of the University of Paris was forbidden to deliver the funeral oration. Nonetheless, even without overt condemnation of Descartes in the 1660s, an organization sponsored by the crown would probably not be among his supporters.[131]

Perrault began to develop his own theory of animal cognition and animal soul in the 1660s, based on the work he oversaw at the academy. Only hints of this theory (which did not appear in print until the 1680s) were included in the academy's publications on the anatomical project, which focused on the exotic animal dissections. As the main compiler of the text of the *Mémoires pour servir à l'histoire naturelle des animaux* (1671–76), Perrault emphasized mechanical physiology rather than animal cognition. But this was not a Cartesian mechanism of particles and vortices—as pointed references to the pineal gland made clear—but a Galilean mechanism of fluids and forces, following the lead of Pecquet. He shared with Descartes, and with virtually every other natural philosopher in the second half of the seventeenth century, the idea of the body as a machine.[132] Mechanistic anatomy did not encompass a single theory, but mechanists agreed that the Aristotelian-Galenist explanations of function based on faculties were outmoded. Many mechanists argued that matter and motion could explain everything.[133] Perrault, however, rejected the Cartesian animal-machine and retained an incorporeal and self-moving soul.

Perrault's most developed exposition of his philosophy appeared in his *Essais de physique* (1680–88), a four-volume work on the mechanical philosophy that, in keeping with the academy's broad definition of *"physique,"* included discussion both of mechanics and of anatomy. The discussion of sound in volume 2, as we shall see in chapter 5, also included a Severino-like (or Casserio-like) comparative anatomy of the ear in different animals, although Perrault's descriptions and illustrations far surpassed Severino's in sophistication and detail as well as in the wide variety of animals, familiar and exotic, he employed. Anatomical examples and topics appeared throughout the *Essais de physique*, but he devoted volume 3 to "la *mechanique* [*sic*] des animaux" and threw down the gauntlet to Cartesians on the first page:

> To impede the bad effect that the equivocation and ambiguity of the title of this work might produce in the minds of those who are understood to say that most animals are purely machines, and who then believe that one has solved

that problem; I warn that I understand by animal a being that has feeling and that is capable of exercising the functions of life by a principle that we call the soul.[134]

He claimed that the soul served the organs of the body—which were indeed machines—as the principal cause of their action. Thus, he attributed to animals what Descartes only attributed to humans. Perrault followed the epistemological modesty of Harvey that, as we shall see, had also governed his work in the *Histoire des animaux*: he would describe only the functions of animals as observed in dissection and vivisection without attributing causes. These causes were, he said, known only to God, "the admirable Worker of miracles who sees himself in the structure of the organs of animals." Perrault added, "I have not followed the sentiments of the new Sect . . . [who believe] that by means of mechanics one can know and explain all that pertains to animals."[135] Plants could indeed be viewed as purely mechanical in their actions (although, Perrault admitted, there were sensitive plants), but animals could not. His work on sound and hearing described a precise and methodical process of reasoning from the anatomical parts to their actions, proceeding from the known to the unknown by analogy "with the parts of other senses, to see if analogically also, they can have the same uses." These analogical particulars along with particulars of the natural history of the organ furnished the materials for induction, "establish a probability, and supply a clarity that all these things have not at all had with each other." Such a method led inevitably to a discussion of animal perception.[136]

Like Cureau de la Chambre, Perrault believed that sensation or the ability to perceive implied consciousness. Not all animals exhibited the same level of consciousness, which he linked to the ability to move: so oysters were less conscious than a lion. Nor did they have the same level of sense experience, and these gradations included the ability to feel pain. This ability to feel could be measured by an animal's voicing its distress. Perrault agreed with Cureau de la Chambre that animals had sufficient intelligence to distinguish between good and bad, at least in regard to their own well-being.[137] But this intelligence, as Cureau de la Chambre had argued, had its limits. "One cannot say that beasts are capable of Science, of Philosophy, or Morals and Politics, because they do not have a formal and reflective knowledge of universals," but they had an "internal reasoning" that guided their actions. Perrault noted that "the inflexions of the voices of beasts are proper to make their intentions known to each other." We could not understand "what they intend to signify" because we could not understand their language, not because they did not possess a language. In the *Histoire des animaux*, Perrault had argued that even though monkeys had larynxes and other anatomical parts that could form human speech, they could not speak because they were not human. Animals may have a form of speech, but they could never have fully human speech.[138]

Although, said Perrault, one could explain quite plausibly (and, indeed, correctly, to a point) how the animal spirits operated mechanically to cause an animal to act, this explanation was incomplete. "Tout le monde" believed that the seat of sensation resided in the brain, and that tubular nerves filled with animal spirit conveyed sensations to the organs. But how could this spirit (or indeed two spirits) move along the same tracks in contrary directions toward the brain and from the brain to the organs? If a horse remembered a road because, when coming one way, a certain configuration of the animal spirits left physical "traces" in the horse's brain (as Descartes would have argued), how, then, did those traces reverse themselves so that the horse remembered the road going back?[139] More dramatically, Perrault cited a snake from which he had removed the head and the heart that nonetheless found the stone behind which it habitually hid in back of the Bibliothèque du roi. These stories, reminiscent of Rorario, proved that "in the functions of animals there is something that cannot be explained by all that we know of the properties of bodily things; so that one must suppose that even in the memory there is some sort of reasoning . . . to make the operations of memory, one must find a means to join things together that are not at all disposed to that which depends on the end and the order that the causes of corporal things have naturally within them." As we have seen, Chanet had defined reason in just this way.[140]

But the story of the viper displayed Cureau de la Chambre's concept that the rational soul was not merely contained in the brain but was somehow spread out through the entire body. Far from being separate, mind and body were intimately intermingled: "Our soul is not in our body as one is in a house, but it is united with it; it ought to be considered as involved in all of our actions." The entire body exhibited consciousness. Therefore, the brain (and the pineal gland) did not direct motion and activity, conveying its commands via the animal spirits to the muscles. The directive function was disseminated throughout the body.[141] Mechanism could not entirely explain the actions of an organism, whether conscious or unconscious.

Perrault's views differed also from the anti-Cartesian ideas of the Jesuit Ignace-Gaston Pardies (1636–73), who had published his *Discours de la connoissance des bestes* in 1672. Pardies had poured scorn on the Cartesian idea of the animal as automaton, saying it would make animals little more than marionettes.[142] Although Pardies spent the first half of his book explaining

Descartes's ideas on animal mechanism, in the second half he refuted these with the customary arguments of the Scholastics, based on Aristotle's distinction between sensory and spiritual knowledge. Animals, argued Pardies, "do not have, in truth, the spiritual knowledge that only belongs to reasoning souls, and to pure spirits; but they have nonetheless sensory knowledge [*connoissances sensibles*], which are able very well to suit all animals that Nature has equipped with diverse sense organs."[143] It would be strange if God had given animals the capacity to feel but not the ability. In the next year, Jean-Baptiste Du Hamel, the academy's secretary, argued similarly that the existence of sense organs must imply the existence of feeling in his *De corpore animato*.[144]

Perrault's views on animal soul developed in the course of his dissections at the academy. His growing conviction of the uniqueness of each animal, coupled with the experimental nature of dissection and vivisection, whose results were not always the same, led Perrault away from a view of animals as identical machines and of knowledge as certain rather than probable. As we have seen, this notion of uniqueness, which ultimately stemmed from Galenic disease theory, was reinforced by the perceived failure of blood transfusion. He filled the *Essais de physique* with examples from the animals described in the *Histoire des animaux* and the academy's minutes.[145] In the *Histoire des animaux*, as we shall see in the next chapter, Perrault and his colleagues recognized what Jacques Derrida referred to as the "unsubstitutable singularity" of the animal.[146]

FOUR

The *Histoire des animaux*

1. The First Exotic Dissections

The last week of June 1667 was hot and humid in Paris. On Friday the 24th, a thresher shark that had washed up on the shore in Brittany arrived at the Bibliothèque du roi, where the academy, specially assembled, awaited its arrival. The effects of the journey and the heat made for a hasty dissection of the huge fish, which measured over eight feet (nearly two and a half meters) long. The meticulous dissection set the pattern for many to follow. It began with a minute external examination, with many measurements. One academician sketched the animal—possibly Perrault, an accomplished artist who would often assume this task—while another took notes of its form, its fins, its skin, its eyes. The academicians used a microscope to examine the shark's tongue, and tasted its flesh to dispute its ancient designation of "renard marin," or sea fox. It tasted good, unlike a fox. Gayant or perhaps Pecquet dissected while others carefully described and sketched all the parts.[1]

As the shark was being dissected, news reached the academy that a male lion had died at the royal menagerie at Vincennes.[2] Following Auzout's suggestion of a few months earlier, the *compagnie* decided to convey the lion's corpse to the Bibliothèque du roi for dissection, which went on through the weekend and was completed on Tuesday, 28 June. The hot weather held, and like the shark's, the lion's corpse deteriorated quickly and soon smelled quite bad, particularly in the confines of the small rooms at the library, although not as bad, according to Perrault, as a dead deer following the hunt. After four days, maggots appeared in the lion's skin. Nonetheless, as with the shark, the academicians took thorough notes and measured everything, and sketched the animal and then its dissected parts. Particular attention was paid to the eyes, in keeping with ongoing debates about vision. The

Figure 4.1 Lion. Claude Perrault, ed., *Mémoires pour servir
à l'histoire naturelle des animaux*, 1676.
(Linda Hall Library of Science, Engineering, and Technology.)

heart seemed large for the size of the animal, but the brain was surprisingly small. There was a tiny pineal gland, an inch and a half in length.[3] As they had done with the shark, they examined the lion's tongue with a microscope. Only two years earlier, Marcello Malpighi and Carlo Fracassati had published their microscopic explorations of the tongue, and a short account had appeared in the *Philosophical Transactions*. Lorenzo Bellini's a work on taste and the tongue appeared in the same year. The academy's observations may have responded to this research, but Perrault did not cite it.[4]

While numerous other anatomical activities continued, the dissection of exotic animals afforded members of the academy a new outlet for their curiosity and a new opportunity to retain the attention of Colbert and the king and to publicize their work in an acceptable manner. Only the descriptions of "foreign" animals were worthy of publication, and such animals were much less readily available to other groups such as Bourdelot's or Denis's. The academy's minutes chronicle dissections of badgers and foxes (and polecats and weasels and many birds), but exotic animals merited a wider audience.[5]

In each case, whether published or not, meticulous dissections affirmed the circulation of the blood, and confirmed particular interpretations of digestion and respiration. As they had with the lion, in most cases the anatomists located the pineal gland, even if, as in the badger, it was only as big as a grain of wheat. In *Traité de l'homme*, Descartes had identified the pineal gland as the seat of the soul and memory. It harbored specifically human functions, a notion he had previously mentioned in his 1649 *Passions de l'âme*. Descartes thought this gland only existed in humans, but Thomas Willis quickly dismissed that idea in his 1664 book on the anatomy of the brain.[6] As we have seen, Stensen's dissections of the brain at Thévenot's house in early 1665 also specifically refuted Descartes on the uniqueness of the pineal gland to humans.[7] But the insistence of the *compagnie* (and particularly Perrault) in pointing out its appearance aimed to make evident the academy's anti-Cartesian philosophy.

2. Faber, Johnstone, and a New Model for Natural History

By the time Cureau de la Chambre died in 1669, the academy's animal program was well under way, with, as we shall see in the next section, a 1667 pamphlet and a more substantial publication of 1669. Although the *compagnie* performed the dissections as a whole, Claude Perrault seems to have been the author of these early descriptions. He drew on the humanist, encyclopedic tradition of Gessner and Aldrovandi and the comparative anatomical study of Severino, but more particularly on recent works by Johnstone and Faber. The Leiden-educated physician John Johnstone (or Johannes Jonstonus, 1603–75), born in Poland of Scottish ancestry, published four folio volumes on the natural history of animals between 1650 and 1653.[8] He aimed to be comprehensive, and as with Gessner and Aldrovandi, each volume treated a different class of animals: fish, birds, quadrupeds, and "serpents and dragons"—that is, reptiles and insects. Within these categories he attempted a classification based on a particular quality; he classified quadrupeds based on the type of foot, although this meant grouping the elephant (which he mistakenly thought had no separate toes) with the horse and the unicorn.[9] Like his predecessors, Johnstone's method might be compared to a commonplace book, in which tidbits from various authors were cut and pasted to form the text. He based his commentary largely on accounts from Pliny, Aelian, and other naturalists, including Gessner and Aldrovandi, adding some original observations and expressing skepticism about some of the wilder stories. His account of the unicorn "rest[ed] on the fidelity of the Relators," and he dismissed some stories but noted the existence of unicorn horns at several locations.[10]

Although Johnstone never left Europe, his reading was wide, and he included many more American and other exotic animals than had his predecessors, relying particularly on Georg Marcgraf's (1610–44) work on Brazil, which had appeared only a few years earlier. Among birds, for example, he described the toucan, the rhea, the dodo, and the emu, none of which had appeared in earlier works. Quadrupeds included the opossum and the coati, both described by Marcgraf. But Johnstone wrote a natural history, not a comparative anatomy: he described only external morphology, making no reference to dissection and giving little or no account of internal parts.

Illustrations were the most original aspect of Johnstone's work. Each animal was illustrated, and the engravings, by the noted Frankfurt engravers Matthäus Merian and his sons Matthäus and Caspar, surpassed earlier works in both beauty and accuracy. The images did not appear alongside the descriptions as in Gessner, but grouped at the end of each section. Several animals of similar kind appeared on each page, each in a separate vignette. The Merians situated each quadruped in a natural setting, often showing the animals in motion. Johnstone's books, originally in Latin, were translated into a number of European languages and reprinted well into the eighteenth century.[11]

Johannes Faber (1574–1629), a German physician and anatomist in Rome and a member of the Accademia dei Lincei, inspired both Johnstone and Perrault. Faber was a prolific dissector of both animals and humans,

Figure 4.2 Hyena, civet cat, muskrat. Johnstone, *Quadrupetibus*, 1650.
(Linda Hall Library of Science, Engineering, and Technology.)

Figure 4.3 Techichoatal. Hernández et al., *Rerum medicarum novae Hispaniae thesaurus*, 1651. (Wellcome Library, London.)

claiming to have dissected one hundred human bodies, and like his contemporaries Aselli, Riolan, and Harvey, he also dissected many different kinds of living and dead animals. He had over one hundred animal skeletons in his home, and the Roman artist Filippo Liagni made engravings of some of his collection in the early 1620s.[12] Faber wrote a work of some four hundred folio pages on animal (with some human) anatomy and natural history in the 1620s in the form of a commentary on the animals described by the Spanish physician Francisco Hernández (1514–87) in the 1570s. David Freedberg has recently described the complex history of Hernández's manuscript, and how it landed in the hands of the Lincei. Although a few copies of Faber's anatomical work were printed in 1628, it remained obscure until the publication of a full edition of *Rerum medicarum novae Hispaniae thesaurus* in 1651.[13]

Perrault and his fellow academicians were most likely familiar with this edition. However, it is possible that they may have known of earlier printings via Bourdelot. His friend and correspondent Cassiano dal Pozzo, a Linceo and secretary to Francesco Barberini (the original patron of the *Thesaurus* project), was a driving force behind the *Thesaurus* project, for which Louis XIII had also been a patron. Faber's work, sparsely illustrated with engravings of individual animals, constituted only a portion of the massive *Thesaurus*, which mainly concerned plants. He followed a format of a "descriptio" that included an image of the animal and a brief description of its external appearance and origins with what he called "scholia." These wildly discursive narratives made Aldrovandi look reserved; Faber's description of a Mexican civet cat, for example, ranged through other odor-producing animals, to spermaceti, sperm whales, then to ambergris, amber, insects in amber, and back to whales, before returning to civet cats and the anatomy of their scent glands.[14] Faber employed the full armory of classical and more recent references, from Pliny to Rondelet, Gessner, and Aldrovandi, and included descriptions of his own dissections of related (or sometimes not so related) animals.

3. The Literary Technology of Dissection

Johnstone's illustrations and Faber's dissections provided Perrault and the *compagnie* with additional models of description and presentation. Perrault probably conceived the idea of publishing descriptions of the lion and the shark. Before the end of 1667, an illustrated pamphlet of twenty-seven pages appeared, in the form of two letters addressed to Marin Cureau de la Chambre.[15] Frédéric Léonard, a royal printer, had already published academician Christiaan Huygens's ten-page description of a ring around the sun.[16] For Perrault, it would have been a welcome distraction from the flurries of pamphlets continuing to appear on blood transfusion. The academy's pamphlet adopted a conversational tone, but in keeping with its rules on anonymity named no one other than Cureau de la Chambre. It included two engraved illustrations. One consisted of two views of the shark: a lateral view of the entire animal and a dorsal view of it opened for dissection. The other displayed several dissected parts of the lion, including a microscopic view of its tongue. Both included scales indicating actual dimensions. Each engraving covered half a folio sheet, which was then folded in half to fit within the quarto pamphlet. Because they appear in different places in different copies of the pamphlet, they seem to have been sold unattached.[17]

With the 1667 pamphlet, the academy began to build a public paper archive of its anatomical research, with multiple functions: it demonstrated its worth to its patrons, Louis XIV and Colbert; it provided a public record of the academy's activities; and it promoted its distinctive brand of research to a broader public at home and abroad. In the latter it certainly succeeded, with notices in both the *Journal des sçavans* and the *Philosophical Transactions*.[18] With the important exception of Huygens's pamphlet, these publications had no author but the academy. They presented a sanitized public image of consensus and self-effacement by its members that the masses of internally generated paper belied. Although the academy's minutes were edited and redacted, the Pochettes de séance were not, and names and credit were liberally distributed in these documents, but only for the benefit of the members and possibly Colbert.[19]

Jean Gallois edited the *Journal des sçavans* from the time of its relaunch in January 1666, and by April 1668 he was also the academy's second secretary. Nonetheless, plans to publish the academy's activities in the journal remained unrealized for some time. In 1667, as we have seen, the journal published a single account of the academy's activities, the letter from Pecquet in April. Work by academicians Auzout and Picard appeared in essays on astronomy and physics by Pierre Petit, the intendant of fortifications but not a member of the academy. The descriptions of the shark and the lion were summarized separately late in the year.[20] Gallois also included extracts on various topics, particularly blood transfusion, from the "Journal d'Angleterre," the *Philosophical Transactions*. He was a great enthusiast of that technique even if the academy was not, publicizing the efforts of Jean-Baptiste Denis as well as Claude Tardy's book promoting human transfusion. He devoted the entire issue of 6 February 1668 to reports on the many pamphlets that had appeared for and against transfusion.[21] But religion and history continued to dominate the pages of the *Journal des sçavans*. Typical was a 1667 review of a book entitled *La présence de Jésus dans le Saint Sacrement* by the Jesuit Jacques Nouet that occupied thirteen pages, far more than any scientific account in that year. As we saw in chapter 3, the physical presence of Jesus in the Eucharist was highly debated between Jesuits and Jansenists, and the review took the Jesuit side. It was followed, however, by a review of the published letter from Denis to Montmor that described his first human blood transfusion.[22]

This situation changed little after Gallois became second secretary. He included a few reports from "the Company that meets in the King's Library" in 1668, including an account of "worms" found in the livers of various animals, and a piece on mathematics explicitly extracted from the minutes.[23] Huygens submitted a few more pieces in 1668 and 1669. But Gallois was not an energetic editor and the *Journal des sçavans* appeared irregularly over the next several years and only revived in 1675 under a new editor, Jean-Paul de la Roque (d. 1691).[24] For the most part, it did not report on the academy's activities but on its publications. Its place was partially taken in the early 1670s by Jean-Baptiste Denis's *Recueil* and the reports of his *conférences* that appeared between 1672 and 1674, but Denis was not a particular friend of the academy and his reports ranged widely.[25] In January 1675, the academy repeated the resolution it had made in 1667 to "send to the author of the

Journal" each "particular that has been written" after it had been read to the *compagnie*.[26]

Without a journal of its own, and with its rule of anonymity, the academy required careful management to publish anything. Direction of the public image of its anatomical work fell mainly to Claude Perrault. He supervised dissections, reported on those that took place outside the *bibliothèque*, and often read to the *compagnie* for their approbation the individual descriptions—many of which he wrote—that later became the *Mémoires pour servir à l'histoire naturelle des animaux*. These reports, which took place alongside Duclos's lengthy readings from Boyle and van Helmont, established what Steven Shapin has called a "literary technology," which could establish "matters of fact" and mobilize assent to these facts.[27]

At the same time, much in the manner of Cassiano dal Pozzo's *museo cartaceo* or the Hernández *Novae Hispaniae thesaurus*, the academy's anatomical publications displayed the wisdom and generosity of its patrons to the wider world. The Accademia dei Lincei, which had counted Cassiano dal Pozzo, Johann Faber, and Galileo as members, could not have been far from Perrault's mind as he organized the academy's publications. Much as Louis XIII had been a patron of the *Thesaurus*, so too Louis XIV, via Colbert, hovered over the academy's works. More proximately, the *Saggi* of the Florentine Accademia del Cimento had appeared in 1667.[28] The appeal to Louis as patron would be explicit in the *Mémoires pour servir à l'histoire naturelle des animaux*. But meanwhile Perrault made sure that the academy's work stayed in the public eye.

The academy's animals were not ordinary animals, and they held a variety of roles in Louis's universe. In representing the gift economy of diplomatic exchanges and indicating the reach of the French imperium, they displayed his power. In addition, animals played important roles in establishing the public image of the king. The taming of powerful wild animals in the menageries seemed to parallel the regulation of the aristocracy by the civilizing process of the court. The transition from the animal fights at Vincennes to the controlled display at Versailles, which featured exotic birds as well as fierce beasts and valiant cows, manifested the transition of the *noblesse d'épée* represented by the Grand Condé to a nobility entirely under the thrall of the court. The menageries and the academy also brought prestige to the court, and animal symbolism entered into many aspects of Louis's image-making apparatus, "le roi-machine."[29]

In 1669, a *Description anatomique d'un caméléon, d'un castor, d'un dromadaire, d'un ours, et d'un gazelle* followed the 1667 pamphlet on the lion and "renard marin."[30] In between these two publications appeared a 1668 pamphlet consisting of two letters written by Pecquet and Mariotte on the seat of vision.[31] Like the 1667 pamphlet, the 120-page *Description anatomique* was the standard quarto size. Instead of the half-folio engravings from 1667, however, the new publication included folio-size foldout illustrations of each animal. It followed the style that the pamphlet had established, but with more detail. The meticulous dissection of each animal was preceded by accounts of its name, provenance, external appearance, and behavior, corresponding to Faber's *descriptio*. But the narrative that followed did not indulge in the humanism-on-steroids excesses of Faber's *scholia*. Myths were debunked or simply ignored; the academicians had little patience for lengthy digressions, and although there was much reference to the ancients and to a variety of other authors, the narrative always returned to the animal itself and its physical presence.[32]

In this text, Perrault and the *compagnie* displayed their awareness of the philological tradition while also situating the animals in their natural, exotic environments and acknowledging the importance of place as part of each animal's identity. The narrative descriptions followed humanist practice in including etymologies, classical references, and despite Perrault's later disclaimers, attempts at classification, at placing the animal into some more general category. But unlike encyclopedists such as Gessner who sought to gather all known information about an animal, or earlier naturalists who sought out the anomalous, the academicians aimed to sift through this information to find the single true fact. Discovering the proper, original name for an animal, therefore, not only established ownership (following the Renaissance tradition) but also established its identity and its place among other animals. The chameleon, perhaps the most exotic of these five animals, commanded half of the volume's pages.

4. The Chameleon

Although ostensibly mortality in the royal collections dictated the choice of animals to be dissected and described, the process of choosing descriptions for publication was far more deliberate, and the academy dissected many more animals than appeared in the pamphlets or later in the volumes of the *Histoire des animaux*. The 1669 volume began, "There is hardly an animal more famous than the Chameleon," and it was indeed an immensely popular animal in seventeenth-century Paris and one that evoked inordinate fascination from naturalists because of its ability to change color.[33] Aristotle had described the chameleon in great detail and apparently even vivisected one. It had appeared in every zoological book since Gessner, and even its

Figure 4.4 Pieter Boel, Chameleon, ca. 1668.
(© Musée du Louvre, Dist. RMN-Grand Palais / [Gérard Blot] / Art Resource, New York.)

illustration had become standardized; Faber's account of it occupied over twenty pages of the *Thesaurus*.[34] A few years later, the novelist Madeleine de Scudéry, who ran a famous Saturday salon in which Claude Perrault participated, had a pair of pet chameleons. When they died, Perrault and Pecquet dissected them, and she displayed their skeletons in her sitting room.[35]

On Thursday, 20 September 1668, "un ordre exprès du Roy" called the *compagnie* together on a nonmeeting day to examine a live chameleon that had been presented to the king by a Capuchin father who had lately been in Egypt.[36] This may have been the chameleon that the royal animal painter Pieter Boel depicted. The group made several "observations and experiments," including the usual measurements, noting that the animal, a male, had the ability to "inflate" itself to make it look bigger, but was otherwise quite thin. Much attention was paid to his skin: its coldness, its texture, and of course its color. It was covered with "petite éminences" the size of a pinhead or slightly larger. His natural color was a bluish gray, but he did not change color as quickly or as readily as had been assumed; the academicians put the animal on different-colored backgrounds, and in the sun and the shade. His skin changed in the sun, slowly becoming brown in the part exposed to the light, while the rest took on yellow and reddish tints. When they wrapped him in a white linen cloth, he became paler but not wholly white. The academicians even marked some of his spots with ink to determine if and how they changed color. But he did not change to the color of his background; the changes rather seemed related to temperature.[37] They also attempted to feed him, notwithstanding the common notion that chameleons did not eat. When they placed some flies nearby, they were startled to see the animal's long, slender tongue flick out and pull the flies into his mouth. While these activities were going on, Perrault made a "peinture exacte" of the animal. The eyes, always a topic of interest, were minutely examined. The *compagnie* carefully recorded his fishy smell and even the color and texture of his excrement.[38]

Although the *compagnie* handled the animal with great care, he nonetheless lived less than a month, dying on 13 October as the weather moved toward winter. Since the academy took its annual holiday from early September to mid-October (the meeting on 20 September was during the holiday period), it is not clear where the chameleon was housed during that period. He was very thoroughly dissected immediately following his death, with more "figures exactes" drawn of every part of interest. The following month, Perrault read to the *compagnie* his description of the chameleon and his dissection. The gathering then approved the account for publication, making it a truly collaborative work, and this became the standard practice for the animal publications.[39]

The drawings had already been entrusted to Perrault to arrange for their engraving. His lengthy narrative occupied over sixty pages of the *procès-verbaux*, and the activities surrounding the chameleon entirely displaced the chemical disquisitions of Duclos for several meetings. Indeed, Gallois, the secretary, rather testily reminded the *compagnie* that they had already agreed to hear Duclos's opinions on Boyle, but the members put this off to another meeting while also resolving to dissect other animals that resembled the chameleon.[40]

Perrault's description of the chameleon was far more elaborate than any of the 1667 narratives, and it drew liberally on the traditions of humanist natural history. The chameleon was an animal exceptionally surrounded by myth and symbolism, indeed "the most celebrated symbol in moral philosophy and rhetoric, to represent the cowardly complaisance of courtiers and flatterers, and the vanity that simple and fickle minds feed upon."[41] Such pointed language distinguished the serious scientific aims of the academicians from other courtiers. After puzzling over why such a "base and disagreeable" ("*vile et laïde*") beast could be called a lion of any sort, a short

discourse on etymology and general classification (with examples from Belon, Faber, and Gessner, among others) led to detailed physical description
of the animal's body and an account of the academy's experiments with
comparison to the standard sources, ancient and modern. Although Perrault
drew on the *procès-verbaux*, he added both examples and explanations. His
frequent references to "our chameleon" emphasized its singularity.[42]

After twenty pages of external description, Perrault turned to the dissection. Here too the detailed and meticulous account elaborated on the
procès-verbaux. The *compagnie* evidently agreed with Perrault's physiological
views; although the lacteal veins revealed themselves readily enough, a part
"in the middle of the Mesentery" might be either the pancreas of Aselli or
the *receptaculum chyli* of Pecquet: he remained skeptical of Pecquet's explanation of the thoracic duct.[43] He spent several pages on the structure of the
chameleon's tongue and the contents of his stomach, exploding the ancient
myth that chameleons did not eat but gained nourishment only from the air
and the sun's rays. Perrault wittily asserted that the loss of the metaphorical
value of the chameleon's diet and skin was made up for by other qualities,
just as his quick tongue and his all-seeing eyes compensated for his lack of
ears and slow movement. The *compagnie* thoroughly examined the eyes and
particularly the structure of the optic nerves, noting that the nerve from each
eye never joined the other: did the chameleon therefore not have binocular
vision? This was a much-discussed topic in scientific circles; Descartes had
concluded that the pineal gland combined images conveyed by the optic
nerves.[44]

The main myth to be exploded concerned the skin and its color changes.
The *compagnie*'s experiments had shown that the chameleon did not adopt
the color of his surroundings. Perrault identified three theories of the color
change: the reflection theory of Solinus, the suffusion theory of Seneca, and
a "change in the disposition of the particles that compose the skin, following
the Cartesians." Perrault inclined toward the second theory, believing that
the change in colors was actually a suffusion of another color into the skin,
either by change in temperature or, he thought more likely, by the action
of the passions. Choler or yellow bile, "with which that animal abounds,"
could mix with the natural gray-blue of the skin to produce a greenish shade,
such as appeared when the chameleon was in the sun, "where he is happy."
When he was bothered, "black and bitter" humors in his blood could produce darker shades. Whiteness, such as was produced by the linen cloth,
constituted an absence of such humors. Although he did not mention the
soul, Perrault's attribution of emotional states to the chameleon fit with his
and Cureau de la Chambre's ideas about animal soul. Perrault admitted

his ideas were speculative and concluded with a more concrete topic, the animal's skeleton.[45] The large foldout engraving showed the chameleon in life (in a characteristic pose gripping a branch) and its dissected parts in a trompe l'oeil drawing that appeared to be pinned to the page. Each part had an "explication" except the skeleton; the "netteté" of its image and the clarity of the textual description eliminated the need for further explanation. But the engraving of the chameleon in life drew far more upon standard iconographical tropes dating back to Gessner and Belon than upon Pieter Boel's painting.[46]

5. Of Beavers and Bears

Perrault's fifty-page essay on the chameleon was a tour de force, elaborating not only the academy's style and methods but also his own philosophical approach. It was moreover written in elegant and stylish French. If the remaining four essays in the volume—on the beaver, bear, camel, and gazelle—were not as extensive, they nonetheless displayed the multiple meanings and motivations behind the exotic animal project.

While the camel and the gazelle could, along with the chameleon, claim exoticism, neither the beaver nor the bear was rare or unfamiliar. Yet both fulfilled important functions in the all-encompassing symbolic universe of "*le roi-machine.*" The beaver had appeared in every standard animal encyclopedia of the era, going back to Gessner, and Johnstone had pointed out its ubiquity across Europe.[47] Nonetheless, it was, after the lion, among the first group of "*étrange*" animals to be dissected by the academy, in December 1667, and its skeleton was the first to be mounted in the academy's rooms. Jean Gallois read the description of its dissection to the *compagnie* on 1 September 1668.[48]

What was so special about the beaver? Gallois claimed that the ancients spoke little of it and that even modern naturalists had neglected to describe its anatomy. It is true that Johnstone's account of the beaver's anatomy is cursory, Gessner's nonexistent. But more important in this case was the provenance of this particular animal. The academy's beaver was not the European variety, but from North America, "taken in Canada along the banks of the St Lawrence."[49] This origin immediately put the beaver into a category not only of "*étrange*" but also, like the chameleon, the more important category of pleasing to the king, in this case serving as a reminder of his expanding North American empire: the Sieur de La Salle reached Canada in 1666 and embarked on his first expedition into the Ohio country as this volume appeared. Moreover, the French had dominated the fur trade in North

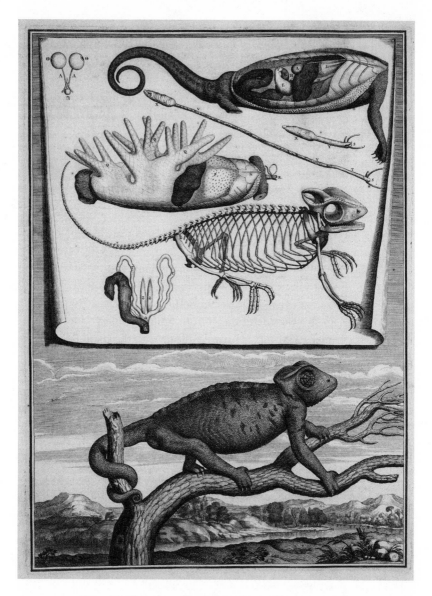

Figure 4.5 Chameleon. Claude Perrault, ed., *Mémoires pour servir
à l'histoire naturelle des animaux*, 1676.
(Linda Hall Library of Science, Engineering, and Technology.)

America since its beginnings in the sixteenth century, although they were under increasing competition from the Dutch and the English. The English founded the Hudson's Bay Company as an outpost for the trade in furs and other North American commodities in 1670. Beaver fur was particularly prized for making felt for hats. While Gallois did not mention the fur trade directly, he described the qualities of beaver fur with care, and the *compagnie* had examined it with a microscope.[50]

The academicians used the microscope sparingly in their anatomical studies, but that they used it at all put them at the cutting edge of research techniques. Faber had described the "Novus & mirabilis tubi opticus seu *Microscopium*" in the 1620s, and Harvey had made a few observations with the aid of magnification for *De motu cordis*. Severino in 1646 mentioned the *microscopia* as a new invention that could be useful for examining small parts.[51] Only in the past decade, however, had the microscope become a regular part of anatomical research. Pecquet had not used magnification, but Marcello Malpighi began to use a microscope in the late 1650s, and Robert Hooke's *Micrographia* appeared in 1665. While Hooke wished to employ the microscope as an alternative to dissection, Malpighi and the Parisians used it as an accompaniment, to see structures dissection could not alone reveal. In theory, the microscope could reveal the micromechanisms that propelled nature, whether Cartesian or some other variety. But the academicians were not immediately concerned with such theoretical questions, and used the microscope rather to examine particular structures.[52] Microscopic examination of the beaver's fur helped to reveal the purposes of Nature, which gave the beaver two types of hair that kept it warm and repelled water. These characteristics also made the fur useful to humans, although Gallois did not draw that conclusion.[53]

Gallois (unlike Perrault) wasted no time in referring to ancient or other more recent accounts, plunging directly into a very precise external and internal examination with a number of measurements. He compared the beaver's morphology and anatomy to several other animals, particularly the otter, but also rats, squirrels, and dogs; indeed, the dog, rather than the human, seems to have been the basis of comparison in many of the animal dissections. Gallois spent considerable time describing the beaver's paws, which showed that "Nature has destined this Animal to live in the water as well as on land." Like Faber, the academicians were fascinated with the animal's musk glands and spent several pages in their examination, skinning the animal to get a better view of its hindquarters, and dismissing the accounts of several who had confused the musk glands with the testicles, in-

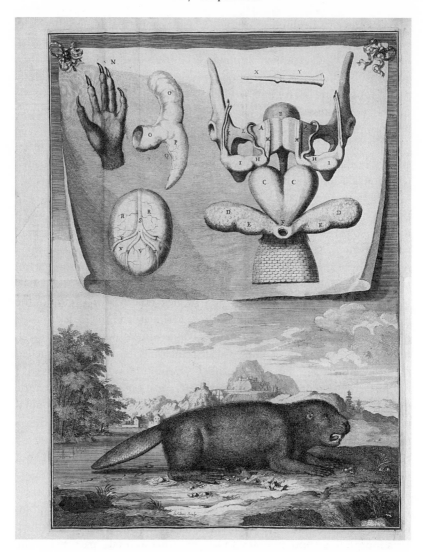

Figure 4.6 Beaver. [Claude Perrault, ed.], *Mémoires pour servir
à l'histoire naturelle des animaux*, 1671.
(Wellcome Library, London.)

cluding Rondelet.[54] During 1668, the academy dissected other animals with
scent glands, including a polecat and two civet cats.[55] Castoreum from the
beaver and other varieties of musk were valuable ingredients in medicines
and perfumes. But they could not dissect the beaver's eyes because they had
been eaten by a rat, an indication, perhaps, of the less than ideal conditions

at the Bibliothèque du roi. Gallois's account was strictly morphological and said nothing about function.

The bear was even more familiar than the beaver: dancing or fighting bears were common street entertainment, and Johnstone had pictured these semidomesticated creatures chained and muzzled. Medieval *jongleurs* and contemporary *spectacles* featured performing bears alongside dogs and monkeys. At Louis XIV's "Grand Carrousel" of 1662 at the Tuileries, performing bears were part of the entertainment, and a bear, a dromedary, and an elephant took part in the Versailles *spectacle* titled "Les plaisirs de l'isle enchantée" two years later. As with the lion, the act of looking at the animal and then dissecting it served further to demystify it, even to domesticate it, or at least "deexoticize" it. Pliny and other ancients noted the many similarities between bears and humans: they stood on their hind legs, had "hands," and ate a varied diet. Johnstone claimed they even had sex like humans, not like other animals. If we are to believe Gessner, bears attracted more myths and legends even than chameleons, among them the notion that their young were born without form and required maternal licking to give them shape.[56]

In contrast to the beaver, Perrault said nothing about the provenance of the bear whose dissection he recounted to the academy in June 1668, nor did he state when or where the dissection took place. Various clues indicate it was a captive animal, most likely from Vincennes. Perrault denied ancient claims of similarities between the bear and the human. Galen and Pliny may have thought their "hands" were similar, but Perrault concluded, with much anatomical detail, that they were not, and bears' paws were in addition highly nutritious and good to eat, further distancing them from human hands. He expressed skepticism but did not entirely doubt a tale told by both Aelian and Pliny that a bear lived for forty days simply by licking its right foot. In fact, Perrault wrote, bears resembled lions in many ways.[57] Here lies a clue to their significance. The lion was not only among the first animals to be dissected by the academy, but it also held enormous symbolic significance at Louis's court. Louis viewed himself as the new Alexander, whose symbol was a lion. As we shall see, Perrault would play on the Alexander theme in the preface to the *Mémoires pour servir à l'histoire naturelle des animaux*. Bears too held considerable symbolic meaning and in ancient times had symbolized royal status. It is not surprising that Louis would want to appropriate this symbol as well.[58]

Although the weather was cooler than when the lion was dissected, the bear, once opened, was particularly pungent, and its entrails were doused with several pints of eau-de-vie. More was distributed on handkerchiefs for the *compagnie* to hold to their noses.[59] Perrault's account of the dissection of "our bear" focused particularly on the kidneys, which in bears, unlike in other animals, have multiple small lobes; Perrault counted fifty-six, describing their structure as resembling a bunch of grapes.[60] Unlike Gallois, Perrault spoke at length about the kidneys' function. He described in detail the three different vessels (the emulgent artery and vein and the ureter) that entered each kidney and "lost themselves in the Parenchyma" of the kidney. He was not convinced by the arguments of Lorenzo Bellini, who had argued in his 1662 pamphlet *Exercitatio anatomica de structura et usu renum* that the vessels branched into "nearly infinite" divisions, therefore becoming imperceptible. Perrault was more inclined to agree with another mechanist, Nathaniel Highmore, who had discussed the structure of human kidneys in his 1651 *Corporis humani disquisitio anatomica*.[61] Highmore described a process whereby chyle traveled to the liver, then via the vena cava to the heart and the arteries. The emulgent arteries were "attracted" to the kidneys, which acted to separate out one kind of "excrement" from the blood. According to Highmore, arterial blood "percolated" through the spongy parenchyma of the kidneys, straining off urine and leaving venous blood. Perrault employed homely metaphors to explain this process, comparing urine to the whey left behind in cheese making, and the filtering process to soapy water passing through fabric without the existence of "*canaux*" to guide it.[62] He did not indicate any use of magnification. Perrault's acceptance of the relative roles of arterial and venous blood in the kidneys assumed (as did Highmore) that the blood was therefore formed in the liver, which remained on its throne.

Perrault also wrote the descriptions of the camel and the gazelle. He presented his account of the camel (or "*dromedaire*") to the *compagnie* in December 1668, with no indication where or when the dissection occurred, or the origins of the animal, but there had been a camel at Versailles at some point. Perrault debated whether the animal in question was a camel or a dromedary, citing numerous observations of other authors as well as his own interview of an Arab ambassador. On the basis of the Arab's testimony, he decided it was a dromedary, defining the dromedary as the variety with one hump, although there was much confusion about this; Johnstone had come to the exact opposite conclusion. If Perrault's dromedary looked more like a llama than a camel, at least it did not have the horse's head of one of Johnstone's.[63]

The origins of the gazelle, on the other hand, were quite clear: it was one of four that had died at Versailles in the past year, all of which had been dissected. Perrault read his description to the academy on 29 December 1668. It compared the four gazelles (three females and a fawn, also female) to one another, but maintained "the gazelle" as a category.[64] But within a few years, Perrault would change his mind about the idea of natural categories, and attempt to view each individual separately.

This volume too earned notice in both the *Philosophical Transactions* and the *Journal des sçavans*.[65] The latter gave the short book extravagant praise, claiming that such work on natural history had not been done since the time of Alexander the Great, who alone had devoted attention to such matters. Now, the author (presumably Gallois) went on, there was another king whose "magnificence does not shine any less on all that can serve the advancement of the sciences." At the "Bibliothèque de sa Majesté" men worked incessantly "to dissect all sorts of animals, to verify what Naturalists have advanced, and supply what they have forgotten." These observations, said Gallois, will one day fill "many volumes."[66]

6. The *Mémoires pour servir à l'histoire naturelle des animaux*

The 1669 volume opened with a short note "from the printer to the reader":

> After having printed last year the anatomical descriptions of a Renard Marin and a Lion, which were extracted from two letters written to Monsieur de la Chambre, I continue to give to the public the Observations that are made in the dissection of all sorts of animals in the King's Library. These five Descriptions that I have put in this collection, are those of animals of which I have found the figures engraved. I hope to give others as soon as the Engravers furnish the plates.[67]

The five described in 1669 were only a fraction of the number of animals dissected in the early years of the academy; a cursory count indicates at least forty different animals between 1667 and 1670.[68] Between 1671 and 1676, the French royal printing office published six illustrated folio volumes of aspects of the work of the Paris Academy of Sciences. Five of them were elephant folios, fifty-five centimeters in height. These included two editions of the *Mémoires pour servir à l'histoire naturelle des animaux* (henceforth referred to as the *Histoire des animaux*), published in 1671 and 1676; Jean Picard's *Mesure de la terre*, on the circumference of the earth, also published in 1671 and often bound with the *Histoire des animaux*; and the *Mémoires pour servir à l'histoire des plantes*, first published in 1676 and reissued in 1679. Another even larger folio (fifty-eight centimeters) from 1676 was the *Recueil des plusieurs traitez de mathématique*, which collected several shorter works by academicians, including some that had previously been published separately, such as Picard's work, as well as Nicolas-François Blondel's *Résolution des quatres principaux problèmes d'architecture*, which had first appeared in 1673. In addition, a description of Christian Huygens's pendulum clock, *Horologium oscillatorium*, appeared as an ordinary folio in 1673.[69] Claude Perrault's 1674 translation and abridgement of the ten books on architecture of the Roman Vitruvius may also be seen as part of this group, although it was printed by Jean-Baptiste Coignard rather than the Imprimerie royale. All boasted the same royal emblem on the title page and followed similar formats, and all were illustrated with detailed, full-page engravings by some of the best-known artists of the time.

By 1669, plans were unfolding for a much bigger publication project. The Imprimerie royale, the royal printing house founded by Richelieu in 1640, had moved from the Louvre to the rue Vivienne, next door to the *bibliothèque*, by the end of 1666.[70] The printing house employed a battery of artists and engravers, including Louis de Châtillon, Sébastien Leclerc, Abraham Bosse, Jacques Bailly, and Israel Silvestre. In March 1667, Colbert moved the king's collection of engravings to the rue Vivienne as well. Also housed at the rue Vivienne was the Académie des inscriptions et belles-lettres. The *"petite Académie"* met on Tuesdays and Fridays to discuss medals, engravings, inscriptions, and other works for the glory of the king. Charles Perrault was its secretary.[71] At the end of 1667, the Council of State allowed Colbert to centralize control of engraving, and only engravers of his choosing could depict the king's objects. Included among these were "les figures des plantes et animaux de toutes espèces et autres choses rares et singulières."[72] Already busily producing engravings of Louis's chateaux, his artworks, views of the towns of France, and images of Versailles, these artists now turned to drawing and engraving plants, animals, and astronomical observations. Among these, Leclerc, Bailly, and Bosse particularly worked on the animal studies. Two years after the 1669 volume, a much different publication emerged from the Imprimerie royale.

By 1670, the atelier on the rue Vivienne had produced hundreds of engravings. A *mémoire* from February 1670 from Colbert to Charles Perrault (as secretary of the *"petite Académie"*) and Pierre de Carcavi, head of the Bibliothèque du roi, discussed the publication of these engravings.[73] Colbert's *mémoire* is speculative and proposes several ideas for publication, but it reveals the place of the Academy of Sciences' work among the other artistic and scientific projects of the crown. One of his proposals was to publish groups of engravings on various topics in annual volumes that would display the work of the past year. So, for example, at the end of 1670, the king would receive a volume composed of ten or twelve engravings of animals and twelve to fifteen of plants, along with a variety of other images to round out a volume of forty or fifty plates, including ancient medals from the royal cabinet, the king's busts and statues, carousels, tapestries, royal houses, "and all other things of the same genre." At the end of ten or twelve years, these volumes would compose an entire history "de toute sorte de science."

This plan was soon abandoned in favor of separate volumes on each topic. But it is clear that Colbert viewed the images as prior to any text. The composer of the volumes should be "a clever man" who can get all the plates in the right order and in a uniform size and quality. As if as an afterthought, the *mémoire* adds, "Choose someone else to make the text and the descriptions." The *mémoire* lists seven animal engravings but does not mention that they had already been published in 1667 and 1669. It does acknowledge that two of the engravings are in quarto format (presumably those from 1667) and would need to be enlarged for a uniform folio.[74] Colbert also suggested that if the number of engravings of animals and plants was insufficient to make two volumes, they could be joined in one, consisting of fifteen or twenty animals and twenty or thirty plants. But the 1671 volume contained only animals, with fifteen descriptions of thirteen different species.

In Colbert's *mémoire*, the images retained priority: "As far as the anatomical dissections Mr Perrault physician who has done the seven that are printed can continue to do them more easily than another because he did or saw done the dissections of the animals or drew the whole and the parts. . . . He also [will take] such care [as] to make understood the observations he has made."[75] Claude Perrault's value lay in his anatomical skill that could produce the dissections that led to better engravings, and secondarily in his literary skill that could explain the meaning of the images. Colbert did not mention Perrault's artistic skill as a draftsman. His (and later Philippe de la Hire's) initial drawings of animals and their dissection served as the basis for the artists and engravers who followed.

The *Mémoires pour servir à l'histoire naturelle des animaux* was the first of several presentation volumes derived from the work of the Paris Academy.[76] It served a number of functions: it was a work of art, printed on fine paper, with full-page engravings of each of the animals described, bound in red morocco with the fleur-de-lis of the crown embossed in gold.[77] The illustrations used a combination of copper engraving and etching, in which the copper plate was dipped in an acid bath to deepen the lines to give certain effects; both techniques were used in each image. This combination, employed mainly in fine art, was labor intensive and not often used in illustrations, indicating the high value placed on the work by the crown. The title page, emblazoned with a large royal emblem encompassing a crown, scallop shells, and fleurs-de-lis, named no author. The illustrations alone cost over four thousand livres.[78] It was a patronage object, too rare and expensive for the open market; the two hundred copies of the book, noted Alexander Pitfeild, who translated the book into English in 1687, were given away by the crown and its agents to "persons of the greatest quality."[79] It provided public evidence of Louis XIV's *gloire* in several ways: as a work of art, a display of his power, an imaginarium of his menageries, and a contribution to the new science. This volume was reissued in a revised and expanded form, with thirty descriptions of thirty-one different species, in 1676. The additional descriptions also appeared as a separate volume.[80]

Publication of the new knowledge gained from the academy's dissections benefited the wider Republic of Letters as well as fulfilling scientific and symbolic functions for the academy. It asserted Louis's control of natural knowledge and its creation in the same way he controlled other forms of cultural production, while declaring the high status of the new science as an activity that the court supported. The front matter included an illustration of a mythical visit of the king to the Paris Academy that not only assumed royal patronage but hoped it would continue; Louis fulfilled this wish with a visit in 1681. Moreover, the *Histoire des animaux* functioned as a virtual menagerie that could extend the reach of Louis's glory beyond those few who could actually visit Versailles or Vincennes and see the animals themselves. It was a "paper zoo" like the "paper museums" that contemporary antiquarians compiled, consisting of engravings of antiquities, or Cassiano dal Pozzo's natural history *museo cartaceo*.[81] At the same time it served as a virtual *spectacle*, an entertainment of the sort featured in fairs that combined puppets, rope-dancers and other acrobats, music and dancing, and not least, animals, both performing and simply on display.[82] Unlike Aldrovandi, who had included "usus in spectaculis" among the categories in his description of the elephant, this purpose was never explicit in the *Histoire des animaux*.[83]

Images of animals from a number of publication projects, including the *Histoire des animaux*, appeared in other works of art celebrating Louis's reign. Colbert's *mémoire* viewed these images as essentially interchangeable. Indeed, as we shall see in more detail in chapter 5, animals were everywhere at Versailles, outdoors in fountains, statues, and architectural details, and indoors in paintings and tapestries. They also featured in the elaborate *spectacles* that displayed the king's glory, not to be confused with fairs and other popular entertainments.[84]

While the 1667 descriptions revealed little of this symbolic role, the narratives from 1669 onward fully acknowledged it. Perrault's preface to the 1671 *Histoire des animaux* effusively praised Louis, employing a comparison to Alexander the Great that Gallois had used in his review of the 1669 volume.[85] This theme had been established in a series of paintings (and later tapestries) by the court artist Charles LeBrun beginning in the early 1660s, and continued in the decoration of various rooms at Versailles in the 1670s and tapestries and engravings in the 1690s.[86] The academy's natural history, Perrault wrote, would "not be unworthy of the greatest King who ever lived." Since Louis was the equal of Alexander in everything else, he would be equal in this too, with the academy assuming the role of Aristotle. As Charles Perrault would proclaim a decade later, this was indeed the "siècle de Louis XIV." No detail was too small for "*le roi-machine,*" and the order of presentation in the *Histoire des animaux* played further on the Alexander theme. It was not the order of the earlier publications, or the order of dissection. Instead, the lion came first, introduced with a headpiece replete with royal symbols. Perrault introduced the comparison to Alexander toward the end of the preface, and the reader would then turn the page to the magnificent full-page image of the lion (see fig. 4.1), followed by the symbolic headpiece. This echoed Jacques Bailly's 1668 presentation volume *Devises pour les tapisseries du roy,* which featured two lion heads among the numerous symbols on its title page, and Jean de la Fontaine's fable "Le Cour du Lion" presented the king himself as a lion.[87] Medieval bestiaries had also most often commenced with the lion, the king of beasts.[88]

7. Perrault's Preface

As well as situating the *Histoire des animaux* within the royal universe, Perrault's preface argued for the credibility and legitimacy of the Paris Academy as a scientific institution. It established principles for animal observation, description, and anatomy, blurring the boundaries between descriptive natural history and analytical natural philosophy by establishing observational data, *historia*, as sufficient in itself to make knowledge, without an overarching causal theory.[89] These distinctly anti-Cartesian principles were in keeping with the Paris Academy's ban on Cartesians or other system builders.[90] Yet Perrault's claim that "what is most considerable in our *Memoires*, is that irreproachable witness to a certain and recognized truth" has a ring of Descartes's clear and distinct ideas. However, these truths came not from within an individual mind, who "only perceives easily what confirms the first thoughts he had." The witnessing of the *Histoire des animaux* was "irreproachable" insofar as it was collective: "The *Memoires* are not at all the work of an individual [*un particulier*], who can let his own opinion prevail. . . . Our *Memoires* only contain facts that have been verified by an entire Company, composed of men who have the eyes to see these kinds of things, apart from most of the rest of the world." Only the trained eye could truly see; the members of the academy were not ordinary "merchants and soldiers" who might make observations by chance, but men with the "esprit de Philosophie."[91]

According to Perrault, the *compagnie* refused to speculate about causes or to generalize from particulars or indeed to state any general principles at all. Its members were less concerned with animal mechanism than with animal structure, the *"particularitez"* of each animal. Although the general tenor of the descriptions was mechanistic, they did not, as we have seen, always conform to current theories, particularly with regard to the formation of the blood. The duty of the natural philosopher, said Perrault, was first simply to describe. Any attempt at generalization was premature, because each animal was unique. It was singular because in many cases it was the only, and possibly unreplaceable, specimen in the royal menageries. It was rare because no other specimen existed in France. Claire Salomon-Bayet, following Gaston Bachelard, has referred to Perrault's "systematic pluralism" as a sign of modernity.[92]

But as he had revealed in the debates over blood transfusion, Perrault remained in many ways the faculty physician. His objections to blood transfusion centered on a Galenic and medical recognition of the specificity of the individual. Blood was not interchangeable, even among members of the same species, because each person (and each disease he or she suffered) was unique and individual, the product of a particular combination of humors and temperaments. Even with multiple specimens of the same animal, such as the lion, dissection revealed many differences, and long experience had shown that dissection itself yielded inconsistent results. Perrault's emphasis on the individual and the collection of discrete facts contrasted with those such as Descartes who sought causal explanations. If it was not as Baconian a program as Fontenelle later painted it, it did share a faith in the unique fact.

Dissection revealed what could not ordinarily be seen. Perrault characterized the *compagnie* as mere "reporters" who held up a mirror to nature and described "avec simplicité et sans ornement" what they had seen, a "peinture naïve," recalling the "peinture exacte" that Perrault had made during the course of the dissection. *"Naïve"* had the particular meaning in this period of "natural" or "without artifice"; the 1694 *Dictionnaire de l'Académie française* added a second definition: "what well represents the truth, what well imitates Nature." A "peinture naïve," then, offered a true description of nature.[93] "Avec simplicité et sans ornement" echoes Descartes's critique of classical rhetoric in the *Discourse on Method* and the *Meditations*. In his reply to Gassendi's critique of the latter, he claimed (perhaps disingenuously) that the clear and distinct language of natural philosophy made the ornamentation of traditional rhetoric obsolete.[94] Although the mirror was a commonplace image in seventeenth-century natural philosophy, simply "to see as in a mirror" held several layers of meaning in the *Histoire des animaux*.

Perrault claimed that the *compagnie* saw "de mesme qu'en vn miroir, qui ne met rien du sien, & qui ne represente que ce qui lui a esté presenté" ("the same as in a mirror, which adds nothing of its own, and only represents what is presented to it"). The use of the word "*miroir*" rather than the more common "*glace*" immediately evoked the etymologically related Latin word "*miror*" ("to wonder"), and the subjects of the *Histoire des animaux* were indeed wondrous to behold. The "mirror of nature" was in any case a venerable trope. Paul the Evangelist, like Plato, rejected sense experience—employing the metaphor of a mirror—as providing only imperfect knowledge compared to spiritual knowledge: "For now we see in a mirror dimly, but then face to face." Similarly, Augustine contrasted the mirror of the mind, which reflected the deceptive world of the senses, to the mirror of the soul, which reflected ideal truth. But by the seventeenth century, mirrors had so improved that they no longer reflected dimly, but clearly, and humanistic imagery echoed this change. Perrault's clear mirror reflected not an analogical truth but an empirical one.[95]

Perrault also recalled to the attentive reader the critical development of mirrors themselves. Costly and beautiful, mirrors were status symbols. Colbert attempted to break the Venetian monopoly on the production of mirrors in the mid-1660s and established a successful factory by 1670.[96] Nicolas Fouquet had possessed a famous collection of them, and construction of the Hall of Mirrors at Versailles began in the late 1670s. It was completed in 1684.[97] Mirrors were also increasingly significant to science, used extensively in optics and astronomy, and Perrault may even have been referring to the practice of some painters (such as Vermeer) of using mirrors or the camera obscura as drawing aids.[98] Mirrors, then, indicated not only a heightened level of verisimilitude, but also a particular way of looking at nature and possibly a veiled reference to Versailles itself.

The *Histoire des animaux* differed from contemporary accounts either of natural history or of comparative anatomy.[99] Unlike most naturalists, Perrault paid much more attention to anatomy than to classification, noting the particular care the *compagnie* devoted to the verbal descriptions, "marking the size and proportion of all the parts which are seen without dissection, because these are as little known as all that is enclosed within."[100] Unlike most comparative anatomists, however, Perrault and company described a wide variety of different animals rather than focusing on a single kind or a single organ or system, reflecting their sources in Louis's menageries, so to a certain extent, this accessibility (or lack) drove their science. The order of presentation was neither that of the previous publications nor the order of dissection, and it followed no discernible system of classification. The chameleon followed the strategically placed lion. The renard marin was seventh. New animals included four catlike animals—the loup-cervier (Canadian lynx), civette (civet cat), chat-pard (probably a serval), and coatimundi—as well as two more North American animals, the otter and the elk.

The 1671 volume comprised fifteen descriptions of thirteen different species—the three lions each merited its own description. Five years later, an additional fifteen chapters described seventeen more species.[101] The *compagnie* did not attempt to classify this motley assortment of animals. Each animal was an individual, and only individuals carried ontological meaning. They were not even exemplars of a type, because it was not clear what the type was. As new individuals appeared, their similarities to and differences from others were carefully noted; each description was "toute particulière." In 1676, Perrault spent considerable time trying to sort out the various species and varieties of monkeys, noting differences in tails, fur, and general morphology, but added confusion rather than clarification.[102]

Perrault placed the hedgehog and the porcupine together, he said, because the ancients classified them together. Johnstone had also pictured them together but was uncertain whether they were of the same kind. His image showed them as nearly equal in size. The academy's image showed a much larger porcupine, and Perrault's narrative debunked any notion that they might be similar. On the other hand, he placed the Sardinian and Canadian deer together because of their physical similarities, despite their geographical separation.[103] Yet when in the text he acknowledged geographical specificity, he went beyond his customary humanist textual references in seeking new empirical knowledge of places. In addition, this acknowledgment of varied origins implicitly demonstrated the monarch's power and the extent of France's colonial empire, which ranged from the "cerf du Canada" to the "grande Tortue des Indes" taken on the coast of Coromandel.

The 1676 volume reflected the academy's intensive dissection activities, which intensified after Duverney assumed the position of anatomist in 1674. It also reflected the court's increased preference for Versailles over Vincennes as a site of animal display. Within Versailles, the emphasis was on ornamental animals rather than fierce beasts, although fierce beasts certainly remained. These ornamental animals consisted mainly of exotic birds, and the 1676 volume included eight birds while the 1671 volume had none. This new emphasis on beautiful birds such as the "Demoiselles de Numidie," graceful African cranes, pointed to the court's modeling of values of self-restraint and civility for the aristocracy to replace the older warlike values symbolized by the battling animals at Vincennes.[104] But the inclusion in the *Histoire des animaux* of predatory birds such as the eagle, the cormorant, and the cassowary complicates an idea that the royal menageries as chronicled by the Academy of Sciences aimed solely toward the promotion of civility. Louis waged war throughout the 1670s, and the warlike values of the lion and the eagle were equally prominent as the feminine qualities of dancing birds.

Perrault therefore navigated a multiplicity of meanings and values attached to the animals of the *Histoire des animaux*, and his and the academy's refusal to classify indicates their unwillingness to privilege one set of meanings over another. The *compagnie* in this way also differentiated its method from the generalizations of other naturalists, both ancient and modern.[105] The "grands & magnifiques ouvrages" of Aristotle, Pliny, and Aelian sought to classify, order, and find similarities, but these aims rested on faulty foundations. In a report to the academy on its anatomical activities at the end of 1669, Pecquet had extolled the dissection of animals alive and dead, "ordinary and extraordinary," to add to Aristotle's *Historia animalium* "the exact anatomy that it lacks." This precision meant that one could not claim, said Perrault, that all bears had fifty-six kidneys on each side, only that the bear they had dissected had this particular conformation. So the gazelle described in 1669 no longer functioned in 1671 as a representative of the three (by 1671, four) other gazelles; each merited its own consideration. The 1671 *Histoire des animaux* also added descriptions of two other lions, a female and another male, to the original 1667 account. The "Avertissement" to the 1676 edition made this policy explicit: as more animals of the same species were made available, their descriptions had been rewritten "because it is important to note as much as possible the differences and the similarities that one often encounters in animals of the same species."[106]

"Species" ("*espèce*") here functioned not as a biological term but as a logical one, a means of organizing knowledge. It acknowledged morphological similarity, but was not a taxonomic category. Thomas Aquinas had written that "in natural things the shape is the sign of species," and external morphology was the main distinguishing characteristic among natural types for Perrault and the academy.[107] But it was not the only one. The ancients had noted the differing sizes of the porcupine and the hedgehog but had concluded that they were essentially the same. Perrault disagreed: their bodies differed not only in size but in shape; their spiny protuberances were quite different in length and quality; and they originated in widely different places. All of these characteristics made these animals distinct species. Yet the extent to which Perrault was caught between humanist etymology and modern empiricism is evident in his straining to reconcile the ancient beast known as the *alce* with the modern elk, despite considerable differences in their descriptions.[108]

The "chat-pard" provides a good example of this open-ended notion of classification. This cat, small but spotted like a leopard, was a puzzle to the *compagnie*. It came from Africa, and might be explained as a "mélange of different species," which was more common on that continent. Aristotle had explained in the *Historia animalium* that the lack of rain in Libya compelled animals to mingle at waterholes and "mate with each other even if not of the same breed" if the animals were of similar sizes and had similar periods of gestation.[109] But what animals could have combined to produce this particular animal? The possibilities were a leopard or a panther and a common cat, despite the differences in size and period of gestation. The *compagnie* was not satisfied with either of these answers, yet found evidence that the animal was a hybrid of some kind in the imperfection of its generative parts: perhaps, like a mule, it was sterile. But it was also very fat, which made some wonder if it had been castrated. They debated for some time during the dissection on a cold January day in 1670. In the end, they came to no conclusion and simply described the animal and its parts.[110]

The coati was a different kind of puzzle. "The *Coati* is an Animal of *Brazile*, which is variously described by Naturalists; and their Descriptions do not exactly agree with what we have observed in ours: which may cause a belief that there are several Species."[111] Perrault noted discrepancies among the descriptions of others such as Johannes de Laet and Georg Marcgraf as well as between their descriptions and the academy's observations. In particular, several artists' depictions of the animals in situ at Versailles diverged from earlier descriptions of the length and shape of the snout, which Marcgraf and the sixteenth-century explorer Jean de Léry had compared to the trunk of an elephant. Nonetheless, there was enough in common among the descriptions and images to assert that the specimens in the King's Library were indeed coatis. Following Marcgraf, Perrault further distinguished on the basis of coloration between a coati and a coatimundi.[112] Confronted with such variety, Perrault and the academicians did not feel they could classify beyond the most basic types. New observations could always arise that could cast doubt on any agreed-upon certainties. This distrust of established principles revealed the fundamental nature of the academy and of Perrault: a profound conservatism underlay its determined modernism.

8. The Illustrations

The illustrations in the 1671 and 1676 volumes of the *Histoire des animaux* asserted the volumes' status as a work both of natural history and of comparative anatomy, and the academy's twofold project of description of the living animal and dissection of its dead body.[113] A full-page illustration accompanied each of the animals discussed, displaying the animal in life as well as some of its dissected parts, which were presented in trompe l'oeil fashion as another sheet pinned to the underlying image. Perrault employed a similar format in his 1674 work on the Roman architect Vitruvius, with the same aim of showing the interior and exterior simultaneously.[114] Although illustrations played critical roles in natural history, they were, as we have seen, much less important to anatomists, who relied more on textual description.[115] Among recent works, Severino's *Zootomia democritaea* illustrated only isolated dissected parts. Johann Faber's portion of the Hernández *Thesaurus* had included an image of each animal at the head of the chapter, and occasional additional images of dissected parts or other animals. Johnstone's volumes were well illustrated with multiple examples of certain animals; but the impact of the illustrations was blunted by their grouping together at intervals in the volumes rather than interspersed throughout the text. He did not include any images of dissection.

The trompe l'oeil format was lost in later reprints, such as the 1733–34 quarto reprint, which separated the two images, but it was retained, though in a somewhat smaller size, in the 1687–88 English translation of the 1676 edition.[116] Like the cabinet, where one might view both a whole specimen and various parts, the illustrations in the *Histoire des animaux* revealed more than the animal in the menagerie, or a single dissected specimen, could do. They showed that the dissection literally superseded the live animal, revealing the *compagnie*'s process of observation followed by dissection. While dissection revealed more than simple observation, it was also destructive. Once the animal was dissected, it was no longer whole, and because it was a singular specimen—often no other examples existed in Paris—the underlying illustration provided the only evidence of its existence. The cabinet might solve this problem by having several examples of the same animals in various states of dissection; the catalog of Duverney's cabinet described such an arrangement. The illustration therefore took the place of these multiple specimens.[117]

Unlike the cabinet, however, the illustrations in the *Histoire des animaux* also depicted the live animal in an imagined natural setting. This style departed from most previous works of natural history. Illustrations in such

Renaissance works such as Gessner's *Historia animalium* were flat and without depth or context, and not separated from the text. The image in no way supplanted the text, but the concrete image served as a counterpoint to the heterogeneous verbal descriptions. Although Gessner claimed that his images were "ad vivum," many were not original to his book.[118] This flat and noncontextual style—and the reuse of images—continued to dominate natural history illustration for the next century. Johnstone's illustrations showed the animals in a natural setting, sometimes even in action in the background, but like Aldrovandi's, the images were stacked on the page in a way that emphasized their factitious quality.

The *Histoire des animaux* encompassed natural history, comparative anatomy, and human anatomy, but the *compagnie* also self-consciously aspired to make and display new knowledge of these new animals gained from direct observation and careful dissection, and the illustrations reflected this approach. The image of the live animal often derived from sketches made while it still lived in the menagerie. The king employed artists such as the Fleming Pieter Boel (1622–74) to document animals in his menageries, particularly at Versailles. Boel's images of animals appeared in tapestries and paintings, and at least a few also formed the basis of the engravings in the *Histoire des animaux*, although, as we have seen, Boel's painting of the chameleon differed significantly from the image in the book. Perrault, or later others such as Philippe de la Hire, drew the animals as they arrived from the menageries and also during and after their dissection; thus, some of the images were from death rather than from life.[119] Perrault claimed a parallel precision between the text and the image. The *compagnie* examined the whole animal and its dissected parts. Then the person designated to write the description was also charged with drawing the animal "sur le champ." But even this was collaborative: this drawing did not serve as the basis for engraving until all those who had been present at the dissection confirmed it. Artistic skill was not the first requirement, "because what was important was not to represent well what one sees, but to see well what one wishes to represent."[120] Perrault minimized the extent of the collaboration between the artist and the scientist to privilege the eye of the natural philosopher above all others and privilege the Paris Academy above all the other tenants at the rue Vivienne.

Nonetheless, the alliance between natural philosopher and artist was key to this interchange among observation, epistemology, and picturing. While Perrault's preface elevated the status of the natural philosopher above the artist in order to establish the position of the Paris Academy in the hierarchy of royal patronage of the arts and sciences, his assertion appears to be more of an ideal than an account of what actually took place. A series of *mémoires* to Colbert throughout the 1670s indicates lengthy and sustained collaboration among artists, engravers, printers, and naturalists, who occupied the same buildings for many years. Later, in a work entitled *Le cabinet des beaux-arts* composed with his brother Charles, Claude Perrault acknowledged engraving as one of the arts that had reached perfection during the seventeenth century.[121]

Perrault noted differences between the published image of the live coatimundi and his own observation of the animal on the dissection table. The implication is that he never saw the animal alive, and that he depended on the artist's observation to depict the live animal.[122] But even if the images were indeed drawn from life (or death), their iconographic presentation often drew upon older illustrations. In the case of the chameleon, the pose closely resembles the image in Pierre Belon's *De aquatilibus* of 1553, an image subsequently used by Gessner.[123] There are also important differences between the two images, particularly the eye and mouth, and the later image more closely resembles the actual animal. Yet compared to Boel's contemporaneous painting of a chameleon, the illustration nonetheless looks posed and static, designed to display certain anatomical characteristics—the eye, the feet, the tail—rather than to give an impression of a living creature. The image of the beaver too appears to have been copied—though slightly modified—from Gessner, including the pose, the shape and position of the tail, and the bared teeth. It is a picture of a picture, and one with which Perrault's audiences would certainly have been familiar. The animals, moreover, appear whole and healthy, not aged or injured as they may have been at the time of their deaths. The organs too are precise and clean, not bloody or mutilated.

Most of these engravings bore the name of Sébastien Leclerc. Born in 1637, Leclerc came to Paris in the mid-1660s and entered the circle around the court painter Charles LeBrun. LeBrun advised Leclerc to devote himself to engraving, and Leclerc quickly rose to the top of his profession; his royal stipend of eighteen hundred livres surpassed that of most academicians. In 1670, Leclerc engraved the designs for royal tapestries (*Devises pour les tapisseries du roi*) that Jacques Bailly had painted in the mid-1660s for the Gobelins tapestry works. Bailly's designs included animals and plants in allegorical settings based on the four seasons and the four elements, and the allegorical *culs de lampes*, or tailpieces, in the *Histoire des animaux* closely resemble those of the *Devises*.[124] Leclerc's appointment to engrave the animal volumes gives further indication of the prestige they were intended to accrue; among his other works were a series on Alexander and one on Louis XIV's battles. Although the 1670 *mémoire* to Colbert about engraving assigned the lion and the renard marin to Bailly, in the printed volumes the engravings bear Leclerc's name.

The emphasis on direct observation by naturalist and artist extended only to the figures of the animals, which were placed in imaginary landscapes that bore little correspondence to the animals' native habitats. Backed by a dramatic sky, the lion stood in a rocky landscape quite unlike its native savanna, but also unlike the enclosure he would have occupied at Vincennes. The camel was situated before a rural and very French-looking landscape with a small farmhouse and deciduous trees. Many animals were depicted on hilltops with cities in the background. The 1676 text identified the tortoise as originating in the East Indies, but its image occupied a temperate rather than a tropical landscape, with a classical temple, obelisk, and bridge. Sardinian and Canadian deer, from widely separated parts of the world, were depicted together as they would have lived in the Versailles menagerie, with the further embellishment of a forest of what appear to be chestnut trees. Other illustrations similarly showed generic landscapes of the sort that were common backgrounds in human anatomical illustration.

Contemporary painters such as Claude Lorrain and Nicolas Poussin as well as the court painter LeBrun mingled classical architectural elements with natural landscapes.[125] A few illustrations showed ruins or tombs, commonly inserted in human anatomical illustration as a memento mori, thus making a statement, probably inadvertent, about parallels between human and animal souls, while following artistic convention. Even if human figures were absent, buildings in the background reinforced a human presence in these images. These landscapes did not acknowledge habitat but ownership: although they were depicted outside the enclosures of the menageries where they had spent their last days, these animals nonetheless resided in domesticated landscapes for denatured animals, human settings rather than animal settings. The inclusion of dissected parts further asserted human hegemony.

In the 1676 volume, the monkeys are depicted in the most humanized setting of all, the terrace of a chateau, with a formal, geometrical parterre in the background. If this was not the actual palace of Versailles, the image was certainly meant to evoke it, and one historian has identified Charles Perrault's labyrinth in the background. The parterre and the potted plant in the foreground portray a nature completely subsumed to human desires, particularly the king's desires: the orange trees at Versailles had been confiscated from Fouquet's Vaux-le-Vicomte.[126] As with the deer, two different but similar species from opposite sides of the world are pictured together: the "sapajou," or capuchin monkey, from South America, and the "guenon,"

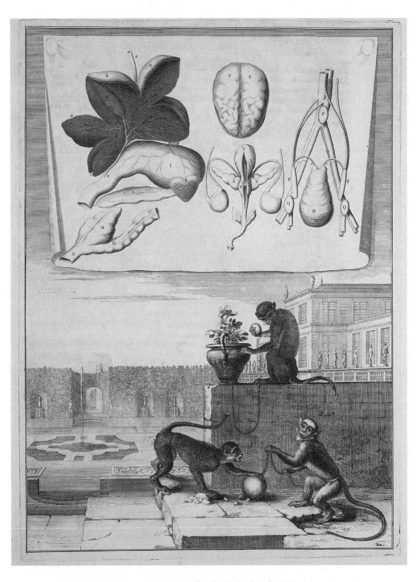

Figure 4.7 Sapajous et guenon. Claude Perrault, ed., *Mémoires pour servir
à l'histoire naturelle des animaux*, 1676.
(Linda Hall Library of Science, Engineering, and Technology.)

or cercopithecus, from Africa. The buildings are not in the background but constitute the habitation for the animals. Among all of the animals, they are the only ones shown under restraint, with a ball and chains. The chains are delicate, and the monkeys appear to be playing with the single small ball, its form echoed in the fruit another monkey holds.

Performing monkeys and apes were frequently seen at fairs and other spectacles, and the ape Fagotin had a large following in midcentury Paris. There were at least two of that name. The first, who flourished in the 1640s in the marionette show of a certain Brioché, wore clothes and was supposedly killed by the playwright and duelist Cyrano de Bergerac (1619–55), who mistook the ape for a man. A second was known as a rope-dancer in the 1650s. "Fagotin" became a standard name for a performing monkey, and one or another of them appeared in works by Molière and La Fontaine; in La Fontaine's fable "Le cour du lion," Fagotin even entertains the king.[127]

Fagotin earned no mention in the *Histoire des animaux*, but the image of the living monkeys acknowledged that similarity that had misled Cyrano de Bergerac. Yet the "explication" that accompanied the figures emphasized instead their differences, particularly the hands and feet (nonetheless "*mains*" rather than "*pattes*"). Galen had argued that hands are uniquely human, "the instruments most suitable for an intelligent animal." Only the monkey grasping the fruit showed a humanlike grip, and even then the thumb was obscured. Another monkey was shown on all fours, and the handlike foot of the third was prominently displayed: this form, said the caption, was also true of other "Brutes."[128]

The text—written jointly by Perrault and Duverney—acknowledged the close comparisons between monkeys and humans while concluding in every instance that the differences between them were in fact profound.[129] The second critical point of identity was the humanlike larynx. Duverney detailed the close anatomical similarities between the organs of speech in humans and monkeys: "They are entirely similar to those of Man, much more than the hand . . . according to the Philosophers monkeys should speak, since they have the instruments necessary for speech." But in this case anatomy was deceptive, because monkeys did not speak, affirming the status as brutes that their hands demonstrated. Duverney did not mention Cureau de la Chambre's contention that animals nonetheless have language.[130]

The *Histoire des animaux* therefore displayed multiple meanings to multiple audiences, and these audiences in turn brought their own expectations and interpretations.[131] To the king and Colbert, the volumes, and particularly their illustrations, contributed to the royal propaganda machine that conveyed prestige to the king and his office by means of their size, expense, beauty, and symbolism. The Paris Academy gained prestige as the recipient of royal favor but also as an intellectual body. The size and accuracy of the images far surpassed those of any predecessors in natural history, natural philosophy, or comparative anatomy. The volumes bridged the gap between comparative anatomy and natural history, and between natural history and natural philosophy, as no other works had done, and represented an unprecedented depth of collaboration between naturalists and artists. The unusual format of the illustrations, displaying both the entire live animal and its dissected parts, highlighted the process of investigation, which included dissections and experiments that explained specific aspects of animal form and function.

Yet the impact of *Histoire des animaux* was less significant than it might have been. The expense and size of the volumes, as well as their status as patronage objects, made them less well known to the broader community of the new science than less elaborate works. Most of the copies were given away; in 1679, Colbert made a gift to each of the twenty-odd members of the academy of three large volumes published in the 1670s: *Histoire des animaux*, *Histoire des plantes*, and the 1676 *Recueil de plusieurs traitez de mathématique*, a compilation of individual works. The 1671 volume earned a long review in the *Philosophical Transactions*, which indicates that the Royal Society somehow obtained a copy, but the review did not appear until 1676. The Royal Society's future president Hans Sloane also owned one. But far more copies existed of Alexander Pitfeild's English translation (reprinted in 1702) than of the 1670s originals. Not until 1733 did the volumes appear in a quarto French edition.[132] By then, their impact on natural history and natural philosophy was muted.

What, in terms of contemporary science, was the purpose of *Histoire des animaux*? The reviewer in the *Philosophical Transactions* did not mention the illustrations but focused on the text and the new observations described therein, "the truth of matter of fact."[133] In size and format it resembled anatomical atlases such as that of Vesalius. Lorraine Daston and Peter Galison argue that "the purpose of these atlases was . . . to standardize the observing subjects and observed objects of the discipline by eliminating idiosyncrasies—not only those of individual observers but also those of individual phenomena."[134] In contrast to eighteenth-century naturalists, whose notion of what Daston and Galison call "truth-to-nature" aimed for a composite "average" or "ideal" representation, *Histoire des animaux* was resolutely individual and particularist. Its images, however much they borrowed from standard iconographic tropes, were of particular and individual animals. The dissected parts might be useful to aid future naturalists in their work on similar animals, but they too belonged to the particular individual under consideration, as the text pointed out. Yet these images do not conform to what Daston and Galison call "the preoccupation of many sixteenth- and seventeenth-century naturalists with what Francis Bacon [called] . . . 'irregular or heteroclite' phenomena."[135] Indeed, the image they use to illustrate this preoccupation, supposedly from Perrault's studies for the *Histoire des animaux*, is for another project with quite different aims.[136] Perrault and the *compagnie* had little interest in the preternatural. The academy occasionally discussed two-headed cats—a drawing of a two-headed calf is among Perrault's papers—but it did not publish anything about them.[137] The animals in *Histoire des animaux* were not anomalous, but they were by circumstance and philosophy singular. As this chapter has shown, not all naturalists sought either regularity or anomaly in nature. Following the method of *historia*, the *compagnie* collected facts without aiming for an inductive causal science. The purpose of *Histoire des animaux*, then, was to add to this collection of verified facts about nature, as the *Philosophical Transactions* acknowledged in its review.

FIVE

Perrault, Duverney, and Animal Mechanism

1. Duverney Comes to the Academy

The siege of Maastricht, the fortified Dutch city blocking Louis XIV's access to the Rhine, took only a few weeks in June of 1673. Among the casualties was the musketeer Charles de Batz-Castelmore, Comte d'Artagnan, who had arrested Fouquet a dozen years earlier.[1] Traveling on with the royal armies was the surgeon Louis Gayant, who had been appointed *chirurgeon des armées du roi* in the previous year as the king began his new war against the Dutch. For Gayant, a surgeon of Saint-Côme who had served as the *prévôt* of the surgeons' company, this appointment was the pinnacle of his career. But the French were in retreat back to Maastricht when Gayant was killed in October.[2] The loss of Gayant, who had performed most of the academy's dissections, was a blow to Claude Perrault's anatomical program. Pecquet, the other active dissector, died only a few months later, in February 1674, likely from the effects of alcoholism.[3] Edmé Mariotte wrote to a friend, "At present we have no more dissectors in our Company."[4]

Perrault therefore needed urgently to find a new anatomist. Although he knew how to dissect, he preferred the role of interpreter and commentator and often, artist. In the headpiece to the preface of the *Histoire des animaux*, which shows a dissection of a fox taking place at the academy's rooms at the King's Library, Perrault is in the foreground, holding an open book and pointing to a page while looking directly at the viewer (see fig. 3.3). The dissection takes place behind him, the dissector most likely Gayant. To the dissector's left, two other men examine a proof page of one of the illustrations of the *Histoire des animaux*.[5] The image subtly indicated that the anonymous text to follow was in fact the work of Perrault. Gayant and Pecquet had dissected close to twenty different animals in 1672, but only half that number was dissected in 1673, mainly by Perrault.[6]

Only one anatomist in Paris was as skilled as Gayant and Pecquet, the young Joseph-Guichard Duverney, who had burst upon the Paris scene seven years earlier at the age of nineteen. The son of a village physician from Feurs in the Massif Central, Duverney arrived in Paris with a medical degree from Avignon (which at this time was still a papal territory) and a letter of introduction to Bourdelot.[7] Although André du Laurens had received his medical degree from Avignon a century earlier, it was currently known as a diploma mill, much like the medical faculties in Rheims and Valence, and it had no chair in anatomy until 1677, a decade after Duverney received his degree.[8] How he obtained his skills in dissection, then, remains unclear.

As we have seen, the precocious Duverney began his anatomical career with a dissection of a human brain at a *conférence* of Jean-Baptiste Denis in 1667. Stensen had performed a similar operation before Thévenot's academy two years earlier.[9] Bourdelot invited Duverney to do regular demonstrations, and he duplicated a number of recent experiments before the two groups. According to Fontenelle, "He soon gained a reputation."[10] Duverney's demonstrations gave him, if Fontenelle is to be believed, a rock-star status among Parisian *salonnières* of both sexes. Fontenelle praised Duverney's "eloquence," which was a matter not only of his technical knowledge and fluency of speech. The "clarity, accuracy, and order" of his speech conformed to contemporary standards of oratory, but it was an "ardor in his expressions, his manner, and even in his pronunciation" that, in Fontenelle's view, made him an orator. But perhaps even more important, he was young and very good-looking, "d'une figure assez agréable."[11]

Perrault regularly attended Bourdelot's meetings and undoubtedly met Duverney there, recruiting him to dissect at the academy around 1674. Duverney had already been entrusted with the education of the Dauphin (born in 1661). Bishop Jacques-Benigne Bossuet (1627–1704), who had been appointed tutor to the Dauphin in 1670, decided that anatomy should be part of his education. Duverney regularly hauled his anatomical tools and preparations from Paris to St-Germain or to Versailles. According to Fontenelle, the Dauphin, surrounded by bishops and aristocrats, was so fascinated by Duverney's lessons that he skipped the hunt to attend, and even recommenced after dinner. Duverney repeated his lessons in greater detail at the house of Bishop Bossuet, with an additional audience of courtiers. Fontenelle claimed that these lessons earned Duverney the title of "the courtiers' anatomist, known by all" ("*l'anatomiste des courtisans, connu de tous*").[12]

The 1682 appointment of Duverney as professor of anatomy at the King's Garden (the actual title was "démonstrateur-opérateur de l'intérieur des plantes"; the title of "professor of anatomy" was not created until 1729) shifted much of the academy's anatomical work away from the library. By the late 1680s he also lectured on surgery at the Hôtel-Dieu, and gave demonstrations at the salon of Mathieu François Geoffroy.[13] His official appointments gave Duverney access to large numbers of human and animal bodies. His role at the garden also gave him public exposure of the sort that the academy shunned, and made him perhaps the best-known man of science in Paris.

2. Collaboration and Credit

The 1676 edition of the *Histoire des animaux* named Perrault on the title page for the first time as the "compiler." In an *"avertissement"* added to the original preface, Perrault distributed further credit, naming Pecquet and Gayant, who had worked on the earlier dissections with "much care and exactness." He reserved greater praise, however, for Duverney, "who has acquitted himself with such success, that one might say that he has provided these descriptions with a good part of the most curious particularities that are reported here."[14] When the academy's minutes resumed early in 1675, Duverney and Perrault often reported together on the *Histoire des animaux* project, with Perrault giving the external morphology and Duverney the internal; Gayant had always left the reporting to Perrault, and Pecquet reported infrequently. Duverney had not yet been admitted to the academy.[15] Unlike Gayant, Duverney was a physician; and although he did not have the scientific achievements of Pecquet, he also did not have Pecquet's personal and political liabilities. Duverney was young, energetic, and fully versed in the most recent concepts of the new science, and the academy's minutes recorded his presence at almost every meeting during the late 1670s and 1680s. With Perrault, he continued the dissection of exotic animals as well as continuing and extending the research on animal function and mechanism that had begun with the transfusion experiments. Although he began his career at the academy as little more than a technician, by 1676 he gained academy membership. Like Robert Hooke at the Royal Society, Duverney used his particular skills to advance his social standing. Like other courtiers, as Jacques Revel explains, he "created a social role for himself. . . . [He] had to invent qualities that would compensate for the deficiencies of [his] birth."[16] In this he was spectacularly successful.

Perrault's willingness to share credit with Duverney suggests he saw the younger man as an equal. Both came from striving bourgeois families for whom family advancement was paramount, although Duverney's background as the son of a provincial physician was humbler than Perrault's.

Cureau de la Chambre's success had been the culmination of several generations of striving; Duverney's advance proceeded much more quickly. The Perrault brothers had done well individually, but leapt forward in society in the 1660s under the leadership of Charles, the youngest and most ambitious, whose relationship with Colbert enabled him to pull his brothers, particularly Claude, into the orbit of the court. When Duverney came to Paris, he opened a door for his brothers and later his son. All three of the Duverney brothers became members of the Paris Academy, and Joseph-Guichard's son Emmanuel-Maurice, a physician of the Paris Faculty, eventually succeeded him as anatomy professor at the garden. By the time he attained his position at the garden, Joseph-Guichard Duverney, still in his early thirties, was generally recognized as the successor to Jean Riolan the younger as the best anatomist in Europe.[17] But did Perrault moreover seek someone who, unlike Pecquet and Gayant, shared his philosophical views? Perrault had trained in the Paris Faculty of Riolan and Gui Patin, while Duverney had gained much of his knowledge of animal form and function by duplicating the work of Stensen and de Graaf and absorbing their mechanical philosophies in the process. Duverney's thirst for anatomical knowledge led to his request to the Hôtel-Dieu in 1677 that he be allowed to "follow the physicians in their visits, and after the death of a patient he had observed during treatment, open the affected part, to examine the causes [of the illness.]"[18]

With Duverney's participation, the *Histoire des animaux* project moved beyond description to demonstration. Perrault's concern had been the structure of particular organs, but Duverney focused on function. Although the *Histoire des animaux* project gained the most attention, the *compagnie* had continued to look at the basic physiological issues that Perrault had outlined in his initial address to the academy in 1667. Apart from vision, which preoccupied Perrault, Pecquet, and Mariotte into the 1670s, another project focused on the mechanism of digestion. While the academy used the term "*méchanique*" simply to mean structure, the vocabulary of the body as a machine could be found throughout the academy's paper archive. Following the dissection of a female cadaver in March 1667 to determine the course of what Pecquet called the "thoracic canal," the academy had experimented on live dogs to determine whether chyle entered the mesentery veins, and whether there were vessels that carried chyle from the thoracic duct to the breasts. In both cases the result was negative. This confirmed Perrault's skepticism about Pecquet's explanation of the function of the thoracic duct, but he accepted the existence of the lymphatic system.[19]

A few years later, Perrault and the *compagnie* injected various substances into the veins of another female cadaver for further determination of the course of this thoracic canal and concluded that it connected with the vena cava below the heart. These experiments indicate that by the 1670s Perrault had abandoned the idea of the blood-making function of the liver and accepted that it was merely a filter, and that the chyle went entirely to the heart. Although there was much contemporary speculation about the bone marrow and its function, its role in the production of blood was not suspected and the process by which chyle became blood remained unclear.[20] Nonetheless, he remained skeptical that all animals possessed this thoracic canal. Such experiments continued, but they remained essentially structural rather than functional, as were the numerous comparisons of the flow of sap in plants to the circulation of blood in animals.[21]

The academy dissected living and dead animals to investigate chemical phenomena as well as anatomy and physiology. Experiments on coagulation in the spring and summer of 1669, for example, used a number of live lambs, and in June of that year Francis Vernon, the secretary to the English ambassador, watched Gayant and Pecquet dissect a live horse in the cramped quarters of the King's Library. The horse's pericardial fluid served as the basis for further experiments on coagulation, and Vernon recounted to Henry Oldenburg, "The reines [i.e., kidneys] & heart they tooke out to make experiments at home."[22]

In the course of a long treatise on the alkahest, Duclos described digestion as a chemical process, citing observations on animals ranging from dogs and chickens to ostriches. But as we have seen, in the spring of 1669 the *compagnie* witnessed the dissection of a young living fox, noting the peristaltic motion of its intestines. This might have been the tableau depicted in the 1671 headpiece mentioned above.[23] Perrault's theory of animal mechanism hinged on this motion; he read an essay on this topic to the academy in 1675, which he later published in the first volume of his *Essais de physique* in 1680.[24] Duverney applied these ideas to the *Histoire des animaux* project, and both he and Perrault employed them to explain the operation of the senses, particularly hearing.

3. Peristaltic Motion and Animal Mechanism

Peristaltic motion, said Perrault, caused all the operations of life. Unlike the *Histoire des animaux*, which had studiously avoided discussion of causes, the four small duodecimo volumes of the *Essais de physique* were all about causes, about *physique*—that is, natural philosophy. From the 1660s onward, Perrault had developed a theory of matter and of the operations of life intended to displace the ideas of Descartes. Perrault firmly believed that all of nature could be analyzed in mechanistic terms. He could make "probable hypotheses able to explain the most unknown aspects of Nature, by the intelligible means that *la Méchanique* furnishes us." The mechanical philosophy—the idea that nature was basically a machine and could be analyzed in terms of the principles of mechanics—was an essential doctrine of the new science, and by the 1680s its application to living things was widely acknowledged. For example, Le Clerc and Manget in 1685 assumed that the body was a machine, but the works they excerpted in *Bibliotheca anatomica* showed that there were many ways to talk about corporeal mechanisms, and no single theory dominated. From Malpighi's Cartesian micromachines, to Borelli's pulleys and levers, to a wide variety of chemical theories, the mechanical philosophy was protean in its applications to life.[25] Contemporaries borrowed ideas from one another, testing them in different contexts. The notion that Descartes dominated the discourse on animal mechanism is a construction of later historians and practitioners. As Dennis Des Chene has recently shown, by the 1680s Cartesians such as Pierre Sylvain Régis fought a rear guard battle against a number of competing theories.[26]

"*Méchanique*" held multiple meanings to Perrault. He viewed it primarily as a method to interpret the world rather than, as Descartes had seen it, an ontological principle. Perrault did not believe that mechanism could explain all phenomena, particularly those of life.[27] He modestly claimed in the first volume of the *Essais* that "my intention is not at all to establish a new system of the world," noting that he spoke only of "the Globe that we live on" and not the universe, an obvious reference to Descartes and to the academy's often stated opposition to philosophical systems. As other historians have noted, Perrault's ideas of mechanism nonetheless owed much to Descartes, as well as to Mersenne and Gassendi.[28] Like them, he conceived of the world as composed of invisible, indivisible particles.

These particles were not merely hard and inert, but possessed of two qualities, "dureté" and "ressort." Perrault defined these qualities as having the same cause: "dureté" was "the power by which bodies resist the separation of the parts" while "ressort" was "the same power by which the same parts are reunited" after being separated: in other words, cohesion and springiness or elasticity. An "internal disposition" caused particles to cohere, and an external force compelled them toward each other. Left alone, particles tended to cohere. This external force derived from the air, which he described as composed of three different types of particles: heavy ("grossière"), light ("subtile"), and ethereal. Only the heavy particles could be compressed; the light particles were very small and incompressible, and the ethereal particles were yet smaller and had no weight but caused the weight of all other bodies. These three types of particles strongly resemble the theory of matter Descartes had proposed in the *Principia philosophiae* (1644), although Perrault elaborated the roles of the particles differently than had Descartes.[29]

Perrault's theory, then, combined mechanical ideas of the corpuscular nature of matter with notions of inherent qualities. He defined the springiness of the air, the quality called "elater" by Pecquet and discussed by Boyle and Mariotte, as a function inherent in its particles. Huygens had introduced the air pump to the *compagnie*, and Perrault followed him in explaining its action as removing the heavy parts of the air but leaving the lighter. These subtle parts of air, explained Perrault, have "a weight equal to their subtlety": "because if the subtlety renders it capable of penetrating a body as solid as the glass of the recipient, in passing between the spaces of the corpuscles of which it is composed, it appears that it makes the effects of compression inside it that can reasonably be attributed to its [i.e., the subtle particles'] weight."[30]

The transcription in the minutes of Perrault's account of peristaltic motion indicated its importance to the academy. He read an additional portion in July 1676.[31] The version published in 1680 was little changed, and perfectly illustrated his concept of animal mechanism. Peristaltic motion, the "cause of all the operations of life," derived from the cohesion and elasticity of the particles, such that the living body was in a constant state of alternate contraction and relaxation. Although digestion retained a chemical component, it was primarily a function of this motion of the particles of matter both in the body and in the food. The motion of the heart and blood vessels similarly was a function of the dilation and contraction of their constituent particles as well as those of the blood.[32] The entire body was a throbbing, pulsating machine in continual motion: "all the functions of living bodies consist in the motion of the particles of which they are composed."[33] Perrault argued that the contraction of the arteries, and not their dilation, impelled the blood, and that therefore the contraction of the heart was followed by the contraction of the arteries—another instance of disagreement with the now-prevailing ideas of Harvey.[34] Perrault established these principles by means of experiments on live and dead animals, but he also applied them to the circulation of sap in plants. Perrault, Mariotte, and Philippe de la Hire had all worked on the question of the circulation of the sap without coming to a clear conclusion. To Perrault, such circulation was above all a function of life.[35]

Perrault devoted the entire third volume of his *Essais de physique* to the general topic of animal mechanism, beginning with the treatise "De la méchanique des animaux" and going on to discuss the senses, movement, and nutrition. He dealt with reproduction in the fourth and final volume, published in 1688. As we saw in chapter 3, this volume also published a belated account of the academy's experiments in blood transfusion in the 1660s. The term *"méchanique,"* as Perrault well knew, carried multiple meanings. In the *Histoire des animaux*, it referred broadly to structure. But he also described a valve in the heart or veins—a specific mechanism for a specific function—as a *méchanique*, and he disabused readers of the *Essais* of the notion that his would be a Cartesian treatise on animals as "pure machines."[36]

The ultimate cause of animal organization and motion—of life—was an inorganic soul. He compared the body to a musical instrument, an organ: its parts gave it the capability of making different sounds, but only an organist could actually make those sounds. What could be known, Perrault wrote, were the parts, not the "Author of this excellent work," the body. Unlike Malebranche, who believed the doctrine known as occasionalism, which declared that God intervened in all motion from moment to moment, Perrault asserted that the only way we could know God ("the impenetrable depth of eternal wisdom") was through his works, and the animal body was both the most wonderful and the most knowable of these. While Perrault disavowed membership in the "new sect" that explained everything about animals mechanically, yet he found himself closer to that way of thinking than to the Scholastics, who "glorify themselves in their ignorance and laziness" and refused to learn anything new.[37]

Dissection provided the foundation of all knowledge of the animal body. It required "patience and a singular dexterity, and indeed a particular genius" that Perrault credited to Duverney. Yet if the facts of dissection were evident to the observer—indeed, the most perfect as well as the most perfectly known—their interpretation allowed several opinions. Perrault's science was not that of the "certain and demonstrative" but of the probable, which might encompass a number of explanations. Nature's beauty lay in its diversity. He moved from historical to philosophical explanation "to discover by reasoning the causes and hidden reasons of all the particularities," using a favorite term for the minute details of dissection. These causes were mainly mechanical but mostly, as yet, undiscovered. As he had acknowledged the imprecision of classificatory categories in the *Histoire des animaux*, so now he acknowledged a diversity of causal categories to explain function. Mechanism could explain most things, but not everything, and dissection and experimenting might reveal causes other than mechanical ones.[38]

4. The Birds of Versailles

Duverney contributed a few examples of "philosophical" explanation to the second volume of the mostly historical *Histoire des animaux*. These included discussions of digestion and respiration in the ostrich, and a long discourse on respiration in the tortoise, the latter accompanied by several experiments on live animals.

Aristotle had not been entirely certain that the ostrich was a bird, but it took pride of place among the birds in the 1676 *Histoire des animaux*. Although, as we have seen, no birds appeared in the first volume, they constituted half (eight of sixteen) of the descriptions in 1676. Ostriches had first made the long journey from Africa to Paris in the early seventeenth century; the young Louis XIII's "volière," or bird enclosure, at Fontainebleau included them as well as storks, cranes, eagles, herons, and cormorants. Duverney had dissected an ostrich at Bourdelot's in the late 1660s, before the academy dissected one of the Versailles birds in 1671.[39] There were dozens at any time at Versailles, although they did not come cheap; Colbert's animal buyer Monier spent 330 livres to buy eleven ostriches in Alexandria in 1679, and another 600 livres to transport them from Marseille to Paris on special carts. Between 1687 and 1694, over one hundred ostriches made their way to the menagerie, and around 1700 they had their own *quartier* there.[40] The *compagnie* even received ostrich eggs from Versailles, which they attempted, unsuccessfully, to incubate.[41] Pieter Boel painted the ostriches and many other birds.

Like the *Histoire des animaux*, the menagerie at Versailles was notable for its lack of classification or order. The six animal enclosures surrounding the central octagonal structure that served as salon and observation site mingled species. At some point, these enclosures were named for birds, although by the time of Madeleine de Scudéry's 1669 promenade through the menagerie, many other animals shared these quarters.[42] Several historians have seen in the menagerie a reflection of Louis XIV's court society at Versailles, by means of which he constrained the aristocracy and prevented a revival of the passions of the Fronde.[43] Yet the courtly birds such as the "Demoiselles de Numidie" shared space, if not quarters, with birds of prey such as eagles, vultures, and cormorants, as well as large and decidedly ungraceful birds, including the cassowary and the bustard. Other animals included, at various times, tortoises and crocodiles, and various small and big cats—from lynxes and servals to lions and tigers. Mlle. De Scudéry described an animal she called a "chapas," "as sweet and eager to please as a dog," no doubt the "chat-pard," or serval, of the *Histoire des animaux*. She used the term *"flateur,"*

Figure 5.1 Pieter Boel, Three ostriches, ca. 1668.
(© Musée du Louvre, Dist. RMN-Grand Palais / [René-Gabriel Ojéda] / Art Resource, New York.)

often ascribed to courtiers; Perrault had used this term disparagingly in his description of the chameleon. The academy dissected a female crocodile from Versailles at the end of 1681. Presumably, she was kept well away from the birds during her time at the château.[44] At the beginning of that year they had dissected an elephant, a gift from the king of Portugal, that had lived at Versailles since 1668.[45] Even before Louis closed the menagerie at Vincennes in 1700 and transferred its inhabitants to Versailles, there were large and fierce animals there.

Between the publication of the two volumes of *Histoire des animaux* appeared another animal-themed addition to Versailles, the labyrinth. Charles Perrault conceived of this "place where one cannot find the exit."[46] Designed, like the rest of the gardens at Versailles, by André Le Nôtre (1613–1700), who had worked on Fouquet's Vaux-le-Vicomte with Le Vau and Charles LeBrun, the labyrinth opened in 1674. Labyrinths had featured in several of the royal *spectacles* of the 1660s, but Perrault's concept was both larger and permanent.[47] Each of its thirty-nine fountains was based on a fable of

Aesop, related in verse quatrains on a plaque at its base. The verses were by a minor court poet, Isaac de Bensérade, rather than by the well-known Jean de la Fontaine. The first volume of La Fontaine's *Fables*, a retelling in verse of tales of Aesop and other classical authors, had appeared in 1668 and was immediately popular. His best-known previous work, the *Contes*, published in several parts beginning in 1665, were, like the *Fables*, retellings of older stories. But the *Contes* were ribald and licentious; in particular, the *Nouveaux contes* of 1674 featured lecherous monks and lewd nuns, and the police commissioner in Paris banned its sale in April 1675. In contrast, the *Fables*, dedicated to the young Dauphin, aimed to instill moral virtues. Mme. de Sévigné was not alone in finding the tales irresistible, and she urged them on her daughter as delightful distractions. Nonetheless. the tinge of moral opprobrium, coupled with La Fontaine's long association with Fouquet, earned him the intense and continued dislike of Colbert and the king.[48] Although La Fontaine had no direct association with the labyrinth, he played at least an indirect role. Twenty-three of the thirty-nine fables could be found in La Fontaine's *Fables*, including two that were not from Aesop but only appeared in La Fontaine's work.[49]

The labyrinth's thirty-nine fountains included 333 brightly colored lead sculptures of animals. Among these animals were many from the Versailles menagerie and several that appeared in the *Histoire des animaux*, including monkeys and a number of birds. The labyrinth too was intended for the education of the Dauphin. What Charles Perrault called the labyrinth's "*moralité galante*" could perhaps counteract the effects of the animal combats at Vincennes.[50] Perrault's first description of the labyrinth was included in his *Recueil de diverses ouvrages en prose et en vers* in 1675, a collection of panegyric poems to the king and other occasional works. Perrault wrote his own versions of the labyrinth's verses, emphasizing courtly love rather than moral lessons. Perrault's *galanterie* deftly skirted the shoals of libertinism upon which La Fontaine had foundered. Perrault's official guide to the labyrinth appeared in 1677 with Bensérade's original verses and engravings by Leclerc. This guide was reprinted and plagiarized dozens of times over the next century. One copy of the 1677 edition survives with Leclerc's engravings hand-colored by Jacques Bailly. Several years later, in the fairy tale "Peau d'âne," Perrault described a "magnificent and powerful King" whose menagerie included Barbary chickens, rails, guinea fowl, cormorants, "and a thousand other birds of strange habits, nearly all different."[51]

Animals were everywhere at Versailles: in the labyrinth, paintings, tapestries, and sculptures, as well as in the menagerie. While some of these may have been intended to model the behavior of the nobility, others reflected

Figure 5.2 Jacques Bailly, Swan and crane, *The Labyrinth of Versailles*, after S. Leclerc.
(© RMN-Grand Palais / Art Resource, New York.)

the *gloire* of the king and his martial prowess or diplomatic successes. These included the elephant and a cassowary that had been a gift to the king from the governor of Madagascar.[52] Mlle. de Scudéry remarked on the animal paintings in the menagerie's pavilion by the Fleming Nicasius Bernaerts that foreshadowed the actual animals, and the lion symbolism that linked Louis to Alexander could be seen in a number of places. As we have seen, another Fleming, Pieter Boel, regularly sketched and painted the menagerie's animals, and the academy as well as artists who made the labyrinth and tapestries often used his work.[53]

Yet the *Histoire des animaux* project imperfectly mirrored the universe Louis and Colbert created at Versailles. Scientific and political goals battled for prominence in the volumes. Claude Perrault and the *compagnie* did not always indicate the provenance of the animals they dissected. While, particularly after the mid-1670s, most seem to have come from Versailles, not all did. The cormorant that appeared in the 1676 volume had been killed by the cook in a "hostellerie" in Sceaux after it wandered into his kitchen and bit him. Colbert, who had recently acquired an estate in Sceaux, may have arranged for the bird's passage to Paris.[54]

The birds that were dissected at the academy and appeared in the *Histoire des animaux* included such ornamental birds as the dancing Demoiselles of Numidia, the *cocq-indien*, and the *peintade* or Guinea hen; the ostrich may be included in this category, but also could be classified with birds such as the cassowary that were objects of curiosity. The 1676 edition also included an eagle and a bustard, neither of which were either exotic or ornamental. The *compagnie* dissected at least eight additional bird species from Versailles, the descriptions of which were not published until 1733; except for the vulture ("grifon"), all of them were ornamental: the *pallette* (spoonbill), the *becharu* (flamingo), the *poule-sultane* (purple gallinule), the ibis and the stork (considered together), the pelican, and the *oiseau-royal* (African crowned crane). The *oiseau-royal*, like the monkeys, was pictured in an exterior courtyard. However, the parrot named "Arras" that Duverney dissected after it died at Versailles in January 1682 never made it into the *Histoire des animaux*.[55]

5. Duverney, Ostriches, and a Tortoise

Eventually eight ostriches made their final journeys in very large carts from Versailles to the King's Library and into the pages of the 1676 *Histoire des animaux*, and Duverney dissected many of them. The description of the ostrich was afforded an extra illustration of its internal parts, an indication of the great interest it held. In November 1675 Perrault read his and Duverney's account of the ostrich to the academy, detailing their concept of animal mechanism.[56]

Without ever employing the term "peristaltic motion," the account of digestion in the ostrich gave a good example of its operation, recreating the tremulous pulsating body out of the dissected remains. Structure displayed function; as the *compagnie* wrote of the cassowary, "But we here treat of a Machine, all the parts whereof are visible, and which need only to be lookt upon, to discover the Reasons of its Motion and Action," and in the *Essais de physique*, the ridged flesh of the esophagus visually displayed *péristaltique*.[57] In opening an ostrich, the *compagnie* discovered that what Perrault and Duverney called the *ventricule*—that is, the stomach—was full of grass and other foodstuffs as well as stones and seventy coins. The coins appeared to be worn away by rubbing, not by corrosion. "This made us to think that in Birds, and generally in all Animals, the dissolution of the Nourishment is not performed only by subtile and penetrating Spirits, but also by the Organical and Mechanical Action of the *Ventricle*, which compresses and incessantly beats the things which it contains."[58] This action—the same "Contraction and Dilatation" that the heart performed—caused the "boyling of the Fermentation" in the stomach as well as the "Concoction and alteration" of the blood.[59] A few years later, Duverney continued this investigation in other birds, feeding them pearls and retrieving them "diminished in weight but more beautiful than ever," proving that the stomach juices were not acidic and that a "mouvement oblique" did most of the work of digestion. Like Perrault, he asserted that birds did not have lacteal veins, and that chyle in them therefore traveled directly through the mesenteric veins to the liver. The promised treatise on this topic, like most of Duverney's treatises, never appeared, so it is not clear if he believed that the liver made blood.[60]

Respiration in the ostrich proceeded by a similar contractile motion, and indeed one of its functions was the "Concoction and Distribution of the Nourishment, by the continual agitation and constriction of the *Thorax*." Such motion occurred in the pectoral muscles, the diaphragm, and the lungs themselves. "These Actions," declared Duverney and Perrault, "are essentially necessary for Life." The "Tension and Relaxation" of the muscles and the diaphragm, the "Reciprocation and Vicissitude of Impulsions" of the lungs, all contributed to a body in constant involuntary motion.[61]

The "grande tortue des Indes" that Duverney dissected late in November 1675 had traveled even farther than the ostriches.[62] He, for it was a male, had been captured in Coromandel, on the southeastern coast of India. The French East India Company, reestablished by Colbert, was opening trading "factories" in this area at precisely this time; the best known, at Pondichéry,

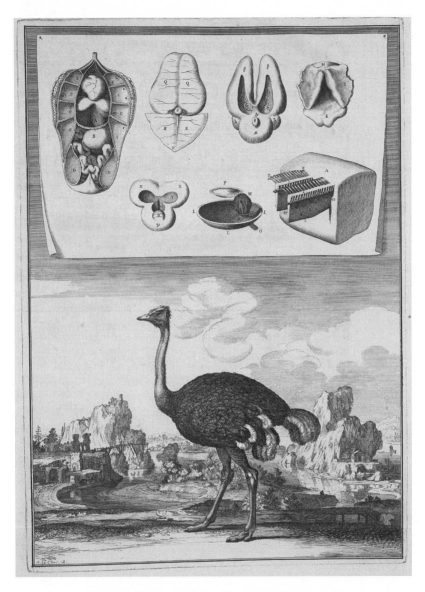

Figure 5.3 Ostrich. Claude Perrault, ed., *Mémoires pour servir
à l'histoire naturelle des animaux*, 1676.
(Linda Hall Library of Science, Engineering, and Technology.)

was founded in 1674.[63] The tortoise therefore had considerable political significance, and had been at Versailles for just over a year when he died and was conveyed to the King's Library. With the description of the chameleon, this was the longest account in the *Histoire des animaux*. But while the chameleon had many myths to untangle, the value of the tortoise—the only other reptile in the book—lay in his anatomy. Duverney and Perrault appear to have jointly authored this account as well. But once the preliminary bow to the ancients had been made, it diverged significantly from the rest of the volume, reflecting Perrault's preoccupations with function as he wrote the *Essais de physique*.[64]

The tortoise was indeed "grand," over four feet long from nose to tail; the shell alone was three feet long and two wide. His dissection interrupted the long series of chemical analyses of plants by Claude Bourdelin that dominated many meetings in 1675 and 1676, although Bourdelin took the opportunity to distill the animal's urine. With twelve pounds of urine in the bladder, he had plenty to work with.[65] Duverney and Perrault followed their anatomical description with a detailed account of the lungs and respiration. They distinguished the lungs of what they termed "amphibious" animals, including what we classify as reptiles and amphibians, from those of quadrupeds and birds, basing this distinction on the "fleshiness" of the lung tissue. Quadrupeds and birds (i.e., warm-blooded animals) had fleshy lungs, imbued with blood vessels, while amphibious (i.e., cold-blooded) animals had "membranous" lungs in which the blood was less evident (but not, as some writers claimed, missing). Aristotle had similarly distinguished viviparous animals, whose lungs contained blood, from oviparous ones (including birds and reptiles), whose lungs were "usually spongy."[66] In his essay "Méchanique des animaux," as we shall see below, Perrault placed birds in a third category between quadrupeds and amphibious animals.[67]

Structure and function could not be separated: the motion of the lungs in birds and quadrupeds was "continual, regular, and periodical," while in the tortoise and the chameleon it was "seldom and unequal." Perrault had previously noted the chameleon's irregular breathing, and he and Duverney saw comparable action in living tortoises, including some they had specifically opened alive to observe the lungs: "We saw that the Lungs remained continually swelled by the exact compression of the *Glottis*, and that it shrunk entirely and suddenly, when entrance was given to the Air by cutting the *Aspera Arteria* [i.e., trachea]."[68] This experiment did not in itself show the necessity of respiration, and they also replicated the Royal Society's 1664 open-thorax experiments. Duverney and Perrault, like Robert Hooke

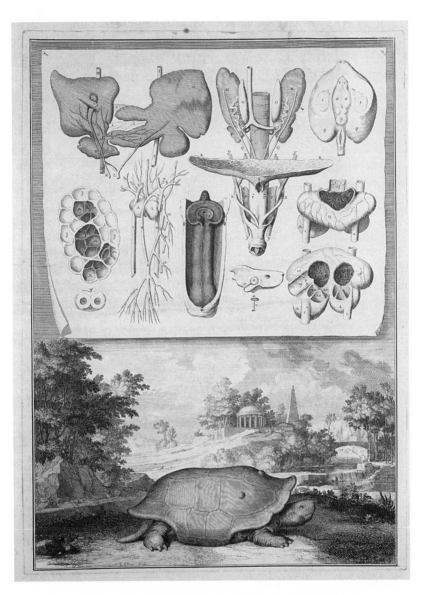

Figure 5.4 Grande tortue des Indes. Claude Perrault, ed., *Mémoires pour servir à l'histoire naturelle des animaux*, 1676. (Linda Hall Library of Science, Engineering, and Technology.)

and Richard Lower, opened the thorax of a dog to observe the motion of its lungs. When the dog's lungs ceased to move, its heart also stopped beating, but it began to beat again when they pushed air into the lungs with a bellows. When a tortoise ceased to breathe, however, "the Circulation and Motion of the Heart do continue so well in their Natural manner, that it was experimented that a *Tortoise* has lived above four days in this Condition" without further intervention.[69] They did not recognize that the tortoise had other means of retaining air.

In keeping with the notion of *"péristaltique,"* Duverney and Perrault believed the motion of the lungs sustained life, and they did not mention the second open-thorax experiment of Hooke and Lower in 1667, which confirmed that a dog could be kept alive by a continuous flow of air from a double set of bellows even if the lungs were motionless. Duverney and Perrault injected a fluid into the right ventricle of a dead dog and showed that if the lungs were made to pump by means of a bellows, the fluid would flow through the lungs and into the left ventricle. This proved, they said, that the dilation and contraction of the vessels—here provided artificially by the bellows—propelled the blood.[70]

But the lungs of "amphibious" animals acted quite differently; they were "useless for the entire Circulation." The porous structure of their lungs duplicated the "spongious" texture of their hearts, so that the blood had no need to pass through the lungs to get from the right to the left ventricle—it simply passed through its "very large holes." Perrault additionally noted in the "Méchanique des animaux" that many such animals had only a single ventricle in the heart. The blood in these animals merely served for nourishment, and the motion of the lungs did not contribute to the motion of circulation. They claimed that circulation in the human fetus operated in the same manner.[71]

Duverney and Perrault demonstrated the "true use" of the lungs in such animals with several more experiments on live tortoises, tying off the blood vessels to the lungs or otherwise disrupting the blood flow between the heart and the lungs. In each case the blood continued to circulate as usual with no contribution from the lungs. But these experiments also went beyond the stated goals of the *Histoire des animaux* of describing and not drawing conclusions. Perrault listed several uses for the lungs in the "Méchanique des animaux": promotion of the circulation, aid to voice (the tortoise had none), "refreshment of the internal Parts," or "Evacuation of their Vapours," but none of these applied to the tortoise. "Compression of the Internal Parts" was a minor use.[72]

It was impossible that such a major organ served no purpose. They weighed several possibilities in reaching the "probability of this Opinion": even these true causes were only probable, continuing Perrault's epistemological modesty. If the lungs had no metabolic function, they had a structural one. They functioned as air bladders that enabled a tortoise to float or sink on water. The Accademia del Cimento had noted such bladders in fish. When a fish was placed in the air pump and the air pumped out, upon removing the fish, its bladder deflated. It spent the rest of its life on the bottom of its tank, unable to rise to the top.[73] Duverney and Perrault reasoned that the lungs could serve such a purpose in tortoises. They argued that the bubbles that issued from a tortoise's mouth as it entered the water released enough weight to allow it to sink. Chest muscles compressed the remaining air; when the animal wished to rise, it relaxed these muscles, and "the Air by virtue of its Spring [*"ressort"*] returning into its first State," the tortoise would float. They further tested this theory with an experiment, enclosing a tortoise in a glass jar filled with water and sealed at the top except for a protruding glass pipe. The rise and fall of the water level in the pipe corresponded, they said, to the changes in density of the tortoise as it rose and fell in the water. The ability of the epiglottis of the tortoise to close completely aided this action.[74] While the release of air bubbles would not have significantly changed the weight of the tortoise, it would have changed its specific gravity, which would indeed have allowed it to sink.

Duverney was not satisfied with the account in the *Histoire des animaux*, and he continued to investigate the role of the lungs in the tortoise for the next twenty-five years. First alone and then with Jean Méry, who joined the academy as an anatomist in 1684, he dissected many more living and dead turtles and tortoises, publishing two more accounts of the heart and lungs in 1698 and 1702.[75] As we shall see in chapter 6, the question of the function of the lungs of the tortoise fed into debates between the two anatomists in the 1690s on the nature of the circulation in the fetus.

Perrault also continued to investigate respiration. A few years after the account in the *Histoire des animaux* appeared, he gave further attention in the *Essais de physique* to the purposes of respiration and particularly the physical and chemical qualities of the air. Again, he relied on his own theory of matter. The subtlety of the particles of air and their innate qualities of *péristaltique* allowed it to penetrate any solid body. In addition, "being composed of the expirations or exhalations of all bodies, this mélange results perhaps in some new quality composed of the union & fermentation of all these different materials, & to which one ought truly to attribute the different effects that the air is capable of producing." The air moreover is influenced by temperature, by seasons, and by the amount of sunlight; Perrault explained that what was once referred to the influence of the stars may now more legitimately be referred to the influences the air receives from the earth. He agreed with (but did not mention) the arguments of John Mayow and others that a penetrating nitrous quality in the air gave the blood its characteristic color and perfection. But Perrault then went on to reiterate the *méchanique*, as opposed to the *physique*, qualities of the air, differentiating uses of the lungs in different kinds of animals. Once more he described the lungs of "amphibians" as membranous rather than fleshy, serving primarily "to sustain their body in the water."[76]

Perrault placed birds in a middle category between *"animaux terrestres"* and *"amphibies,"* attributing to them both fleshy and membranous lungs, seemingly moving back toward an Aristotelian characterization. But the membranous part of avian lungs served not to support the animal in water but to supply the compressive movement for digestion that would normally be supplied by muscles of the lower torso. These were lacking in birds because of the size of the breastbone and flight muscles. Perrault did not support this ingenious argument from anatomy to function with experimental evidence, and did not explain how flightless birds such as the ostrich might differ in this regard. He compared the action of avian lungs to a bellows, noting that air in both cases came from two sources, above and below, from the vessels of the chest and those of the stomach. His illustration showed the lungs of an ostrich alongside a bellows.[77] While Perrault and Duverney demonstrated their awareness of contemporary work on respiration and circulation, their particular beliefs about innate motion and structural variation among animal kinds, as well as their particular version of the mechanical philosophy, led them to significantly different conclusions from other European naturalists.

6. Sense and Sensibility

In 1676, Perrault and Duverney had examined the eyes of the tortoise and wished to dissect the head to see the ears. But the *grande tortue* in the end belonged to the king:

> The Necessity that there was of keeping the Remains of this rare and extraordinary subject, for an Ornament of the *Aviary* of *Versailles*, having hinder'd us from persuing any farther the Enquiry of the Organs of sense in the Head of our *Tortoise*, we have supplied this defect with the dissection of several other *Tortoises*.[78]

In his 1671 preface to the *Histoire des animaux,* Perrault had extolled the evidence of the senses, particularly the eye, above all other ways of gaining knowledge. In the *Essais de physique,* Perrault praised "la noblesse de l'organe de la vue," although, as we shall see, hearing also played a significant role.[79] Work on the nature of vision and the structure of the eye continued investigations on optics and anatomy that Descartes, Mersenne, and Thomas Hobbes as well as Peiresc and Gassendi had pursued in the 1630s. The *compagnie* examined the eyes of every animal dissected in the *Histoire des animaux* project (except for those of the unfortunate beaver), and dissected many more human and animal eyes. The descriptions in the *Histoire des animaux* included the marvelous structure of the chameleon's eyes, debates about bipolar versus binocular vision, even the color of the lions' eyes (a gray-yellow shade known as "isabelle"). Perrault employed all of this evidence in his detailed and comparative anatomical description of the eye in volume 3 of the *Essais de physique* and in his discussion of the mechanism of vision in volume 4.[80]

The controversy over the seat of vision, which occupied the *compagnie* in the 1660s and 1670s, revealed the academicians' deep knowledge of the smallest parts of the human and animal body and their use of the microscope to probe even further. But it also revealed the limits of Perrault's descriptive anatomical method. It began with the blind spot. In the early 1660s, Edmé Mariotte had noted a peculiar phenomenon. He fixed two small round pieces of paper to a wall, one at eye level and the other a few feet to the right and slightly lower. When he closed his left eye and backed away from the wall, keeping the central paper in his vision, the paper to the right at some point disappeared from view. He attributed this phenomenon to a blind spot in the eye, identifying its location as the place where the optic nerve entered the eye, based on numerous dissections of human and animal eyes. He repeated this experiment with several friends and also demonstrated it at the Bibliothèque du roi in the spring of 1667, always with the same result. He had established the fact of the blind spot beyond dispute.[81]

If the academy—and the Royal Society as well—accepted the fact, it did not accept Mariotte's explanation of its cause. In an essay he read to the *compagnie* in June 1667, Mariotte argued that this experiment showed more than simply the existence of the blind spot. It revealed that the seat of vision was not the retina, as most natural philosophers had believed, but the choroid, the vascular membrane between the retina and the sclera.[82] In 1604, Kepler had argued that the retina was the seat of vision, based on his reading of

Felix Platter's anatomical study of the eye. Subsequent work on vision assumed this as true, and Descartes gave further demonstration with the eye of an ox in his *Dioptrique*.[83] But if one assumes that the blind spot, argued Mariotte, was where the optic nerve entered the eye, then the retina, which covered the optic nerve, would also cover the blind spot. The choroid did not cover the blind spot, so that transferring the site of vision to the choroid made perfect sense. He had proved it experimentally and anatomically.[84]

Although Mariotte based his reasoning on anatomical evidence, he read his essay on a Tuesday, a "mathématique" day; optics was a mathematical rather than a physical topic. The portion of the essay concerning the blind spot appeared in a pamphlet in 1668 together with a reply by Jean Pecquet on the seat of vision.[85] These works were not anonymous and received a substantial review in the *Journal des sçavans* a few months later.[86] Pecquet praised Mariotte's discovery of the blind spot, but defended at length the established view that the retina was the seat of vision, relying on his own numerous dissections of animal and human eyes, including lions, camels, and bears as well as cattle, deer, sheep, cats, and dogs.[87] The variety of animals confirmed the uniformity of the retina both within the eye and across species. He added that the choroid, in contrast, was irregular, making it a poor surface to receive visual images. Only the glossy, mirrorlike surface of the retina could receive all of the luminous particles; the eye was literally a mirror of nature. Pecquet also discounted Mariotte's claim that the retina was not attached to the optic nerve, employing microscopic evidence of its structure.[88]

Pecquet argued entirely from anatomy. He had dissected a human eye with Huygens in February 1667, and their notes detailed the location of the optic nerve.[89] The 1668 pamphlet gained a further audience when it was appended to the 1669 *Description anatomique*; and Mariotte replied to Pecquet in the summer of 1669, reading another treatise to the *compagnie*. He too argued from anatomy, noting differences between the eyes of living and dead animals. In response, the assembly resolved to dissect yet more eyes, and charged Couplet with "finding a bird of prey such as a kite to dissect its eyes, that species having very piercing vision." At the next meeting the eyes were extracted from a living kite and Pecquet dissected them, predictably finding support for his theory over Mariotte's. Once the eyes had been examined, the *compagnie* went on to dissect the remainder of the bird.[90]

Following Pecquet's death in 1674, Perrault took over the role of defender of the retina against the resilient Mariotte. The latter's second letter had been published in 1671—a condensed version in English had appeared in the *Philosophical Transactions* the previous year—but Perrault's reply did not appear until five years later. The elephant folio *Recueil de plusieurs traitez de mathématique de l'Académie royale des sciences* included all of the letters published thus far, adding Perrault's and Mariotte's final response, making five letters in all. Perrault too argued that the glassy surface of the retina was essential for vision, and attributed the blind spot to irregularities in the optic nerve itself, or to disturbances in the animal spirits.[91] He and Mariotte continued to argue from anatomy, and Mariotte's final letter ceded nothing. The issue remained unresolved for another two centuries, until Hermann von Helmholtz, with a fuller understanding of the nervous system and how vision occurred, declared the retina to be the seat of vision in 1867.[92]

Perrault did not mention this controversy in the *Essais de physique*, and assumed the retina was the seat of vision. But the five letters were reprinted in 1682 and again in the posthumous 1700 *Oeuvres de physique* of Perrault, which included the *Essais* as well as the work of Claude's brother Pierre, and many continued to take Mariotte's position, including the academy anatomist Jean Méry. The debate never descended to acrimony, although Mariotte, in his final letter, commented that "*sectateurs*" of the new philosophy would always be opposed to him. His active career at the academy was crowned, one might say, by Duverney's description of his dissection, read to the *compagnie* in 1684.[93]

The debate on the seat of vision was a momentary distraction from Perrault's main pursuit in the 1670s: his anti-Cartesian physiology based on the idea of an embodied soul. In Paris in 1675–76, the German philosopher Gottfried Wilhelm Leibniz commented that "the opinion of Monsieur Perrault" was that "the soul is equally throughout the entire body."[94] In the fourth volume of the *Essais*, Perrault returned to vision with a lengthy description of the structure of the eye based on multiple dissections, many of them performed by Duverney. Perrault even built a mechanical model to show how various muscles moved the eye.[95] The "sensation of sight consists of a nearly incomprehensible delicacy," whereby "the imperceptible movement of particles of a distant object is capable of communicating itself to the particles of the organ; the movements by which all of the organ is turned toward these objects, and which make the figure of its different parts, are diversely changed according to its needs, and so quickly and precisely that one cannot admire it enough." While this description seemed simply mechanical, Perrault hastened to add that "the system I employ to explain the movement of animals . . . [requires] the presence of the soul united to all of the parts of the body."[96]

The motion of the eye therefore partook of the general vibratory motion of the body, originating in an embodied soul, which he had explained in the

"Méchanique des animaux." Although animal structure predicated function, the soul governed all action. Humans had binocular vision and chameleons and rabbits did not because the soul in each case acted for the greater utility of the organism as a whole. Descartes had argued that the pineal gland combined images from the optic nerves.[97] Perrault assumed a mechanical and particulate transmission of images but did not identify the exact mechanism by which visual impressions traveled through the optic nerve to the brain: "whether that communication happens through the vibration of the nerves, by channels full of spirits . . . [or] by the course of spirits between the fibers" was not as important as the two sets of vibrations—of objects on the retina, and of the muscles of the eye—that were thus communicated. Even the structure of the optic nerve itself, which differed in different animals, made no difference to the final result.[98] Perrault distinguished sensory from motor nerves, but as we have seen in chapter 3, he claimed that all nerves acted by "the soul joined and united to each part" and not from the central direction of the brain.[99] Unlike Descartes's notion of the brain as the seat of memory and sense experience, these functions existed in every part of the body. The brain played a secondary role as the generator of the animal spirits required for sensation.[100]

Perrault said nothing about the nature of light and surprisingly little about the actual mechanism of vision. He spent far more time on the mechanics of muscular motion and the limits of a purely mechanical explanation. As he stated in the "Méchanique des animaux," the ultimate goal of natural philosophy was to glorify God, and his outline of what we might call a Christian vitalism fulfilled this goal.[101]

7. Music, Modernity, and the Ear

Unlike other senses, wrote Perrault in the "Méchanique des animaux," vision and hearing dealt with heterogeneous sense impressions that required "assembly" to be felt entirely, in contrast to the uniform impressions of taste, touch, or smell.[102] The nerves in vision and hearing were therefore more delicate and sensitive. Although vision was the critical sense for the practice of natural philosophy, Perrault believed that hearing, even more than vision, demonstrated the operation of the soul within a mechanical framework. Music gave a particularly good example of this operation, and modern polyphonic music, in contrast to the modal style of the ancients, expressed a new subjective aesthetic that relied on complexity rather than simplicity. Perrault devoted the entire second volume of the *Essais de physique* to a study of sound and the ear (*Du bruit*), ending with a treatise on ancient music.[103]

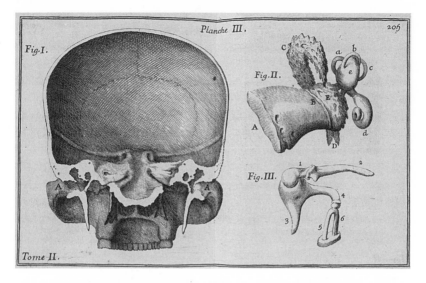

Figure 5.5 Dissected parts of the human ear. Claude Perrault, *Essais de physique*, vol. 2, 1680. (Linda Hall Library of Science, Engineering, and Technology.)

Perrault read the long work on sound that became *Du bruit* to the academy over several months during the winter of 1677–78. Much of this work concerned the nature and causes of sound; only in the last third did he turn to the structure of the ear and the mechanism of hearing. Perrault pointedly distinguished *bruit*, usually translated as "noise," from *son*, or "sound." *Bruit*, he said, was the genre; *son*, the variety. This is the opposite of the definition given in the *Dictionnaire de l'Académie française* of 1694, which listed *"bruit"* as *"son assez fort,"* a loud sound. It defined *"son,"* on the other hand, as *"the object of hearing, that which touches or strikes the ear."*[104]

Based on his mechanistic deconstruction of the air, Perrault identified sound as the vibration of airy particles, another example of *péristaltique*.[105] Various forces caused the particles to vibrate, and their *ressort* or springiness created an "impulsion" that agitated nearby particles in a chain reaction. The vibrating air entered the ear and caused the flexible eardrum to vibrate and then the inner bones, particularly the "snail shaped bone"—the cochlea— that was the seat of hearing. In much the same way, particles emitted by an object struck the retina to cause vision (although, as he noted, the speed of light far exceeded that of sound).[106] Duverney's animal dissections supported Perrault's conclusions. While the structure of the ear in terrestrial animals (meaning quadrupeds) resembled the human, birds differed and fish differed even more. In volume 3 Perrault had noted that sense organs and their uses differed in each animal "according to their different capacity"; the tortoise and the chameleon, for example, could not hear at all, as he had previously explained in the *Histoire des animaux*.[107]

8. Duverney's *Traité de l'organe de l'ouïe*

Duverney dissected many animals in the late 1670s, including exotics such as the civet cat, and cows and birds for projects on the egg and on animal nutrition. The academy's minutes mention him working on ears in June 1677. In February 1678, near the end of his months-long reading of *Du bruit* to the *compagnie*, Perrault demonstrated the ear and its operation.[108] Beginning in the mid-1670s, initiated by Charles Perrault in his capacity as deputy *contrôleur des bâtiments*, the academy began to draw up reports on its activities to send to Colbert.[109] Included in the report for 1678 was "Plusieurs écrites de Mʳ Du Verney" on the sense organs. He described the structure of the ear, the eye, and the nose in animals that included dogs and cats, wolves and foxes, and calves and sheep, but also the civet cats and lions of the *Histoire des animaux* project.[110]

The academy's policy of collaboration and anonymity began to break down almost immediately after its foundation. Stars such as Huygens and Cassini were never anonymous, and the letters in the controversy over the seat of vision were signed. Although Perrault's 1673 translation of Vitruvius was not, strictly speaking, an academy project, it resembled the academy's work in many respects, including the lack of an author on the title page. However, Perrault signed his dedication to the king. A few years later, Perrault's name appeared on the title page of the 1676 *Histoire des animaux*, and he figured prominently in the *Recueil* of the same year. At the end of that year, the *compagnie* resolved to propose to Colbert a new volume of individuals' works in an ordinary folio, but that did not happen.[111] Perrault's *Essais* of 1680 definitively broke the rule of anonymity, although it too was not an official academy work. Much of Duverney's dissection work on sense organs appeared in the *Essais*, but at some point Duverney decided to publish on his own. While Perrault credited Duverney as a skilled technician in the *Essais* and the *Histoire des animaux*, he did not credit him as a coauthor.

By 1683, when Duverney published his treatise on the ear, the academy no longer entirely circumscribed his identity. While the title page of the *Traité de l'organe de l'ouïe* identified him first as a member of the academy, it added the titles of "Conseiller" and "Médecin ordinaire du Roy," as well as professor of anatomy at the King's Garden. Duverney did not avail himself of the royal publishers, preferring instead the academic publisher Étienne

Michallet. Although he credited Perrault in his preface as having treated the subject "with more exactness than anyone"—echoing Perrault's praise of him—the rest of the volume ignored him.[112] Duverney conceived of this volume as the first in a series on the senses, but no others appeared in his lifetime.

Duverney placed particular emphasis on the illustrations in this work. Perrault's illustrations in the *Essais* were for the most part sparse and schematic. Although both works were small duodecimo volumes—approximately 19 × 13 centimeters—Duverney included sixteen foldout engravings of various sizes, at least some of them by Sébastien Leclerc, detailing everything from the bones of the inner ear to the structure of the brain. Some of the drawings and engravings were by Louis de Châtillon (1640?–1734), another royal artist, who had worked on the *Histoire des plantes*.[113] Duverney declared in his preface that his work was not only complete and exact, but "clear and intelligible," with "true and faithful" images that "remove all ambiguity." Although this claim to plain language was a rhetorical gambit intended to set his work apart from others, including Perrault's, the *Traité* reveals the clear and logical exposition that made his reputation as a lecturer. The reader followed his journey from the exterior of the ear layer by layer down to the tiniest bones.[114]

Perrault had employed Duverney's dissections in his own account of the structure and operation of the ear, but *Du bruit* concentrated on the nature of sound rather than the ear. Duverney used the same examples and many more. He revealed the structure of the ear, and the surrounding blood vessels, nerves, and bones, including the jaw and the skull, as well as the anatomy of the brain. Only in the second part of the *Traité* did he move to functional considerations:

> After having given an exact enough description of all the parts of the ear, I believed that to make it more useful, I should add some reflections, and draw some consequences from the *méchanique* of these parts by which one can explain their uses & the ways in which we perceive different sounds and noises.[115]

Duverney asserted that his conjectures, "*assés vraisemblables*," owed much to Mariotte; he did not mention Perrault, although he adopted the latter's concept of sound as the vibration of the particles of air.[116]

Such vibration first struck the skin of the eardrum. It adjusted itself to "the proper tension, so to speak, to represent the different tones of the resonating body." As was well known, when two lutes were placed side by side,

Figure 5.6 Dissected parts of the human ear. Duverney, *Traité de l'organe de l'ouïe*, 1683.
(Bibliothèque interuniversitaire de santé, Paris.)

plucking a string on one provoked a vibration in the same string on the other, or on an octave, a fifth, or a fourth. The royal engraver Abraham Bosse had used lutes and viols to represent hearing in a 1638 series on the senses.[117] Duverney attributed different sounds and tones to the "different nature & different collisions [*choqs*] of the resonating body," whose vibrations were then communicated to the air and via the air to the eardrum. These motions were mechanical but imperceptible and therefore difficult to explain.

Figure 5.7 Abraham Bosse, *Hearing*, 1638.
(Wellcome Library, London.)

However, he denied that the air played a major role in the transmission of sound beyond the eardrum. Such a small quantity of air that might exist in the inner ear could hardly move enough to dislodge the bones. Rather, the vibration communicated directly from the eardrum to the bones. Returning to the lute analogy, he claimed that the sympathetic vibration of the second lute resulted mainly from the table on which the lutes lay, which communicated the vibration of the first to the second through its wood, since if the lutes are held in the air, "the experiment does not succeed." He did not mention Galileo's observation that the sound of a viol could travel through the air and cause a glass to vibrate. Duverney continued, "Nature, mechanics, & the articulation of these three bones seem very favorable to this conjecture; they are dry, they are hard, they are slender, & therefore they are very capable of being shaken."[118]

The most important of the bones in the inner ear, as Perrault had asserted, was the cochlea, the snail-like spiral bone. Duverney argued that the shape of the bone, including its diminishing width from the bottom of the spiral to the top, meant that it resonated differently according to different tones: the thicker end vibrated with lower, slower tones, and the thin top end vibrated with faster, higher tones. So "according to the differing vibrations of the spiral bone, the spirits of the nerve, which spread themselves in its [the bone's] substance, receive different impressions that represent in the brain the diverse appearances of tones."[119]

Perrault explained vision and hearing as acts of the embodied soul, rather than a rational brain, and the "judgment" of that soul that enabled the senses to distinguish truth from error was the same in humans and animals. Although Duverney did not explicitly address Perrault's concept of the corporally dispersed soul, his explanation of tinnitus, as Veit Erlmann has explained, expressed a similar bodily subjectivity. In tinnitus (*"le tintement"*), "the ear hears sounds that are not, or at least are not external." The ancients had believed this was caused by the agitation of air entrapped in the ear, but Duverney rejected this notion. It was rather "a sound that is internal." If hearing was the vibration of the ear (as vision was the vibration of the retina), then that vibration could come from more than one source. Rubbing the eyes produced lights that do not exist, and various ailments of the ear could cause its bones to vibrate. But in addition, Duverney argued, the animal spirits could themselves vibrate the auditory nerve and cause the sensation of sound where no sound existed; "[the ear] is indifferent whether the fibers of the nerve are agitated by the ear or by the brain, it results in the same sensation."[120]

9. Animals and Humans, Ancients and Moderns

Perrault's embodied soul mediated individual experience. As we saw in chapter 3, the soul that is embedded in each part of the body has two kinds of judgment: "habitual," which is formed by habit and is fully unconscious, and "express and distinct" which is an object of will. The habitual judgment of the soul guaranteed the accuracy of our sense impressions, regardless, Duverney would add, of the source of those impressions.[121] The express judgment entered particularly into the aesthetics of sound and hearing, what would come to be called taste. Perrault took it for granted that animals felt pleasure, and much as the chameleon enjoyed being in the sun, so animals such as singing birds took pleasure in music. Other birds listened with attention to the human voice, and even learned to speak. Some authors had claimed that deer and boars liked music. Mersenne, in contrast, had argued that because the ear merely transmits sounds to the brain, animals did not have the consciousness to comprehend sounds, "but only the representation, without knowing if that which they apprehend is a sound or a color or something else."[122]

Perrault attributed more consciousness to animals than had Mersenne, yet he did not believe that birds understood music, since their own songs lacked harmony and diversity. "One sees," he added, "that they show as much pleasure in hearing a confused noise and discordant voices, as in the best music." This was no different from most humans who claimed to love music but could barely tell if the voice was good or the instrument in tune; such "simple amateurs" of music should not be confused with the "true connoisseurs capable of discerning perfection, which is something that touches the mind [*l'esprit*] and the imagination more than the ear."[123] Only humans understood music; only humans had the capacity for taste. And only particular humans had good taste.

Perrault distinguished several varieties and modifications of sound. One modification was tone, and several chapters explored the sounds of particular musical instruments.[124] Perrault closed *Du bruit* with a treatise on ancient music, *De la musique des anciens*, claiming it was relevant to the study of sound because it concerned the meeting of different tones that constituted harmony. Duverney's *Traité de l'organe de l'ouïe* soon superseded Perrault's work on the anatomy of the ear. But Perrault's concerns lay far beyond anatomy; the intersection of music and mechanical philosophy of *Du bruit* fed directly into the opening sallies of the Parisian culture wars of the *querelle des anciens et modernes*.

The production of harmony had concerned many natural philosophers. In his early work *Compendium musicae* (1618), Descartes had argued that music consisted solely of the mathematical qualities of duration and tone. It gave pleasure insofar as it followed mathematical structures. In *Two New Sciences* (1638), Galileo discussed the mathematical proportions of harmony and the correspondence between music and the physics of sound. The pleasing quality of a particular chord corresponded to its mathematical ratios, so that the effect of the fifth "produces a tickling and teasing of the cartilage of the ear drum so that the sweetness is tempered by a sprinkling of sharpness, giving the impression of being simultaneously sweetly kissed, and bitten." A few years earlier, Mersenne's *Harmonie universelle* (1636) had likewise determined the frequency of particular tones and their mathematical ratios. Mersenne viewed such harmony as a paradigm for the order of the universe.[125]

Perrault's focus on physiology and aesthetics—on that tickling of the eardrum—rather than mathematics made no mention of the work of his predecessors. He argued instead that the ancients knew less about sound than the moderns because they lacked knowledge of harmony; their modal music was confined to single notes. Descartes had told Mersenne in a 1630 letter that beauty "signif[ies] simply a relation between our judgment and an object." He quoted from his 1618 *Compendium musicae*:

> Among the objects of the senses, those most pleasing to the mind are neither those which are easiest to perceive nor those which are hardest, but those which are not so easy to perceive that they fail to satisfy fully the natural inclination of the senses towards their objects nor yet so hard to perceive that they tire the senses.[126]

In other words, the perception of beauty is always mediated by the mind. Descartes connected the purposes of music to those of rhetoric—to teach, to delight, to move—but its perception was subjective.[127]

Perrault disagreed with Descartes about perception much as he had about micromechanisms and the soul. His *Ordonnance des cinq espèces de colonnes selon des méthodes des anciens* (1683) argued that musicians always agree that a particular tone is accurate because "there is a certain and evident Beauty in the Exactness of it, of which, the Senses are easily, and even necessarily convinc'd." Although certain architectural elements gave pleasure because of their symmetry, there is not "so precise an Exactness" as in music, where discord is immediately evident to the ear without the intervention of the mind. This was also quite unlike Mersenne, who had emphasized the mechanical precision of musical intervals as a reflection of divine wisdom.[128] In *Du bruit*, Perrault asserted that ancient music lacked that beauty; it "consisted of very little, if one compares it with the essential beauty of our harmony that is particular to it, and that was unknown to all of Antiquity."[129] Moderns such as Perrault differed from the ancients, who had not been true connoisseurs.

Perrault discovered the "essential beauty" of harmony in the new Italian opera, which differed in significant ways both from previous music and from previous dramatic performances. Operatic music was polyphonic and in addition employed several frequently overlapping voices; it often included dance. Unlike classical drama, its plots were fanciful and extravagant, involving the direct influence of supernatural events and beings that were moreover depicted onstage by means of elaborate stage machinery. After attending a performance of Robert Cambert's opera *Les peines et les plaisirs de l'amour* in 1672, Perrault wrote an essay titled "Sçavoir si la musique à plusieurs parties a esté connüe et mise en usage par les anciens" (Whether multipart music was known and used by the ancients). This essay was not published in Perrault's lifetime, but apparently he intended at one point to use it as a preface to *De la musique des anciens*.[130] This preface, however, has more in common with the burlesques that Perrault wrote with his brothers in the 1650s than with *Du bruit*; it opened with Perrault braving a crowded *parterre*—in this case not a garden but the back of a theater—to witness Cambert's opera.

Most of the essay consists of a dialogue the author supposedly overheard before the performance between two music enthusiasts he called "Palae-ologue" (student of the ancients) and "Philalethe" (lover of knowledge). Philalethe argued that artistic taste was not absolute but dependent on con-text, individual temperament, and contemporary consensus. It was a matter of good sense rather than learnedness. What seemed good to the ancients may not seem so to us; we may admire them only because we are told to. Perrault had already expressed his skepticism about the superiority of the ancients in his unpublished essay "Mythologie des murs du Troye" two de-cades earlier. Perrault did not use the term *"honnête homme"* in either of these works (although he did use *"honnéteté"* in the preface), but that term might define his man of good sense. As the discussion moved from art and poetry to music, Philalethe began to argue that the polyphonic music of the moderns was superior to the single-tone modal music of the ancients just as the curtain rose and the orchestra began its overture.[131]

Palaeologue and Philalethe united in their annoyance at the crowd, who continued to talk over the music, gossiping about the female singers. Both felt strongly the beauty of the harmonies of different voices and instru-ments. Palaeologue even "entered in a way into the feelings [*sentimens*] of the music."[132] He and Philalethe remained in their seats while the crowd dis-persed, and he pointed out to Philalethe "all the things that made the piece that has been played so marvelous and so much superior to the spectacles of the ancients with regard to imagination [*esprit*] and ingenuity [*invention*]." Paleologue went on to list the virtues of the performance he had just seen: the felicity of the dance and "machinery"; the variety of characters and sub-jects ("sometimes happy, sometimes sad"); the use of two familiar genres of song, the drinking song and the love song; the way the music carried the story; the representation of conversation by having several voices sing to-gether; and finally, that it went on for three hours without being boring.[133]

Perrault employed the words *"sentimens"* and *"sensible"* several times in this essay. In contemporary usage, these terms referred specifically to sense experience, to a physical rather than an emotional feeling. His concept of the embodied soul meant that the body experienced such feeling at the level of the perceptive organ, not at the level of the brain: the ear itself felt sound. The experience of the *"merveilleux"* that the opera and all its components evoked (music, dance, words, and theatrical machinery) bypassed the brain for the eye and especially the ear. Reason was not involved, and each eye and ear experienced the spectacle differently, subjectively. The radical individualism of the *Histoire des animaux* here translates to a radical subjectivity. The sense of wonder that the term *"merveilleux"* invokes is the wonder of admiration and surprise. While the gods and goddesses of the plot demanded supernatural effects, theatrical machinery connected the supernatural to natural causes.[134] When, a few years later, Perrault defended the mise-en-scène of another opera in a letter to Colbert, he argued that the fact that you could see the ropes that made the supernatural effects was immaterial to the experience of the effects: they were no less marvelous for revealing the material causes of their operation.[135] The experience itself imparted knowledge, just as natural history did. The causes were irrelevant. Indeed, the illustrations to *Histoire des animaux* might be seen as stage sets, with trompe l'oeil effects emphasizing their artificiality. The monkeys in particular recalled acrobats in an opera, and they played where an opera might have been staged at Versailles, in an outdoor courtyard overlooking the gardens of Le Nôtre. Until the late 1680s, the court was the center of opera production.[136] As in the menagerie, so in *Histoire des animaux*, animals were actors in a larger production about the glory of the king.

Delayed by the crush of carriages at the entrance to the theater, Palaeologue and Philalethe continued their conversation, with the addition of one Aletophane, who related his essay on ancient music—Perrault's essay that was to follow this one. Perrault left Palaeologue unpersuaded of the merits of the moderns over the ancients. But by 1680, when Perrault's essay on ancient music appeared in *Du bruit* (without this rather frivolous preface), the culture wars surrounding the opera had heated up, an opening round in a larger debate about science, the arts, and taste.

One evening at the end of January 1674, the Perrault brothers—Claude, Pierre, and Charles—joined the crowds at the Palais-Royal with their friend the court painter Charles LeBrun to see a performance of Jean-Baptiste Lully's *Alceste*, an *opéra-tragique* in the Italian style, with a libretto by Philippe Quinault based on an ancient Greek play by Euripides. The king and Mme. de Montespan had viewed a performance at Versailles a few months earlier, but both Quinault's interpretation of the classical work and Lully's music proved controversial. By the summer, following several more performances, Charles Perrault issued an anonymous *Critique de l'opéra* that defended modern music. Between 1671 and 1673, Charles had been elected to the Académie française and Colbert had named him one of the *contrôleurs général des bâtiments, jardins, arts et manufactures du roi*, a position of considerable power. He soon dropped the façade of anonymity, reprinting the *Critique de l'opéra* in his *Recueil de divers ouvrages en prose et en vers* in 1675. The dedication of that work, by Louis Le Laboureur, claimed that Perrault here did not address the issue of the ancients versus the moderns; he simply expressed his own ideas. Perrault's essay, appearing among a number of panegyrics to the king, tacitly acknowledged the role opera played in the *gloire* machine of Louis's court.[137]

The playwright Jean Racine (1639–99) and the critic Nicolas Boileau-Despréaux (1636–1711) fiercely attacked Lully's opera and Charles Perrault's defense. Although, as we shall see in chapter 6, Boileau had penned a critique of the Paris Faculty in 1671, his *L'art poétique* (1674) staunchly defended ancient literature and the classical style of Racine and Pierre Corneille. He also took the time to direct a few blows toward his former physician Claude Perrault, "*célèbre assassin.*" The Perrault brothers had satirized the ancients twenty years earlier in their verse burlesque *Les murs de Troye*. Claude had at that time written a second chapter of that satire, prefaced by his essay "Mythologie des murs de Troye," that characterized ancient poetry as primitive and its modern imitators as ridiculous. Although this work remained unpublished, Boileau certainly knew Perrault's opinions.[138] Racine too defended the honor of the ancients in the preface to his play *Iphigénie*. Charles Perrault argued that the success of the opera among "*honnêtes gens*" proved the superiority of the modern version, and indeed the *tragédie en musique* soon overtook in popularity the *tragédie classique* of Corneille and Racine.[139]

Claude Perrault's *Du bruit* and *De la musique des anciens* situated sound, and the particular kinds of sound that constituted music, at the forefront of cultural debates as well as scientific debates about the nature of the senses and therefore of epistemology and knowledge. Perrault's *avertissement* took up the theme of "Mythologie des murs du Troye" that the ancients were overrated; the "essential beauty" of modern harmony was simply unknown to them.[140] Much of *De la musique des anciens* is a humanist examination of ancient texts, and it summarized work Perrault had already presented in his translation of Vitruvius, which had included a large chart explaining the music theory of the ancient Greek philosopher Aristoxenus.[141] *De la musique des anciens* sought to prove that ancient music was not harmonic in the modern sense; in particular, the ancients had no knowledge of counterpoint, the combination of two or more melodic lines. Their music was essentially primitive, and its beauty was the beauty we perceive in ancient ruins. Perrault even compared it to the music of the *"nations barbares"* of the New World.[142] Ancient music, he said, sufficed to satisfy the heart and the senses, but it did not engage the mind or imagination (*l'esprit*). So too only those with sufficient knowledge and intelligence could fully appreciate modern polyphonic music. This was what constituted taste, which varied over time as well as among individuals; Perrault gave the example of the Roman fermented-fish condiment *garum*, which the ancients "loved to excess" but which recent centuries viewed as "an abomination." Is it indeed so strange, said Perrault, that the moderns have invented a new and more beautiful music, knowing what has also recently been discovered in natural philosophy and mechanics? But its beauties were not open to all, and Perrault's hierarchy of passion, sense, and reason echoed the Galenic distinction between sensitive, animal, and reasoning faculties; only the moderns, apparently, possessed the latter.[143]

Like his brother Charles, Claude as courtier glorified modern music to glorify the king, whose era, as Charles would write a few years later, surpassed even that of Alexander. But his appeal to modernity and subjectivity was not mere sophistry; it was of a piece with his science, and flowed from his deep conviction of the unity of reason and spirit. Their works on the ear show that Duverney and Perrault shared the Parisian devotion to comparative anatomy as the primary mode of investigation of both structure and function. But their goals differed significantly. Duverney's primary concern was correct anatomy and the derivation of function from structure. Perrault's concerns were twofold. One had to do with mechanics: the nature of sound and its transmission, within a larger causal mechanism of peristaltic motion and an immaterial soul. Anatomy provided illustration and in some cases proof for Perrault, but it was not the focus of his attention. Duverney and Perrault used the same anatomical evidence, largely provided by Duverney; but Duverney was less concerned with proving the mechanical philosophy. To him, that battle was won; Le Clerc and Manget said as much when they published *Bibliotheca anatomica* two years later. But in addition Perrault aimed toward a redefinition of aesthetics and pleasure, based on his anti-Cartesian mechanical philosophy and entering cultural debates well beyond the confines of the Paris Academy.

SIX

The Courtiers' Anatomist:
Duverney at the Jardin du roi

1. The Elephant and the Tiger

During the academy's annual vacation in 1679, which usually took place in September and October, Colbert sent Duverney to Brittany to seek out Philippe de la Hire (1640–1718) and Jean Picard, who were surveying the coastline for a new map of the kingdom.[1] La Hire had been named to the academy a year earlier as an astronomer, although most of his training was in art; he was a protégé of the painter and engraver Abraham Bosse.[2] Colbert wanted Duverney and La Hire to seek out all the fish they could find on the Breton and Norman coasts, dissect them, and draw them. This task would be "most agreeable and most curious, and will even be useful." Two months later, as the expedition drew to a close, Colbert assured La Hire that "there is no time better spent than this, because the work will be very useful, being joined to other dissections that have been made and continue to be made at the Academy."[3] This was the closest Colbert ever approached to a policy for the collection of animals. Monier, as far as can be ascertained, simply bought what he could find on forays to the Levant and, later, South America.[4]

Duverney and La Hire dissected dozens of fish of seventeen different species, and La Hire drew them. They showed the drawings first to Colbert and then to the *compagnie*, before turning them over to Perrault to write up the descriptions. But this task does not seem to have been done.[5] Duverney accompanied La Hire the next year on another mapping and fishing expedition to Bayonne in southwest France, and they again presented their sketches and descriptions to the academy. Duverney demonstrated that, contrary to common belief, fish had organs of hearing; he was in the midst of his work on *Traité de l'organe de l'ouïe*. Such ventures helped to fulfill Colbert's expectation that the academy's work be useful to the nation.[6]

Shortly after their return in 1680, Duverney and La Hire took on a very different animal. Instead of showing their drawings of fish, as they had promised at the previous meeting, they set out for Versailles in the dark early morning of Wednesday, 22 January 1681. With them was Claude Perrault, along with several other academicians and many tools and crates. When they reached the château, a dead elephant had already been hauled up onto a platform, "a kind of theatre," as Fontenelle described it, ready for dissection.[7] Couplet was later reimbursed over one hundred livres for unspecified expenses associated with the elephant's dissection, which may have included the construction of this stage.[8]

The African elephant had been a gift from the King of Portugal thirteen years before, and had survived many Parisian winters before finally succumbing the previous day. Four years old when she traveled from the Congo to Paris, she was therefore seventeen at her death. She was not the only elephant in Paris; a young Asian elephant had been on show when she arrived at Versailles. But by the time of her death she was certainly the best known. In the summer, her many visitors could see her in an open pen; in winter, they could view her through the glass of her heated chamber. Artists came to draw her. She ate twenty-four pounds of bread and twelve pints of wine each day, supplemented by two buckets of "potage" or sometimes cooked rice. During her summer promenades through Versailles, she pulled up grass with her trunk and ate it. Generally very gentle, she knocked to the ground an artist who teased her; another she soaked with water from her trunk. Her trunk was a marvel: she could untie knots with it, and one night opened the door of her enclosure without waking her keeper and wandered around the menagerie. Duverney told his students that she could pick up a glass of water and empty it into her trunk without breaking it.[9]

Elephants had a properly royal history in France: the Caliph Harun al-Rashid had sent Charlemagne an elephant named Abul-Abbas, and Henri IV sent an elephant he owned as a gift to Queen Elizabeth in 1591. During the reign of Louis XIII, an elephant made a progress through France, and Peiresc saw it in Toulon in 1631.[10] An elephant figured prominently in the fourth of five paintings in Charles LeBrun's series *The Triumphs of Alexander*. The painting, *Triumphal Entry into Babylon*, was completed sometime before 1670, and depicts an African elephant like the one at Versailles rather than, as the subject might have demanded, an Asian one. The series, which eventually hung in the Louvre, was copied in tapestries and engravings; one of the tapestries commissioned for the Gobelins tapestry works in 1690s was "un éléphant."[11]

Such a wondrous and enormous beast as the Versailles animal (eight

Figure 6.1 Elephant, *Mémoires pour servir à l'histoire naturelle des animaux*, 1733. (Linda Hall Library of Science, Engineering, and Technology.)

and a half feet long, seven and a half high) held great interest for Perrault and the academy, and her dissection was at once a spectacle and a serious scientific investigation. Although Galen had famously dissected an elephant before his critics in Rome, few other dissections had been made, and much about the anatomy and even the external morphology of the elephant was unknown. Perrault and the *compagnie* carefully examined and measured the elephant's exterior, and scrutinized her skin with a microscope. The printed account, only published in 1733, spent more than twenty pages on external morphology.[12] Duverney dissected slowly and methodically, removing individual organs and parts, including the trunk, to be transported to the academy for further examination. Perrault took detailed notes and La Hire sketched. They discovered that the elephant, which had been thought to be male, was in fact female. At some point in the proceedings, when Duverney was literally immersed in the beast, King Louis made an appearance and demanded to know where the anatomist was; presumably he knew Duverney as one of the tutors of the Dauphin. In Fontenelle's words, Duverney "rose from the flank of the animal, where he had been, so to speak, engulfed," and greeted his king.[13]

Even with winter weather, the parts of the elephant would soon have begun to deteriorate, and members of the *compagnie* met the following Sunday and Monday as well as their regular Wednesday meeting to witness Duverney's dissection of the head and other parts, accompanied by Perrault's explanation. Dissection of the trunk extended into the middle of February.[14] The reading of Duverney's *mémoire* of the dissection and Perrault's account of the exterior occupied several more weeks. The *compagnie* discussed the elephant into the following summer, and she featured prominently in the academy's annual report to Colbert. There it was noted that Duverney's discovery of "many singularities" about the elephant "contribute to clarify many structures of man that remain quite obscure." The list of "volumes ready for the press" included Perrault's description of the elephant as well as the next volume of the *Histoire des animaux*.[15] A year later, the *compagnie* met at the King's Garden to look at the elephant's skeleton, which had been assembled there, and a year after that Perrault showed the *compagnie* several pieces of the elephant's skin that he had preserved.[16]

The trunk and the skeleton received the most attention. Perrault and Duverney described the action of the trunk in mechanical terms of cords and pulleys. Aristotle had believed that the trunk was cartilaginous, but Duverney and Perrault found it consisted of "nervous and tendonous" membranes with strong muscles overlaying them. Duverney located the organs of smell and identified each muscle and its functions in the trunk's ingenious "*méchanique*." The skeleton's interest lay mainly in its enormous size and the fortuitous structure of the elephant's defenses that made her such a deadly foe, and that made her gentleness at Versailles all the more remarkable.[17]

Animal mortality at the menageries was high. Between August 1680 and June 1681, apart from the elephant, Duverney dissected a tiger and three tigresses, a panther, a marmot, a dormouse, several different birds, and a salamander. The name "tiger," explained Perrault in his description (read to the *compagnie* in the summer of 1681), was promiscuously applied to a number of different large cats.[18] "It appears in truth," Perrault commented sternly, "that most of the Moderns who have spoken of the Tiger, have never seen one, and that they can only say what they have read in books."[19] But he nonetheless surveyed descriptions of so-called tigers from Aristotle to Faber and Johnstone, without settling on what exactly a tiger was. Just as the academy had used common names for other animals—"loup-cervier" rather than "lynx," "Demoiselle de Numidie" rather than "Otus"—so Perrault referred to this animal as a tiger. In fact, the image of a spotted rather than striped animal as well as its description indicated it was a leopard or possibly a jaguar.[20] The following year, a "tigre Royal" was killed at Vin-

Figure 6.2 Elephant skeleton, 1681, Muséum national d'histoire naturelle, Paris. This was the elephant dissected by Duverney. (Photograph by the author.)

cennes during a combat with elephants staged before the Persian ambassador. The animal was tossed into an adjacent field (where the locals pulled out all of its whiskers, believing them to be a potent poison). When the *compagnie* received news of the animal's death, "some of us," wrote Perrault, "went to see it." It was too far decayed to dissect, but they made careful observations and measurements, noting its large size and its striped fur.[21]

Over a weekend in December 1681, almost a year after the elephant's dissection, Duverney dissected a crocodile that had lived at Versailles for only a few months.[22] On the previous day the king, accompanied by his son and brother, had made his first visit to the academy. The visit had entailed weeks of preparation. Colbert directed the king's attention to the academy's publications, including engravings for the *Histoire des animaux*, both published and "about to appear." The king also showed interest in La Hire's drawings of fish and Dodart's of plants, and praised the academicians for their diligence. He witnessed some chemical demonstrations, looked over the engraving facilities, and ended his visit with an examination of some "machines" devised by Ole Rømer and Cassini to calculate eclipses.[23]

2. Duverney at the Jardin du roi: *Anatomie à la mode*

Much of the dissection of the elephant and assembly of her skeleton took place at the King's Garden, where Duverney had established his dissection activities by 1680.[24] About the time that Duverney entered the academy, the anatomical specimens and skeletons that had graced the academy's rooms at the *bibliothèque* were transferred to the garden, where there was a large empty room to receive them, a room that became known as the *"salle des squelettes,"* the skeleton room, and which came to hold a sizable collection of animal and human specimens. The skeleton room and the relationship between the academy and the garden that it implied thus preceded Duverney's tenure at the garden; Colbert directed both royal institutions toward common aims.[25] When Marin Cureau de la Chambre died in 1669, his position of "demon-strateur de toutes les opérations de chirurgie" at the King's Garden, to which he had been appointed at its foundation in 1635, fell vacant. As we have seen, it is unlikely that Cureau de la Chambre ever lectured at the garden, and there is no evidence of anyone giving anatomical demonstrations there during his tenure. From the 1650s onward, the governance of the garden fell into disarray: the position of *intendant*, or director, first held by founder Gui de la Brosse, was eliminated after the third director, the Scottish chemist William Davisson, left in 1651. The position of *surintendant*, who by a decree of 1642 oversaw the teaching function of the garden, was occupied by the first physician of the king. This position was also eliminated after the death of royal physician Antoine Vallot in 1671, and as part of his plan of consoli-dation of royal institutions, Colbert took over the garden's administration. But he reestablished the position of *intendant* in the following year, to be held by the king's first physician, while the administration of the garden remained directly under the crown. Antoine d'Aquin (1620–96), the first physician and a Montpellier graduate, held what seems to have been a role with very little power until he lost the post of first physician, and therefore his position at the garden, in 1693. His successor as both first physician and *intendant* was Gui-Crescent Fagon (1638–1718), the great-nephew of Gui de la Brosse. Thirty years earlier, Fagon had successfully defended before the Paris Faculty the first thesis that recognized the circulation of the blood, and he had been named professor of botany at the garden in 1671, where he regularly lectured. He was a popular medical practitioner despite his un-prepossessing physical appearance (he was very short, with bad teeth and a taste for eccentric wigs) and thus came to the attention of Mme. de Main-tenon, the king's mistress. Fagon took advantage of a power vacuum in the late 1690s and assumed the position of *surintendant* in 1699, leaving him in exclusive charge of the garden.[26]

This complex history reveals lengthy periods when no one exercised direct control over the garden, and teaching was spotty at best. Upon the death of Cureau de la Chambre, his position remained vacant for two years until his son François (1630–80), a physician of the Paris Faculty, was appointed "démonstrateur-opérateur de l'intérieur des plantes médicinales" in July 1671. Colbert and his *commis* Charles Perrault saw the potential of the garden as a counterbalance to the Paris Faculty, which was much disliked by court physicians for its antichemical stance. A series of royal declarations between December 1671 and January 1673 completely changed the teaching of anatomy in Paris. First, they reestablished the position of *intendant* for d'Aquin, the king's first physician—who was not a Paris graduate—and second, they declared that anatomical demonstration and surgical instruction were central functions of the garden.[27]

The younger Cureau de la Chambre may have lectured on anatomy during the 1660s.[28] But by 1671 he was physician to the queen as well as "*medecin ordinaire des bâtiments,*" with responsibility for the medical care of all royal employees, and had no wish to teach anatomy. However, unlike his father, he was not allowed to ignore his anatomical duties while collecting his pension.[29] He relinquished these duties to two surrogates, and anatomical instruction at the garden at last began in the winter of 1672–73. He delegated the "*discours*" to Pierre Cressé, a physician of the Paris Faculty, thus keeping anatomy teaching under the thumb of the faculty, or so he thought. Cressé, whom the faculty had nearly ejected over a scandalous affair with a surgeon's wife, seems an odd choice, but perhaps it was a way for him to regain favor.[30] But as the garden's charter instructed, a *sous-démonstrateur* was also hired, a surgeon of Saint-Côme, Pierre Dionis (1643–1718), to give the "cours"—that is, the dissection. Cressé, a Galenist, opposed the circulation; in 1674, he engaged in a three-hour discourse on this topic that attracted an audience of four hundred to the garden. Dionis, following the command of the court, demonstrated the circulation of the blood and other new discoveries.[31] He noted that he used a human body rather than an animal for demonstration not only for its practical value in teaching surgery, but because "the human body . . . is the masterpiece of nature, and therefore the most perfect of all bodies."[32] But the natural theology that this comment implied played little role in his lectures. The "*cours*" and the "*discours*" obviously had little in common and appear often to have been delivered separately.

A royal declaration registered at the Parlement of Paris in March 1673 summarized the measures of the previous year, asserting that lectures at the garden in medicine, surgery, and pharmacy would be performed by competent instructors and would be free and open to the public. In order to perfect instruction in surgery and anatomy, the garden's anatomists would receive preference over all others for the bodies of the executed, "even the Dean and Doctors of the Faculty of Medicine of Paris."[33] One can imagine the howls of dismay not only from rue de la Bûcherie but also from the surgeons at Saint-Côme, who only a year previously, as we saw in chapter 1, had battled the faculty over a corpse. A "*Salle des écoles*" (Dionis called it an "*amphithéâtre*") to accommodate teaching was established at the garden over the fall and winter of 1672–73.[34] The year 1673 also witnessed the opening of the new cemetery at Clamart where the dead of the Hôtel-Dieu henceforth were interred; its proximity to the garden may have provided an additional source for bodies for Dionis as it later did for Duverney. (See fig. 1.1.)

The team of Dionis and Cressé continued through the 1670s. Besides demonstrating all of the newest theories, Dionis salted his dissections with stories of the rich and famous he had autopsied, and he soon drew hundreds of spectators; he estimated four or five hundred regularly attended his courses, which consisted of eighteen demonstrations of general human anatomy (eight on osteology, ten on the rest of the body) and a further ten lessons on surgical operations.[35] The order of presentation, which began with the bones and ended with the muscles, required at least two and probably more bodies, and Dionis took advantage of the garden's newfound priority in obtaining corpses. Although in 1660 the Paris Faculty had successfully subdued the surgeons of Saint-Côme after their union with the barbers in 1656 by enforcing a set of new rules, the king's challenge to its teaching function indicated that the faculty of Riolan could no longer ignore the new science.[36] The faculty's 1660 rules had enjoined the surgeons to use animals rather than human cadavers for their demonstrations at Saint-Côme; not surprisingly, surgical students flocked to the demonstrations of Dionis at the garden, to the extent that they were issued tickets so that they could get priority to enter before the barbers or the merely curious. Because the lectures were free, anyone could and did attend.[37]

Two satires in the 1670s indicated the continued weakening of the faculty's grip on medical and anatomical instruction, as well as a cultural shift, largely but not entirely instigated by the court, from the older values of humanism to those of the new science. The ghosts of Riolan and Gui Patin, who had died in 1672, could not express their outrage at the playwright Molière's Doctor Diafoirus in the *comédie-ballet Le Malade imaginaire* (1673), who was, in the words of his father, "blindly attached to the opinions of the Ancients" and could "never understand nor listen to the reasons and experiments of the pretended discoveries of our century concerning the circulation of the blood and other opinions of the same stamp." The younger Diafoirus indeed carried his anticirculation thesis with him. Yet, following the example of Riolan, he also asked his prospective fiancée "to amuse yourself by assisting at the dissection of a woman upon whose body I am to give lectures."[38]

Already in 1671, the poet and critic Nicolas Boileau-Despréaux had penned an *Arrêt burlesque*, a mock ruling of the Parlement of Paris that forbade the University of Paris to teach anything but Aristotle. The university was in fact preparing such a request to the Parlement, directed particularly against the work of Descartes. Boileau's satire focused on the medical faculty and its opposition to the circulation: Patin had presided over yet another anticirculation thesis in 1670, referring to the circulation as "this famous and murky doctrine . . . confusion in all aspects."[39] Although Boileau, as we have seen in the previous chapter, would soon champion ancient literature over the modern opera, he was not above engaging in a good satire. He had learned of the university's request from royal councilor Guillaume de Lamoignon, president of the Parlement of Paris, whose salon he frequented in the 1660s. Boileau joined forces with the physician François Bernier (1625–88), formerly Gassendi's secretary, to compose the *Arrêt burlesque*, and Bernier contributed the mock request of the university that preceded the mock ruling of the Parlement.[40] Louis XIV did eventually act against the doctrines of Descartes, forbidding their teaching in 1685 as he further displayed his devotion to Catholicism with the revocation of the Edict of Nantes.

At the King's Garden, Dionis continued to teach about the circulation from a mechanical but not Cartesian point of view. His lectures, delivered first at the garden and later at Saint-Côme, were first published in 1690 as *L'anatomie de l'homme* and much reprinted and translated. His listing of the great discoveries of the seventeenth century included Harvey, Aselli, Vesling, and Pecquet but not Descartes. The anatomy that Dionis described in his lectures was structural rather than functional; the configuration of the parts dictated the function. He inclined toward the views of the chemical physicians who dominated court practice. The blood, he said, was a congeries of particles that were strained out by the appropriate glands throughout the body. Digestion was a form of fermentation, occasioned by the highly acid digestive fluid and aided by heat.[41] This was quite unlike Descartes, who characterized digestion as a purely mechanical process in which the digestive fluid disengaged the particles of food by insinuating itself between them, and also unlike Perrault, who, as we have seen, attributed the fermentation of digestion to peristaltic motion.[42]

Dionis agreed with Pecquet that the lacteal veins emptied chyle into the *cisterna chyli*. It then made its way via the thoracic duct to the subclavian vein and thence to the heart, where blood was formed. The lacteals were difficult to see in a dead body but could be viewed in a well-fed dog opened alive four hours after being fed; and if some of his auditors could not believe that what they saw in an animal could also apply to humans, he offered a case very similar to that witnessed by Peiresc and Gassendi in Aix forty years earlier. Dionis lived not far from the intersection known as the Croix du Tiroir, at the corner of the rue Saint-Honoré and the rue de l'Arbre Sec, close to the Pont Neuf. The Croix du Tiroir had long been a place of execution, particularly of counterfeiters, probably because of its proximity to the rue de la Monnaie, where the Paris mint was located. Knowing that a counterfeiter was going to be executed, Dionis sent him food about four hours before the execution. He waited at the scaffold with a coach and whisked the body off to his house. When he opened the corpse, he found the lacteal veins full of chyle, "which convinced me that this was distributed in man in the same way as I had seen it in several animals." Writing in 1690, Dionis said this took place "around eighteen years ago"—that is, around 1672.[43] He scornfully dismissed those who continued to believe that blood was formed in the liver.

The death in 1680 of François Cureau de la Chambre ended the arrangement with Cressé and Dionis. Dionis, named surgeon to the Dauphine, abandoned lecturing for a time, although he later resumed similar lectures at Saint-Côme, in a new amphitheater constructed in 1691.[44] Although Duverney soon assumed the anatomy lectures, the title again fell vacant, this time for two years, until his appointment as "démonstrateur-opérateur de l'intérieur des plantes" in March 1682.[45] Duverney took his position seriously, uniting the "cours" and the "discours," the theoretical and the practical, and he did not delegate his duties. While Dionis focused on structural human anatomy, Duverney's courses, like his chapters in the *Histoire des animaux*, revealed function as well as form.

But in addition, his demonstrations revealed the skills in both performance and dissection that had made him so popular among Bourdelot's *salonnières*. He filled his lectures with action, events, and latent moral meanings. Animals, both alive and dead, and carefully presented anatomical preparations as well as many human bodies created an atmosphere that induced certain audience reactions, including riots. Dionis's lectures were popular, but Duverney far surpassed him both in numbers and in impact.

He appeared in poems, plays, and popular literature, and the numbers he attracted were so great that Louis XIV built him a new anatomy theater in the early 1690s. This audience, which reached 600 spectators, came from all parts of society, and included as many as 140 foreigners.[46]

In assuming the position at the garden, Duverney moved to a new and more expansive arena. He had already successfully negotiated the competing demands of the anonymity of the academy versus his public performances and reputation at the salons. The audiences at the salons of Bourdelot and Denis were limited to those with sufficient knowledge or interest, as well as by the physical confines of their homes. The audience at the Academy of Sciences was even more constrained to its members and the very occasional visitor, who also had sufficient knowledge to understand the proceedings. At the garden, in contrast, Duverney lectured to the wide and varied audience that Dionis had attracted of medical and surgical students, natural philosophers, fashionable members of the court, and whoever else wished to drop by. Guillaume Lamy complained that the "canaille" of the neighborhood, "attracted by a foolish curiosity to see a body dissected," attended one of Cressé's lectures, illustrated with a dissection by Dionis. Such "canaille" impeded the "honnêtes gens" such as him from gaining a view. It ended, he said, with foot stamping and rock throwing.[47]

Such behavior often spilled over into a riot. As at public executions, the close presence of the human body was emotionally disturbing, and the surgical students, mainly teenage boys, acted out their discomfort. In addition, the garden was quite close to the university quarter, and many other students (medical or not) frequented its lectures. An ordinance from 1681 forbade the "students and auditors" who attended anatomy lectures at the garden from wearing swords or "batons," and intermittent complaints about violent disorders continued at least up to 1735. Several declarations complain about disruptive auditors and note that the Swiss guards charged with keeping order were systematically corrupted with money and gifts.[48] This was not the decorous and gentlemanly demonstration of abstract physical laws, or the protocol-bound science of the academies.

3. Duverney's Teaching at the Garden

Like Dionis, Duverney lectured on both general anatomy and surgery. He also gave more specialized courses in osteology. Unlike Dionis, Duverney did not publish his lectures. According to an account of his osteology lectures published in the *Journal des sçavans* in 1689, he was simply too busy to publish. This "Lettre à M. le président Cousin" and a further *Lettre à M.* ***

published as a pamphlet constituted the only published record of his teaching in his lifetime.[49] In addition to these accounts, several sets of student notes survive, dating from the 1680s to 1715. With his posthumous collected works, these notes allow us to glimpse his methods and his style. Duverney attracted a number of foreign students, and two sets of notes are in English. The notes vary widely in completeness and style. Some are carefully copied out minitreatises; others consist of a single day's notes on a particular malady from his surgery lectures.[50] Several sets of notes are on osteology, which may show that Duverney's expertise in this area made these notes particularly worth keeping and handing on. One set of notes on osteology appears from its context to have been given at the Hôtel-Dieu, where Duverney lectured from the late 1680s onward.[51] An anonymous editor, possibly his son Emmanuel-Maurice, issued a two-volume *Traité des maladies des os* (Treatise on the illnesses of the bones) in 1751, based, he said, on a manuscript treatise. The *Traité* closely resembles extant student notes.[52]

ˑ The two sets in English are relatively complete and were written at about the same time in the late 1690s. The Irish medical student Patrick Mitchell, later a prominent medical practitioner in Dublin, attended a number of Duverney's courses in 1697–98. The Scot James Douglas (1675?–1742) attended lectures around 1698, before receiving his medical degree from Rheims.[53] In both cases, the notes appear to have been recopied as a treatise for further reference. This was a common practice; the anatomist William Hunter later advised his students to make a fair copy of their notes each night, and even to attend his lectures twice to ensure the notes were complete. Douglas's notes are in two hands, indicating that the effort of recopying was shared, possibly with his younger brother John, who became a surgeon. A manuscript economy of copying and exchange was well established among medical students, and notes taken at Duverney's lectures were recopied and passed down through generations of students; a set of notes for Duverney's lectures from 1685 were recopied as late as 1730.[54]

Mitchell's notes, which amount to over three hundred closely written pages in a small notebook, are the most complete of these sets, and outline what was probably a typical course of study for a foreign student in Paris. Beginning most likely in the autumn of 1697 and extending to the end of the following summer, Mitchell took three courses from Duverney—in general anatomy, surgery, and osteology—as well as other courses in chemistry and pharmacy. In the spring and summer of 1697 or 1698, he studied pharmacy with a "Mr Charras," possibly Moise Charas (1619–98), an apothecary and academician, who had been the unofficial chemistry demonstrator at the garden in the 1670s.[55] He also studied medical chemistry during that period. But he spent most of his time in Duverney's courses.

As Bossuet had noted, Duverney's lectures were marvels of organization and concision. After a general description of the body and its structure, working from a skeleton, he proceeded from head to toe. He had sufficient bodies not to be concerned about the need to work on the abdomen first, but he was also perhaps conscious of the marquee value of the dissection of the brain, the operation that had launched his career.[56] Proceeding methodically from the skull to the brain to the organs of sense, he frequently compared humans to living and dead animals even as he emphasized that his topic was the human. For example, several experiments on the dura mater of live animals showed that it conveyed the sensation of pain when touched with acid without showing any external signs of disturbance, and the eye of a goose showed the differences between bird and human eyes. He compared the structure of the human brain to that of birds and even of fish, which have "big heads and very little brains."[57]

Animals were constantly present in Duverney's dissection room. Their use varied according to the audience; the general anatomy lectures were more explicitly comparative, while the osteology lectures, aimed at medical and surgical students, emphasized human structures. Mitchell referred to the osteology lectures as a "private course," and it lasted for about a month, serving as a prelude to the general anatomy course, which took three months to complete.[58] The species of the animals were not always named; when they were, they included dogs, sheep, and horses. Sometimes animals acted as stand-ins for humans, as when Duverney showed the lacteal veins in a live dog, or when he looked at the subcutaneous muscles in a human "and even in a dog and a cock."[59] His use of live animals was matter-of-fact. Mitchell described a demonstration of the pain caused by an injury to the periosteum (the vascular tissue that surrounds the bones):

> When a caries comes to the periostium or the membrane investing the marrow the pain is intollerable. . . . We sawed the bone of a live creature, put a probe into the marrow, it excited a horrid pain upon drawing it out, it ceased, and returned upon repeating it.

Duverney and his students concluded that bones themselves "have no sense, for having removed the periosteum, you may scrape without exciting the least pain."[60] Douglas witnessed a demonstration of the lymphatic vessels in a live horse, which "he prepared thus having put each of his feet in the loop of a cord he caused [to] knok him in the head" and then opened the thorax. Duverney proceeded methodically, tying off each vessel as he showed it.[61]

These notes portray a preternaturally calm and competent lecturer. Duverney was never at a loss for words, keeping up a patter as he deftly tied off vessels or injected wax into them, mopped up blood, or made an incision. He sprinkled in anecdotes and everyday observations: rope dancers were a favorite example of agility, and Douglas recounted Duverney's story of a boy in Paris who ruminated.[62] Duverney often employed examples from the *Histoire des animaux* project, particularly the elephant, whose skeleton was on display. In his discussion of teeth, he had at hand a number of teeth and jaws from lions, bears, leopards, and tigers, as well as the more usual dogs and cats, all in the *salle des squelettes*.[63]

Duverney continually returned to two themes: utility and design. God had designed the body for the utmost utility; everything in the body had its best purpose. Mitchell noted:

> The extremitys of the bones are very big for strengthening the motion of the muscles; in the greatest animals that are not so big, as in man; the elephant comes very near man in this; providence appears mightily in this; because this saves a great deal of spirits in moving this great body.[64]

The "admirable méchanique" of the body also gave aesthetic pleasure to the anatomist, and Duverney often expressed how "satisfying" he found the structure of the body, ascribing its teleology to "nature" or "providence" as much as to God.[65]

In content, Duverney's lectures were up-to-the-minute, citing the most recent authors including Thomas Bartholin, Niels Stensen, Raymond Vieussens (1641?–1715) and Charles Drelincourt (1633–97). His critiques spared neither ancients nor moderns. "We know the falseness [*fausseté*] of many of the opinions of the ancients," he said, condemning "certain men who make it an honor and even a kind of religion to conserve as a fertile heritage the errors of their fathers," an obvious reference to the Paris Faculty of Medicine.[66] He held to particular scorn the idea of a "*faculté formatrice*" for the bones, and accepted the circulation of the blood and the lacteals and lymphatic system without question, firmly demoting the liver to a mere filter. Following current views, Duverney was a preformationist with regard to reproduction, rejecting Aristotle's notion of two "seeds."[67] Although he does not seem to have used a microscope in his lectures, he cited the discoveries of those who did, such as Malpighi. A convinced mechanist, he nonetheless embraced chemical explanations. The cause of animal heat was the intestine motion of the particles of the blood, but he explained the formation of bones as a chemical process of precipitation, although the substances that formed bone were mechanically strained out of the blood.[68] Douglas criticized Duverney for ignoring chemical explanations, but he then listed Duverney's four "instruments" of digestion: (1) muscular motion of stomach and (2) diaphragm, (3) heat of neighboring parts, and (4) "dissolvents or levains of ye stomach." In discussing gangrene as a corruption of the blood, Duverney compared it to wine that goes bad, whether because of acid salts or saline sulfur, not specifying which might be the explanation.[69] Such enumeration and weighing of alternative explanations was characteristic of Duverney's style.

Among his disagreements with the moderns was Leeuwenhoek's observation that "the seed is full of animals." This could not be correct, he said, because it reduced the role of the female to simply providing a material substrate. Moreover, Leeuwenhoek had also observed such animals in pepper water, "yet none could claim that these are the eggs of pepper"; insects and their eggs could apparently breed in the air, so the seminal principle might be in the air, an apparent reference to Perrault's ideas of panspermia, the notion that life exists everywhere in the form of seeds or spores; and finally, "they pretend that in a clap, you see none of the animals in the seed; I answer that as in water, when these animals, can finde no more nourishment, they fall to the bottom, so the seed being corrupted they fall to the bottom, it being very unfit to nourish them." He also expressed doubts about the existence of the nervous fluid, since he did not observe it when he cut a nerve.[70]

Duverney's emphasis throughout the lectures was on seeing and showing, and the demonstration of techniques. He demonstrated before the students how to inject wax into a vein rather than simply showing a preparation. If he was not always successful the first time, that too was instructive. He showed how to clean a spleen, and how to press the blood out of a cadaver.[71] He gave his students instruction in handling living and dead animals. By 1690, he was also lecturing to medical and surgical students at the Hôtel-Dieu, and lecture notes on specific diseases may have been for that audience.[72]

Duverney received yearly pensions of fifteen hundred livres each for his work at the academy and the garden, a total of three thousand livres per annum, and he was given an apartment at the garden. The crown reimbursed him for materials and tools. In addition, the royal accounts show payments to various assistants, including a "*garçon*" who assisted with dissections and acquired corpses from the cemetery or the executioner. As we have seen, the academy *élève* Claude-Antoine Couplet purchased or otherwise obtained animals and also reported on live animals he had seen. The purchase and transportation of human corpses had a budget line, with prices ranging from thirty to seventy livres per body, including transport.[73] The crown paid the cutlers André and André-Guillaume Gérard for making instruments of dissection.[74]

André Colson, described variously as an "*ébéniste*," or cabinetmaker; an "*empailleur*," or taxidermist; and an "*anatomiste*," served Duverney in several capacities for over twenty years and may be counted as one of the "invisible technicians" of early modern science. He first appeared in the Versailles accounts for 1672 as a taxidermist who "stuffed diverse animals dead at the Menagerie." By 1676 he made taxidermy specimens and animal skeletons to be displayed at the garden in the new *salle des squelettes*.[75] He became less invisible over time, appearing occasionally in the academy's minutes; in 1682 he is mentioned as charged with "cleaning certain skeletons," most likely the removal of flesh to prepare the skeletons for assembly, a lengthy and unpleasant task. Colson constructed the skeleton of the elephant that still resides at the garden as well as many others. Earlier skeletons, including the first, of the beaver in 1668, had been constructed by academician Nicolas Marchant, who held a medical degree from Padua.[76]

The costs of materials and tools for dissection came out of Duverney's pocket, to be reimbursed later. But the schedule for reimbursement, like the payment of pensions, was highly irregular. Dionis was still being reimbursed in 1685 for expenses from the 1670s. In 1681, for example, the crown reimbursed Duverney for the previous three years of expenses incurred at the *académie*, the not-inconsiderable sum of nearly 400 livres. In 1687 and 1688, he was reimbursed promptly for the previous twelve months' expenditures, but the sums were also much greater: 926 livres in 1687 and 1,346 livres in 1688.[77]

Memoranda survive from between 1687 and 1691 that indicate the kinds of expenses Duverney incurred. A memorandum to the academy from 1688 detailed his expenditures directly related to its dissection projects. These included the purchase of syringes, a microscope, and two "very strong" magnifying glasses, as well as towels, sponges, and basins; coal and wood to heat the dissection chamber; and candles to light it. Duverney purchased animal entrails from a butcher in the Faubourg St. Victor and paid a man to haul off and bury the used animal parts. He paid for several other kinds of animals as well. He also paid a man to clean the dissection room as well as make "several trips" to the cutler and the place of execution. These expenses amounted to 327 livres, about a quarter of what he spent that year.[78] Another memorandum related his expenses for his anatomy courses at the garden in 1687–88. These ranged from five human bodies (one for each of five demonstrations) to seven hundred broadsides to advertise the course—three hundred in Latin and four hundred in French, indicating a mixed audience—as well as someone to post them. He hired soldiers "to impede disorder," bought animals ("4-footed, birds and fish"), and hired a surgeon to demonstrate specifically surgical operations such as bandaging. Candles, instruments, and injections were lumped under one category. In addition to the bodies and animal parts, an executioner was paid 60 livres for "other parts of human subjects."[79] Not mentioned, but perhaps included in this category (or perhaps simply supplied from the chemistry labs at the garden) were the copious quantities of eau-de-vie or spirit of wine commonly used in human dissection to wash out the putrefying cadaver and arrest the process of decay.[80]

In the year 1687–88, therefore, Duverney spent much of his pensions on operating expenses. Although he was reimbursed within a year, this would not be true in the 1690s and later. Even costs for a new dissection room, used both for his own demonstrations and for the academy's work, came initially out of Duverney's pocket. The "petit bâtiment" at the foot of the garden served as a place for preparation and demonstration. In 1692 Duverney was recompensed 400 livres toward its construction. However, that covered only a fraction of his costs, and Duverney was not reimbursed for the remainder—1,933 livres, more than his yearly pension from the *jardin*—until 1697.[81]

The accounts indicate a merging of Duverney's roles as anatomy instructor and academy dissector at these two institutions of the crown, although the roles served different purposes. While some dissection continued at the academy, what became standard practice was described in its minutes early in 1682. The secretary announced the arrival of several birds (presumably dead) from Versailles, including the parrot "Arras." Couplet was instructed to bring them across the Seine to the garden after the academy had examined them and had written up descriptions of their external appearance.[82] In some ways an annex of the academy, the garden also served as an educational institution and an exhibit space—since the *salle des squelettes*, which belonged to the academy and displayed the work of Duverney and Colson, was open to those who attended lectures. Demonstrations at the academy were repeated at the garden, and vice versa. Duverney informed the academy of experiments he performed at the garden; for example, in 1683 he reported the dissection of a woman, and the following year he described four separate operations—two of them on live animals—that he had performed at the garden over the previous week.[83] He completed his work on the elephant at the garden. Accounts from 1688 acknowledged these merged roles, when Duverney was reimbursed for "demonstrations of anatomy, dissections of animals, at the aforementioned Garden and at the Academy of Sciences."[84] Other academicians, including Perrault, also dissected at the garden.

4. Rhetoric, Tragedy, and Moral Anatomy

In a tribute to Duverney's rhetorical skills, Bishop Bossuet claimed that Duverney taught him the secret of organizing a presentation: high praise from one of the great orators of his age.[85] As the most popular lecturer of his era, and one noted for his eloquence, Duverney played an important role in the linguistic changes of the late seventeenth century.[86] Several of Duverney's contemporaries commented on his eloquence, and Fontenelle's eulogy of him compared him to an orator. He detailed Duverney's talents:

> This eloquence was not only about clarity, accuracy, and order, all the cold perfection dogmatic subjects demand; there was a fire in his expressions, in his turns of phrase, and even in his pronunciation, which nearly made him an Orator. He could not indifferently announce the discovery of a vessel, or the new use of a part; his eyes burned with joy, and his entire person became animated. That warmth communicated itself to his audience, or at least kept them from that involuntary languor which was apt to overcome them.

Even actors and actresses attended Duverney's lectures to observe his technique.[87]

Fontenelle's description of Duverney outlined two roles for rhetoric in the discourse of the new science: the "cold perfection" of accuracy and the warmth of conviction. Classical rhetoric referred to the ability of speech to persuade. Plato had contrasted rhetoric as a method of argument to dialectic, defined as a method of discovery. Aristotle, who remained the authority on this topic, instead viewed rhetoric as a subset of dialectic, both of which, he said, dealt with probable truth rather than with the absolute demonstration of philosophical reasoning. Rhetoric therefore was not a method of finding truth, but a method of conveying it to an audience. It encompassed three modes of persuasion: logos, or logical reasoning; pathos, or emotion; and ethos, the character of the speaker, which gave him credibility.[88]

. As literary criticism became established as a genre in the seventeenth century, this classical model of rhetoric came under attack.[89] Descartes claimed (perhaps disingenuously) that the clear and distinct language of natural philosophy made traditional rhetoric obsolete. The novel as developed by

Mlle. De Lafayette and Mlle. De Scudéry challenged the rules—also strained by dramatists and, as we have seen, the opera—of classical tragedy, while Charles and Claude Perrault questioned the value of the classics as a model not only for literature but also for music and of course science. Language was central to the debate between the ancients and the moderns. Duverney's lectures entered this crisis in letters on all three fronts: rhetoric, the redefinition of tragedy, and the rejection of the ancients.

Especially influential to early modern ideas about rhetoric were the works of the Romans Cicero (106–43 BCE) and Quintilian (ca. 39–96 CE). Cicero's discussion of ethos diverged from Aristotle in two important ways: character, he said, was innate rather than learned, and the character of the audience was as important as that of the speaker. The duties of the orator, he said, were *probare, delectare, flectere* (to prove, to please, and to move emotionally).[90] Quintilian (ca. CE 39–96), writing one and a half centuries after Cicero, slightly changed these duties to *docere, movere, delectare* (to instruct, to move, to please).[91] He replaced proving with showing, using the verb "to teach" rather than "to demonstrate." Quintilian placed rhetoric at the center of education, and it was thus adopted as part of the medieval *trivium*. The orator's assets, said Quintilian, included the personal impression he made upon his audience, the possibility of appealing to the emotions, and the organization of proof, always with the understanding that such proof was probable rather than certain.[92]

The rediscovery of a complete text of Quintilian's *Institutio oratoria* by the humanist scholar Poggio Bracciolini in 1416 held particular significance for the humanists' reform of education. Quintilian's emphasis on the moral basis of rhetoric proved extremely influential.[93] Nonetheless, by the seventeenth century, classical rhetoric, as filtered through Scholasticism, had become synonymous with insincerity. But those who so argued had been imbued with Aristotle and Quintilian in their youth. The neoclassical revival that led to the foundation of the Académie française in 1635 aimed to discipline and reconfigure the French language with the goal of *"clarté,"* clarity. Descartes and later the Port-Royal school of Antoine Arnauld and Pierre Nicole criticized even the reformed rhetoric of the academy, claiming that, on the model of geometry, clarity of thought and transparent language would communicate truth better than the persuasive arts of rhetoric. Descartes contrasted method to rhetoric:

I valued oratory and was fond of poetry; but I thought both were gifts of the mind [*esprit*] rather than fruits of study. Those with the strongest reasoning, and the most skill at ordering their thoughts so as to make them clear and intelligible are always the most persuasive, even if they speak only low Breton and have never learned rhetoric.[94]

To Descartes, only mathematics possessed the clarity and precision to be truly persuasive. But he continued to employ the tropes of classical rhetoric in his work; he wrote in prose, not in equations. Pascal's *L'esprit géométrique et l'art de persuader* (1658) similarly argued that geometry could best obtain the truth. Arnauld and Nicole attempted to codify this new interpretation of "method" in their *Logique de Port-Royal* (1662). Yet Pascal, for all his admiration of geometry, concluded that persuasion could not simply be a matter of superior logic, and he left room for intuition, sympathy, and natural human desires and emotions, thus returning to the traditional concerns of rhetoric, citing Montaigne's humanistic *L'art de conférer*. As we have seen, Renaudot in the 1630s and later Denis and Bourdelot adopted the term *"conférence"* as a public discourse on an intellectual topic.[95]

By writing particular works in French rather than Latin, Descartes consciously reached beyond the narrow audience of the Scholastics. He addressed not the scholar but the *honnête homme*.[96] Such a man frequented the new salons and academies, and his knowledge of the classics might be filtered through French translations. As a translator, Boileau presented classical authors as models for modern writing. But the *honnêtes hommes* also read modern novels, and were self-consciously members of a new reading public that included both scholars and laypeople, both men and women.[97] These individuals, who included courtiers but were not identical with them, came to see Duverney's lectures at the King's Garden along with medical and surgical students.

In the garden, Duverney joined the logic of the new science with the emotion of rhetoric, demonstrating and teaching while moving and delighting the emotions. Duverney's credibility as an orator centered on his expertise: both his manual skill and his specialized knowledge.[98] His reputation for eloquence rested on his oral performance, and much of his verbal and physical technique is inevitably lost. Yet what remains suggests the impact of his anatomy lectures on contemporary observers.

Quintilian had compared a good speech to a body, stating that it was important to follow "the natural order, which demands that after dealing with [the principal question], he should then proceed to introduce the subsidiary questions, thereby making the structure of his speech as regular as that of the human body."[99] The body, therefore, gave an inherent structure to the lecture, while the subject matter was both instructive and moral. Quintilian also included a chapter on gesture in *Institutio oratoria*, and the body of the anatomist, his *habitus corporis*, was as important as the body on the table. Because Duverney, like the Perraults, occupied what Jacques Revel called the "difficult middle ground" between court and salon, their presentation of self and their individual qualities determined their position in society.[100] The inherent distastefulness of dissection made this task more difficult. Duverney's self-presentation in his lectures—his voice, his language, and his gestures—identified him as one of the *honnêtes hommes*. What Antoine Courtin referred to as *contenance*, or bearing, "the accord between the inner and outer man," defined civility. Courtin's *Nouveau traité de civilité* (1671) expressed the mores of the salon within a Christian moral frame.[101] Unlike Faret forty years earlier, Courtin described not only how to please but also how to behave morally. Duverney successfully negotiated the challenging terrain of self-fashioning before several different audiences: the court, the academy, the garden, the hospital. His success as the "courtiers' anatomist" was hard-won, and as we shall see, Duverney did not brook challenges to it.

Public anatomical demonstrations such as Duverney's lectures at the garden constituted a ritual with images and metaphors that its audiences understood. These lessons were staged performances intended not to discover new things about the body but to demonstrate what was known. As Le Clerc and Manget confidently declared in 1685, anatomists knew all that was to be known about the body, and a contemporary described Duverney as "knowing anatomy perfectly, not only the human body, but animals and plants as well."[102]

The 1689 pamphlet *Lettre à Monsieur ***, contenant plusieurs nouvelles observations sur l'osteologie* offered an account of one of Duverney's public lectures at the garden.[103] The lecture followed a logical progression in its discussion of the articulation of bones, with two examples: the suture, such as appears in the skull, and the articulation of teeth. Most of the twenty-six-page pamphlet is devoted to the skull. Its suture was not like a comb, Duverney explained, but tightly interlaced, and the location and nature of the suture varied in different animals. He employed the animal skeletons in the *salle des squelettes* for illustration. The audience also experienced a certain proximity to royalty when Duverney frequently returned to the skull of the king's elephant.

The lecture covered a remarkable range of topics under the general theme of the bones of the head and its "*admirable méchanique.*" Duverney proceeded smoothly from skulls to teeth to a classification of animals according to their diet. He progressed from carnivores (displaying the teeth of a lion) to ruminants, to equines, rodents, and finally primates and humans, demonstrated on the skeleton of a monkey. Perhaps responding to the audience's interest, Duverney returned again to the elephant and the use of its trunk before closing with miscellaneous observations.[104] This logical, clear and eloquent lecture presented its underlying theme of teleology and natural theology with subtlety: the assembly of the skull made it less easily damaged; animal teeth were designed for different diets; rumination was "a very ingenious mechanism."[105] In speaking of the teeth of carnivores, he noted, "The *méchanique* of which nature serves itself in these defenses [of] canine teeth is so Ingenious that art cannot copy it [or] imitate it."[106] The lecture's tone was entertaining, enlightening, and morally uplifting. Duverney used metaphors sparingly, and in keeping with mixed nature of his audience, he employed no classical allusions. But the use of animals and comparative anatomy allowed him a much greater range of teleological explanation than could be demonstrated by human anatomy alone.

We can assume that the teleology and moralizing anecdotes of Duverney's lectures at the garden were also delivered to the Dauphin. Bossuet's *Traité de la connaissance de Dieu et de soi-même*, written to instruct the prince, included a chapter on anatomy. Wisdom, wrote Bossuet, consists in knowing God and in knowing the self, and knowledge of the self includes knowledge of the body. Patrick Mitchell noted that "providence appears mightily" in the relationship between the bones and the muscles, and "the wisdom of God appears mightily in the inosculation of the carotid and cervical arterys [*sic*]."[107] Claude Perrault had expressed similar sentiments in the *Histoire des animaux*.

Duverney's public lectures at the garden were performances that, like dramatic performances and the opera, provoked emotional as well as intellectual responses in their audiences. Like dramatic performances, anatomical demonstrations took place in theaters, were accompanied by music, and often required tickets for entry.[108] The most provocative issue in seventeenth-century French drama concerned the use of reason to express emotion. Critics such as Boileau argued that opera and the novel valued feeling over reason. Order and reason, the model of classical drama, gave way to disorder and passion. Despite the anatomist's rational exposition of his subject, public anatomical demonstration had more in common with opera than with Racine's tragedies. A witness to the dismemberment of a human body, as well as the dissection of living and dead animals, might experience horror rather than the healthy catharsis of classical drama. The many riots that followed anatomy lectures at the garden could be seen as the unruly consequences of such disordered emotion.

When Charles Perrault read his poem *Le siècle de Louis le Grand* to the Académie française in January 1687, Boileau shrieked so fiercely and for so long that he lost his voice. Reigniting the debates sparked by the opera a decade earlier, Perrault argued that the era of Louis XIV far surpassed the ancients in every way, singling out the new science, art, literature, and music.[109] Claude Perrault had already asserted that the *Histoire des animaux* offered "an irreproachable witness to a certain and recognized truth"; the ancients, said Charles, who only could see "vain phantoms" and "simple qualities," could not compete.[110] Public opinion was on the side of the moderns: novels and operas were increasingly popular, and by the late 1680s Duverney was the toast of Paris: his auditors at the garden and at Geoffroy's salon took home samples of his anatomical preparations to show their friends, perhaps even to wear.[111]

Yet the category of taste, which Claude Perrault had claimed for the moderns in *De la musique des anciens*, was not easily relinquished by Boileau and his fellow supporters of the ancients. They insisted that the new public who read novels and attended operas and dissections did not necessarily judge these cultural productions by the proper classical criteria.[112] Even though Duverney followed proper rhetorical standards, his subject matter did not. In 1694, Boileau published his notorious attack on learned women in *Satire X*. He criticized these "*précieuses*" for their blind addiction to the new, and their lack of taste and propriety. They valued Roberval and Cassini, looked through telescopes and microscopes ("admirable glasses" to Charles Perrault), and then went to view "a dead woman with her fetus / they must see the dissection at Duverney's / nothing escapes the gaze of our curious woman."[113] This spectacle offended Boileau on several levels: the idea of a woman doing science was absurd and indecent, and in this case a woman was both the audience and its object, which made it even more prurient. Courtin had criticized what he termed excessive curiosity, exemplified in looking through the books of one's host. To Boileau, books were the least of the problems of the moderns. He argued that they sought continual stimulation of the senses rather than invoking reason; while the sublime elevated the spirit, the moderns debased it. In his 1674 translation of the *Treatise on the Sublime* attributed to Longinus, Boileau had asserted that only the ancients could achieve the "elevation of the spirit" and "true nobility of expression" that literature required. Molière, and Dr. Diafoirus, who had taken his fiancée to a dissection, would have disagreed.[114]

Although Antoine Arnauld arranged a public apology in 1694 between Charles Perrault and Boileau, the debate continued. Perrault issued the first of his interpretations of folk tales, *Peau d'âne*, in the same year, and his *Histoires ou contes du temps passés* appeared in 1697. He intended these tales as another blow against the ancients, since they came not from the classics

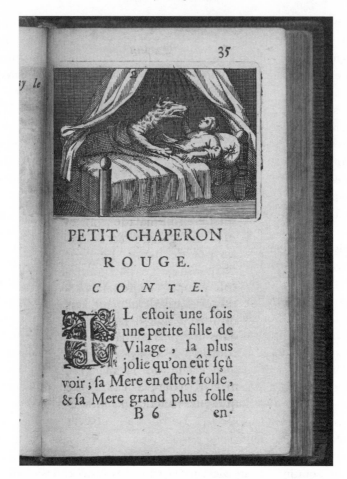

Figure 6.3 Charles Perrault, "Le petit chaperon rouge," 1697.
(Department of Special Collections, Charles E. Young Research Library, UCLA.)

but from folk tradition, and their modern retelling originated in the salons. His retelling of them, moreover, partook explicitly of the modern novelistic style; in the words of folklorist Jack Zipes, these tales were "gallant, natural and witty, but inventive, astonishing, and modern."[115]

Both Perrault's tales and Duverney's dissections depended on an inversion of categories based on the interchangeability of human and animal, a literal or potential metamorphosis.[116] In the stories, a beautiful princess is disguised in a donkey's skin, the puss in boots is smarter than his master, the wolf outwits Little Red Riding Hood. In dissection, animals regularly stood in for humans. Even when animals did not play central roles in these stories, they continued to turn things upside down on the sidelines, much as animals inhabited the edges of representations of human anatomy. In "Cinderella," the lizards and rats were transformed into footmen and coachmen, and in both Vesalius's famous title page and Riolan's less famous one (fig. 1.3), animals occupied the margins.

Cruelty, or the idea of it, was central to both genres. Perrault's tales did not always have happy endings. In the Grimm brothers' version of "Little Red Riding Hood" from the early nineteenth century, the huntsman took the girl and her grandma out of the wolf's stomach alive. In Perrault's version, the wolf ate them both, and there was no huntsman.[117] Duverney matter-of-factly stated that "animals which one opens alive . . . suffer much," but went on to note what particular physiological phenomena were thus produced.[118] He did not comment on the suffering any more than did Perrault. Cruelty made the moral happen; pain led to truth.

Public anatomy, in the hands of a master of eloquence such as Duverney, consciously aimed toward rhetorical goals: to teach, to move, to delight. Duverney demonstrated the new facts of the new science such as the circulation of the blood; he delighted his audiences with the amazing contrivance of nature, and moved it to contemplate the nature of life and death. As a protagonist in a drama about the meaning of life and death, he explained the most ancient of themes in a display worthy of the most ancient forms of sacrificial ritual. Duverney knew how to be eloquent, and as a man deeply immersed in his time, he knew his audience. He gave anatomy a cultural significance that it only partly had before or since. In his hands, human and animal bodies told a deeply felt story.

5. A New Regime and a New Anatomist

Jean-Baptiste Colbert, the godfather and overseer of the academy, died in September 1683, during the academy's annual vacation. The academy was flourishing and active, with ties to the garden and the Paris Observatory as well as to the Collège Royal (where Gallois, La Hire, and Blondel were professors). Carcavi and Charles Perrault maintained ties to the court. Colbert had vested his position as controller of the king's buildings in his fourth son, the Marquis de Blainville, thus ensuring a smooth succession. But upon Colbert's death, the minister of war, François-Michel le Tellier, Marquis de Louvois (1641–91), moved quickly to assume Colbert's powers, much as Colbert had seized Fouquet's powers two decades earlier. He purchased the office of controller from Blainville for the enormous sum of five hundred thousand livres. Louvois made his new regime clear at the first meeting of the academy after Colbert's death, removing Charles Perrault and Carcavi from their positions and replacing them with his own man, Henri Bessé de la Chapelle.[119] Gallois and Blondel reported that "the intention of Monseigneur de Louvois was that [the academy] work particularly on matters that can be useful to the public and contribute to the glory of the King." As Fontenelle pointed out, this had always been the academy's stated goal.[120] Yet a *mémoire* a few years later in which Louvois rejected any research based on "pure curiosity" signaled a change in attitude. One of Colbert's last gifts to the academy was a bird of paradise, which the *compagnie* had drawn (noting, contrary to popular belief, that it had feet); it seems unlikely that Louvois would have made such gestures.[121] As minister of war, he had successfully concluded the Dutch war with the Treaty of Nijmegen in 1679, but he supported Louis's belligerent foreign policy and the hardening of royal religious views that led to the revocation of the Edict of Nantes in 1685 and renewal of war by the end of the decade. Huygens, a Protestant, had left France in 1681 for reasons of health but did not return, although his treatise on magnetism appeared in the academy's minutes in 1683.[122]

Historians have generally agreed that the academy stagnated during the period of Louvois's control from 1683 until 1691, although likely more from lack of funds and a lack of attention than from a deliberate policy of repression.[123] Few new academicians were appointed during this period, although Thévenot received belated recognition of his contributions to science and his intellectual standing with his appointment as royal librarian in 1684 (succeeding Carcavi) and to the academy in 1685. His deep pockets seem to have been especially appreciated at the library.[124] But Bessé regularly attended meetings, and the minutes show a flurry of activity from 1684 onward. The Wednesday and Saturday meetings merged, with anatomy and astronomy at the same sessions, and there was considerable pressure to complete the *Histoire des animaux* project. But payment of stipends lagged; the academy accounts show only partial payments between 1688 and 1690.[125]

By the early 1680s, Duverney, still in his thirties, had established secure niches at the academy and the garden and was able to fulfill his duties for both posts from his apartments and dissection room at the garden. Well known at court, he gained wider attention with his book on the ear, and his collaboration with Claude Perrault on the *Histoire des animaux* had evolved into one of equals. He contributed regularly to the academy's sessions, and he may very well have continued to perform for private academies such as those of Bourdelot (who died in 1685) and Denis; by the 1690s he performed before the *salonnières* at Mathieu Geoffroy's gatherings as well as teaching at the Hôtel-Dieu. Even for someone as energetic as Duverney, it was a punishing workload, and the academy's appointment of another anatomist in 1684 may have acknowledged this. In appointing the surgeon Jean Méry (1645–1722), it even reproduced the earlier duo of Pecquet and Gayant.

Like Duverney, Méry hailed from the provinces. He came to Paris to study surgery at the Hôtel-Dieu at the age of eighteen and rose quickly in his profession. His diligence was such, reported Fontenelle, that he kept a purloined corpse in his bed to practice on.[126] In 1681 he was appointed surgeon to the Hôtel-Dieu as well as to the queen. When the queen died two years later, Louvois named Méry surgeon to the Hôtel des Invalides, which had been founded by Louis XIV in 1670 as a home for wounded veterans. The next year he entered the academy. Martin Lister described Méry in 1698 as "a most painful and accurate Anatomist," as well as a "free and communicative Person." But most viewed Méry, who preferred the solitude of the dissection room, as the polar opposite of the gregarious Duverney. According to Lister, he lived at the rue Princesse, not far from the surgeons at St. Côme. An extensive collection of anatomical specimens, including skeletons and wet and dry preparations, filled two rooms of his house.[127] Three years Duverney's senior, Méry saw him as a rival. According to Eloy, Méry published his work on the anatomy of the ear in 1681, appending it to Guillaume Lamy's collected works, in order to get the advantage of Duverney, who, as was well known, worked on the same topic.[128]

Unlike other new members, who appeared in the minutes without fanfare, Méry's entrance to the academy in April 1684 was trumpeted loudly:

> The Company being assembled, M^r l'Abbé De L'anion [Pierre Lannion] presented to the Company on the part of Monseigneur de Louvois Mr Merri [*sic*] to be one of the Academicians and work at the Academy.[129]

Yet the new economic regime of Louvois manifested itself in Méry's appointment, for his annual stipend was only six hundred livres, the same amount as Duverney's *garçon*.[130] The very next week, Méry plunged into activity, reading his observations on the skin of a frog, and making sure his comments were inserted into the minutes.[131] Unlike the concise and elegant Duverney, Méry's prose was verbose and argumentative. Although they often worked on similar things, the two anatomists did not, from the evidence of the minutes, work together at first. But each stimulated the other to greater productivity. Within a few years, they were dissecting and presenting jointly to the *compagnie*, as when they dissected a civet cat together in November 1686.[132]

6. Authorship and New Publication Policies

Before the 1690s, individual academicians reported their work in the *Philosophical Transactions* and the *Journal des sçavans*.[133] Large presentation volumes like the *Histoire des animaux* fell out of favor, although plans continued for the publication of a third volume on a smaller scale. The 1669 *Description* was reprinted in 1682 by the academic publisher Laurent d'Houry, and Pitfeild's translation appeared in 1687–88, both indications of scholarly interest in the academy's anatomical work.[134] Before his death, Colbert had pressed the academy to advance its publication plans for the third volume of the *Histoire des animaux* and the second of the *Histoire des plantes*. In March 1683 Claude Perrault drew up a publication schedule, and the annual report in June noted that Leclerc had completed several new engravings for the *Histoire des animaux*.[135]

Like Colbert, Louvois promoted publications as a way to glorify the king. Despite a royal interdict of further publication in 1684, owing most likely to lack of funds, the *Histoire des animaux* project continued, with a number of payments between 1684 and 1687 to Leclerc and to the artist and engraver Louis de Châtillon.[136] Duverney reported to Bessé in March 1684 on the revisions and additions that he and Perrault proposed to make. Although it had become their project, the academy still considered it a collective enterprise, and it had to approve any revisions. In order to allow all members to comment, Perrault began the laborious process of reading the entire text to the assembled *compagnie*, and this continued intermittently for years.[137] Yet anonymity was collapsing with the publication of Perrault's *Essais* and Duverney's work on the ear, and in 1688 the academy acknowledged this in a new policy that required it to approve any publications based on work done at the academy. Its approval would then appear in the publication. This marked a shift in the academy's role from a collective producer of knowledge to an adjudicator of knowledge.[138]

Duverney and Méry, whose relationship was civil if not cordial, dissected new animals as well as additional specimens of animals that had already appeared in earlier volumes, adding to the amount to be revised. Duverney and Perrault wrote revisions on copies of the proof sheets from 1676, and the revision had progressed to the point that the *compagnie* examined proofs of the preface in the summer of 1687.[139] The reading was still ongoing, and new animals added, at the time of Perrault's sudden death in October 1688, like Colbert's during the annual vacation. The result of an infection contracted while dissecting a camel at the garden, his death seemed oddly ignored for a time. He was not mentioned in the minutes of the meetings that followed, and the anatomical programs, which he had led since the academy's founding over twenty years earlier, continued without missing a beat: at the first meetings in November, Duverney and Méry discussed their observations on the dissection of yet another ostrich from Versailles.[140] The readings from the *Histoire des animaux* continued, if ever more slowly; in December, the *compagnie* heard more of the description of the sea calf, which they had begun reading the previous July. Duverney continued to bring in new animals from Versailles, and he and Méry dissected them and wrote new descriptions, which were then shared with the *compagnie* for approval. Du Hamel, the academy's secretary, took over Perrault's role as director of the project; in 1690, he was reimbursed for expenses he incurred between 1688 and 1690 for "the continuation of the Histoire des animaux."[141]

Sometime in the late 1680s, an attempt was made to assemble the new engravings and descriptions into a volume. The engravings are listed in the 1691 inventory of the Imprimerie du roi. This volume, an ordinary folio rather than an elephant folio, included fourteen descriptions, some of them newly rewritten, such as the one describing seven lions rather than the four of 1676. The engravings also appear to be new, lacking the trompe l'oeil effect of pins and curled edges, and include new animals such as the pangolin. But the surviving copy was hastily thrown together, and includes images without descriptions, descriptions without images, and images with the wrong descriptions. The images were simply cut down to fit the slightly smaller format. This compilation was never published, but the images, as we shall see, reappeared over forty years later.[142] Another, pirated version of this work was published in 1700, under the title *Mémoire pour servir à l'histoire naturelle du Lion, De la Lionne, du Caméléon, de l'Ours, de la Gazelle, du Chat-Pard, du Renard Marin, du Loup-Cervier, de la Loutre, de la Civette, de l'Élan, du Veau Marin & du Chamois.* The images, "engravées après Séb. Leclerc, par de bons Artistes," were the same as the 1688 versions, but in the correct order; they were not, despite the claim of the title page, reengraved, but more likely somehow stolen. New animals, including the pangolin (which had appeared without a description in 1688), were omitted, and the animals were those from 1671. The author is given as "Charles Pérault [*sic*]," the publisher a mythical "Imprimerie du Louvre."[143]

The major publication of Louvois's era, though it was not published by the royal printers, was *Observations physiques et mathématiques pour servir à l'histoire naturelle, & à la perfection de l'astronomie & de la géographie* (1688), addressed to the academy. It mainly consisted of the scientific observations of a group of French Jesuits in Siam in the 1680s. Edited by the Jesuit Thomas Gouye (1650–1725), it included, among the astronomical and geographical observations, descriptions of several animals that the Jesuits had dissected during their journeys, particularly several crocodiles and a "toc-kaie" (a variety of gecko). But also inserted were a few of the academy's animal observations, including a description of the Vincennes tiger that academy members had examined, but not dissected, in 1682. In addition, Duverney's comments were interspersed amid the Jesuits' account of the crocodiles. The academy also contributed "Clarifications of several doubts about the Camel," which attempted to sort through the confusions Perrault had expressed in 1671 by means of an interview with the Persian ambassador.[144] The small octavo volume included three foldout engravings of the largest crocodile and its parts.

When Louvois died in July 1691, the new volume of the *Histoire des animaux* had yet to appear. The minister of finances, Louis Phélypeaux, comte de Pontchartrain (1643–1727), assumed the role of protector of the academy, and the king transferred its accounts from the King's Buildings to the Maison du roi, the royal household. Pontchartrain inherited from his predecessor the straitened finances of the 1680s, a financial crisis compounded by the War of the League of Augsburg that had broken out in the fall of 1688. The war, instigated by Louis, dragged on for nine years, and a famine in the mid-1690s only added damage to an already weak treasury. As Alice Stroup has detailed, Pontchartrain had good intentions toward the academy, but he had little money to give it.[145]

An indication of his intentions was the appointment of his nephew the Abbé Jean-Paul Bignon (1662–1743) to the academy in a new position as its president. Bignon's background was more literary than scientific, and he harked back to the humanist *littérateurs* such as Chapelain, Carcavi, and Charles Perrault who had advised Colbert. Upon the death of Thévenot in 1692, Bignon became royal librarian, and he soon also was named to the "Petite Académie" of inscriptions and belles-lettres—formerly Charles Perrault's fiefdom—as well as to the Académie française.[146]

With his literary interests, it is not surprising that Bignon emphasized publication. Philippe de la Hire had hoped to publish a collection of short works by academicians in 1687; he had been holding the papers of deceased members, including Roberval and Picard, for some time, not incidentally establishing a precedent of individual members rather than the academy taking charge of such documents.[147] But La Hire's publication plans had foundered, mainly for lack of funds. His *Mémoire de mathématique et de physique* finally appeared in 1694. Before that, Bignon orchestrated the publication in 1692 and 1693 of monthly collections of works of academicians "taken from the registers" (i.e., the minutes), although one might question his wisdom in placing Gallois in charge of this project, who had accomplished little during his editorship of the *Journal des sçavans*. Although these *Mémoires* appeared only for two years, they presaged the yearly *Histoire et mémoires* that began to appear under Fontenelle's editorship after the academy's reorganization in 1699. The majority of the items in the 1692–93 *Mémoires* concerned physics and astronomy. Duverney contributed only one item, Méry five; in contrast, La Hire had fifteen, Cassini sixteen.[148] In addition, two folio volumes appeared of mathematical works and of geographical and astronomical observations of the academy's voyagers, including Richer on Cayenne and La Hire and Picard's mapping expeditions.[149]

The last publication of the academy before the reorganization was Du Hamel's history, which Bignon had commissioned soon after his arrival. Despite the academy's emphasis on the development of French as the language of science, Du Hamel insisted on writing in Latin, although, as he admitted in his preface, many of his fellow academicians did not agree with this choice and felt Latin was "outmoded."[150] The 1699 reorganization and Fontenelle's appointment as secretary affirmed the academy's participation in the Republic of Letters. The façade of anonymity was formally abandoned and the model of the 1692–93 *Mémoires* of signed articles was adopted in new annual publications edited by Fontenelle. The first of these, the volume for 1699, appeared in 1702 and they appeared yearly thereafter, with delays that varied from one to seven years.[151] The reorganization gave the academy the right to print the works of its members without previously obtaining the approval of the royal censors, taking on the role of censor itself. Any work read and approved at its meetings could be published without further scrutiny. It also named an official printer, although academicians could publish their work with anyone with the same lack of censorship.[152]

Yet even with Bignon's support, the *Histoire des animaux* languished. Duverney and Méry continued to dissect new animals. In 1694–95, yet another attempt was made to bring the project to fruition. In January 1695, Bignon signed his approval on each of eighteen folders of new illustrations, constituting nearly two hundred pages, for the new volume of the *Histoire des animaux* by "the late Mr Perrault." The folders, or *paquets*, contained a number of detailed drawings of animal and human parts, including an extraordinarily beautiful set of counterproofs for new engravings. The drawings, signed by Philippe Simonneau the younger and someone identified only by the initials "L.D.C.," illustrated new dissections by Duverney and Méry, and their own much rougher sketches are occasionally included. "L.D.C." was almost certainly Louis de Châtillon.[153] At the beginning of the first folder was a list of animals to appear in the new volume, including the elephant and the "grand lezard écaillé," or pangolin.[154]

Each *paquet* was devoted to a single species and included a number of pencil drawings, notes, and at least one and often several red chalk counterproofs that occupied half of a folio-sized page. The drawings consisted of very detailed explorations of a single organ or part: the paw of a lion, the leg of a bear, the head of a viper, eyes, hearts. A few of the notes are in the hand of Claude Perrault, but most are in the distinctive hand of Duverney, and it is clearly his project; Méry appears only as a dissector, much as had Gayant over twenty-five years earlier. *Paquet* V, with material on the camel, includes a counterproof signed by Duverney with the remark, "I commit that these drawings have been made by order of the Compagnie."

While these are clearly revisions to the *Histoire des animaux*, with comments and rewritings that eventually appeared in the long-delayed third volume, these folders also indicate that more ambitious plans were afoot in 1695. The drawings are far more detailed than any that had appeared in the 1670s, and the notes include extensive comparisons of particular parts across many species; for example, a sheet headed "Glandes analogues aux Prostates nouvellemt [sic] observées dans plusieurs animaux" is followed by a list of animals and the dates of their dissections. Animals include the lion, tiger, gazelle, civet cat, monkey, Canadian deer, muskrat, and porcupine, but also sheep, pigs, boars, and bulls. In addition, the drawings and notes include many examples of human body parts, particularly the heart but also the lungs and liver, as well as the dissection of a human fetus. One folder includes material on a badger, a European animal that never appeared in the *Histoire des animaux*. Did Duverney plan to write a broader work of comparative anatomy that would include the human? Bignon approved of this project, to the extent of paying artists and making preliminary engravings.[155]

But nothing happened. The first folder included a sheet stating, "Premier pacquet remis a Monsieur Du hamel Secretaire de l'Academie Royale des Sciences, par M. l'Abbe Bignon par ordre de M. de Pontchartrain, le 8ᵉ Janvier 1695" (first packet sent to Monsieur Du Hamel secretary of the Royal Academy of Sciences, by M. l'Abbé Bignon by order of M. de Pontchartrain, 8 January 1695). These drawings may have gone to Du Hamel, but at some point they reverted to the possession of Duverney; a similar folder in the academy's archive is labeled "Drawings that match manuscripts of Mʳ Duvernet [sic]." If Du Hamel held these manuscripts, he did not move forward on their publication (with, remember, proofs of a new preface that had been printed in 1687) before his death in 1706.

Figure 6.4 Left paw of lion, 1695.
(© Académie des sciences—Institut de France.)

In 1698, Louis XIV gave the Versailles menagerie to the young Duchess of Burgundy, who had married his grandson the previous year. Under the direction of the architect Jules Hardouin-Mansart, the menagerie and its buildings were renovated and expanded. New enclosures were built for the fierce beasts of the now-closed menagerie at Vincennes. Monier, who had been hired by Colbert, continued to travel every year to bring back animals for the menagerie. But the dissection of exotic animals from the royal menageries ceased, and the *Histoire des animaux* project vanished from the minutes.[156] It is likely that Duverney gained possession of *paquets* and other papers related to the project after Du Hamel's death, following the precedent La Hire had established, and held them until his own death in 1730. As we shall see, efforts began to be made in 1727 to wrest the manuscripts from his grasp.

Why didn't Duverney complete the task of publication? For over twenty years, he held the engravings and proof sheets. His son later attributed Duverney's failure to publish this, or indeed very much of his own research, to his extreme punctiliousness: "Never was he fully satisfied with a topic," and his scribbled-over manuscripts evidence repeated revising.[157] Duverney was not a prolific author. His plans for a series of books on the senses resulted in a single book on the ear, and by the time the academy established a regular outlet for publication, he had largely withdrawn from it. Even the works of others that he purchased with the intention of editing them for publication, including manuscripts of Jan Swammerdam, gathered dust in his study at the garden. But other reasons, both personal and institutional, also intervened. By 1706, Duverney's relationship with the academy had ended.

7. The *Trou ovale* and the Great Divide

Duverney and Méry worked together on a number of dissections from the late 1680s onward, and continued to work together after Perrault's death in 1688. However, beginning around 1692, Méry increasingly focused on a topic Duverney felt was his own: the circulation of the blood in the human fetus in comparison to other animals, particularly the tortoise. Duverney, perhaps seeing a repeat of Méry's anticipation of him on the ear, disagreed vehemently with Méry's conclusions, and the rift that developed between them led to Duverney's retreat from the academy to the garden and the Hôtel-Dieu. The gap in anatomical work caused by Duverney's withdrawal was partially filled by the appointment of Daniel Tauvry (1669–1701) in 1698 and Alexis Littre (1654–1725) in 1699. Tauvry in particular had already shown himself to be as precocious as Duverney had been three decades earlier, publishing a well-regarded textbook of human and animal anatomy when he was barely out of his teens.[158] But the dissection program and animal work did not regain their prominence in the academy until Daubenton's appointment in 1745.

Duverney had compared the circulation of the blood in the lungs of the tortoise to that in the fetus in 1676. As we saw in chapter 5, he explained that the motion of the lungs was not essential to the circulation in the tortoise as it was in mammals. He demonstrated that circulation continued in the tortoise even if its lungs were not functional, which did not occur with a dog. This, he said, was the same as what occurred in a fetus:

> Because in the *Foetus*, as in these Animals, the Lungs receive the Blood only for their Nourishment, and not for the intire Circulation, so that it sends to the Heart only the remainder of what it has not consumed: And in fine as the intire Circulation is not performed but by the *Anastomoses* of the Heart in the *Foetus*; it is done also in the other Animals which we treat of, only by particular Apertures which the *Ventricles* of their Heart have one into the other.[159]

Although the pulmonary circulation had long been accepted as fact, the question of its function remained much debated. Hooke's second open-thorax experiment showed that the motion of the lungs was not the critical element in maintaining life (at least in warm-blooded animals), but it would be another century before the composition and role of air in life was understood.[160]

In 1685, Méry dissected a sea turtle and noted the "vesicles" in the lungs that Duverney had seen in 1676, comparing them to a honeycomb. Duverney and Méry again dissected turtles in 1688.[161] Four years later, Méry presented his treatise on the fetal circulation to the academy, and the next year he presented three additional treatises on the topic. Méry argued that since the fetus had no need to breathe, the circulation largely bypassed the lungs. He claimed that the opening known as the foramen ovale functioned to transmit venous blood directly from the left to the right side of the heart. The fetus in the womb, he said, resembled the turtle in being able to live without breathing.[162] During this period, the academy artist Philippe Simonneau drew dozens of images of hearts at Méry's behest, with the *trou ovale* prominently displayed.[163]

The three *mémoires* from 1693 gave further detail. Although his evidence came mainly from dead humans and animals and dried specimens in his collection, Méry also cited several experiments on live turtles as well as his observations of human births, where he noted that the fetus could suffocate,

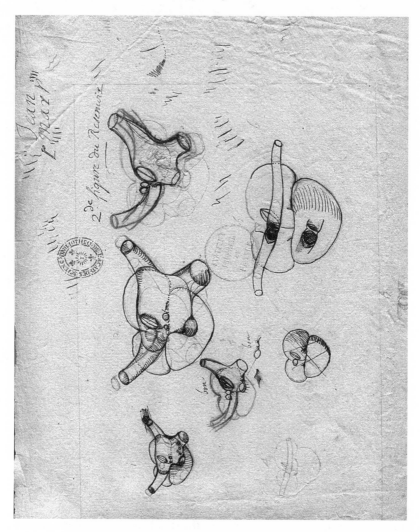

Figure 6.5 Philippe Simonneau, sketches of fetal hearts for Jean Méry
showing the *trou ovale*, late 1690s.
(© Académie des sciences—Institut de France.)

even if it were outside the womb, if the umbilical cord was still attached to
the mother but somehow compressed, indicating that it still received air
from the mother.[164] Blocking the nostrils and mouths of several turtles with
wax, he showed they could live for thirty days without breathing. Although
they needed air, the structure of their lungs allowed them to conserve it,
and their hearts did not require it to make the blood circulate. In postnatal

humans and other mammals, however, the action of the lungs forced air into the blood, propelling it toward the heart; it made the blood lighter and more readily moved; and "the infinity of little balls of air" increased the volume of the blood, increasing its pressure and its motion. The retention of air in the differently structured lungs of the tortoise was sufficient to maintain its much slower circulation, but humans needed the added boost of constant respiration.[165]

Méry seemed unaware of the work that had been done at the Royal Society more than two decades earlier by Richard Lower on the relationship between respiration and the circulation. In 1694, Méry argued that the differing colors of venous and arterial blood had nothing to do with "the subtle parts of air," based on his further observations of turtles and fetuses.[166] As we have seen, Duverney had also ignored Lower's work in his discussion of the tortoise in 1676, and his criticisms of Méry focused on anatomical evidence rather than on physiological conclusions. Working with Pierre Varignon, Duverney found that the foramen ovale was in fact a valve that could not let blood into the heart as Méry described. The debate over several years involved not only the academy but the London Huguenot anatomists Peter Sylvestre and Paul Buissière, as well as the Flemish anatomist Philippe Verhuyen. The academy's new recruit in anatomy, Daniel Tauvry, proved to be a particularly skilled opponent.[167]

Méry submitted what he called his "22 facts" about the fetal circulation to the academy in 1699 for its approval prior to publication, reflecting the *compagnie*'s new role as an arbiter of knowledge. A committee consisting of the botanists Dodart, Tournefort, and Morin approved the twenty-two facts and the publication of Méry's book even as Tauvry demonstrated that some of these facts were incorrect. Later in the year, Duverney read a long paper on the circulation in the tortoise. Nonetheless, Méry's *Nouveau système de la circulation du sang*, with his replies to Duverney, Tauvry, Verheyen, Sylvestre, and Buissière, appeared in 1700, appended to a treatise on the extraction of bladder stones. Méry dedicated the book to Fagon, the new director of the garden (and therefore Duverney's supervisor), and included the approbation of the Paris Medical Faculty, various members of the academy (including Dodart and Morin), as well as a separate testimonial from the academy's new *élève* in anatomy, Alexis Littre.[168]

Mery's bitter reply to Duverney reflected a feeling of betrayal. He attacked Duverney's experimental techniques, ideas, and even his character. He continued his attacks in 1703 in a lengthy "Traité physique" published in the *Histoire et mémoires* of the academy, accusing Duverney of plagiarism and theft as well as incompetence in his dissections of tortoises, an animal Duverney had dissected dozens of times in the previous two decades. Méry escalated the debate far beyond the original issues, and Duverney, apparently, wanted none of it. He did not reply and largely withdrew from the academy, taking the *Histoire des animaux* papers with him. By 1703 he had secured positions at the academy for his two brothers: Jacques-François-Marie, nicknamed Christofle (1661–1748), was a surgeon, as was Pierre (1650?–1728).[169] His son Emmanuel-Maurice entered the Paris Faculty of Medicine. The "Duverney le jeune" who now appeared in the *Histoire et mémoires* was Christofle. "Duverney l'aîné" appeared only once more after Méry's diatribe, with a description of conjoined twins in the volume for 1706, and he disappeared from the minutes. While maintaining an active career at the garden and at the Hôtel-Dieu—where he must have encountered Méry from time to time—he did not return to the academy until after Méry's death in 1722, and then only for a few years until his retirement.

The academy remained well supplied with anatomists even without Duverney, but there were no new undertakings on the scale of the *Histoire des animaux* project. Funding continued to be tight. New members such as botanist Joseph Pitton de Tournefort and chemist Guillaume Homberg were not interested in anatomy; the age of collaboration across disciplines when Auzout could assist at dissections had ended with the death of Perrault. Among the anatomists, the precociously talented Tauvry died in 1701, aged only thirty-one, and Méry, Littre, and Christofle Duverney published on small details rather than large theories. An era was over.

Conclusion

1. The End of the *Histoire des animaux* Project

The collapse of the *Histoire des animaux* project revealed the underlying tensions and weaknesses of the pre-1699 academy. Louvois had already made clear the academy's utter dependence on the goodwill of the crown, and the supply of animals from the royal menageries disappeared abruptly when the king's interest waned in the late 1690s. Like the volumes of the *Histoire des animaux*, the menagerie at Versailles maintained a delicate balance between scientific and courtly purposes. The octagonal pavilion at the center of the menagerie hosted many royal visitors and state dinners.[1] In retrospect, it is clear that the dissection of the elephant in 1681 represented a peak confluence of courtly and scientific goals. As we saw in chapter 6, by 1698, when the king gave the Versailles menagerie to the Duchess of Burgundy, its scientific significance had dwindled, and under the duchess it soon disappeared. Jules Hardouin-Mansart's additions to the menagerie emphasized its human visitors rather than its animal inhabitants.[2]

In addition, the 1670s publication model of large, elaborate books could not be sustained in times of fiscal contraction. Brilliant examples of collaboration between artists and naturalists, the volumes failed to reach a broader audience and, while giving credit to the king, gave much less credit to the people who wrote and illustrated them. The *Histoire des animaux* project had always coexisted uneasily with a new science based on openness, publication, and individual credit. Bignon recognized this in the 1690s with the 1692–93 *Mémoires* and regular publications after 1699. By then, members of the academy had already published works on their own. In addition, the *Journal des sçavans* began to live up to its promise as a publication outlet for the academy under the editorship of the ex-Jesuit Abbé Jean-Paul de La Rocque from 1674 to 1687 and the historian and courtier Louis Cousin

(1627–1707) from 1687 to 1701, although it was never devoted exclusively to science. But Claude Perrault's death in 1688 also made it clear that the overall project of the *Histoire des animaux* owed most of its success to him: to his skills in organization and patronage as well as his talents for dissection, writing, and drawing. Duverney was a great anatomist but a terrible manager, and the collapse of the 1695 publication plans must be laid at his door.

Duverney's departure and Tauvry's death in the early eighteenth century slowed anatomical work at the academy. Duverney had long had other outlets for his talents, and in turning more fully to the garden and the Hôtel-Dieu, he foreshadowed the trajectory of anatomy in eighteenth-century Paris. Anatomical instruction at the garden continued throughout the century, where it was taken up in succession by Duverney's son Emmanuel-Maurice, François-Joseph Hunauld (1701–42), Jacques-Benigne Winsløw (1669–1760), Antoine Ferrein (1693–1769), Antoine Petit (1722–94), and Antoine Portal (1742–1832). The first three of these had studied with the elder Duverney. Ferrein, who had studied at Montpellier, contributed to the reflorescence of anatomical instruction at the Collège Royal in the 1740s.[3] Hospitals, including the Hôtel-Dieu, also gained in importance as sites of instruction. But the eighteenth century was in particular the era of the surgeon in Paris. Having definitively separated from the barbers in 1692 after the forced marriage of 1656, the surgeons organized themselves into the Royal Academy of Surgery in 1731 and steadily raised their standards of training and practice, reestablishing the role that the surgeons of St.-Côme had held a century earlier.[4] There continued, therefore, to be multiple sites for anatomical investigation and instruction in Paris in the eighteenth century.

2. Comparative Anatomies

After his departure from the academy, Duverney dissected animals at the garden only to illustrate human anatomy, and the comparative anatomy of the *Histoire des animaux* project lay in abeyance until the 1740s, when Louis-Jean-Marie Daubenton (1716–99) entered both the garden and the academy.[5] With the important exception of Edward Tyson (1651–1708) in London, whose anatomical studies of individual animals culminated in his *Orang-outang* of 1699, few original works of comparative anatomy appeared. In his *Phocaena, or the Anatomy of a Porpess* (1680), Tyson drew a parallel between the new discoveries afforded by voyages and those revealed by anatomy: "In every Animal there is a world of Wonders," he wrote. Like Perrault, he outlined a method of natural history based on close observation, with anatomy at the forefront. Tyson explained that his work on the porpoise was but one instance of this method, and that "I could therefore wish that at least for the present we had an Account of the most *Anomalous* and *Heteroclite* sorts of Animals, or such whose *Species* are most different."[6]

Tyson, an active member of the Royal Society, may have been among those who promoted an English translation of *Histoire des animaux*. Alexander Pitfeild's translation of the 1676 edition, with new engravings by Richard Waller based on Leclerc's, appeared in 1687–88 under the society's auspices. The translation was an ordinary folio, smaller than the original, and Waller's engravings were about three-fourths the size of the originals.[7] Among the works on comparative anatomy between Tyson and Daubenton was the massive 1720 compilation of German physician Michael Bernhard Valentin (1657–1729), *Amphitheatrum zootomicum*. An animal version of Le Clerc and Manget's *Bibliotheca anatomica*, it included sections of the *Histoire des animaux*.[8]

The many and varied comparative anatomies of the seventeenth century contributed to the new science of the era in multiple ways. Beginning with Harvey and Aselli in the 1620s, dissection became an essential practice of the new science. Never fully confined to medicine, anatomical study in the seventeenth century became a science in itself, developing new experimental methods. Familiarity with human and animal anatomy and some acquaintance with techniques of dissection became essential knowledge for the practitioners of the new science, whatever their philosophical orientation. Although the politics of the court and the academy prevented the *Histoire des animaux* from having a greater impact in the seventeenth century—as we shall see in the epilogue, its influence reached far into the eighteenth—the larger program of animal and human dissection at the academy and elsewhere was a resounding success.

Harvey's insistence that his descriptive "historical" method led to new knowledge freed anatomists to dissect and experiment on living and dead animals with the goal of discovering the workings of the animal and human body, beginning with the circulation of the blood. Pecquet's subsequent discovery of the thoracic duct was one of the most dramatic of these new discoveries, but it was not the only one. Animals also provided ample interest in themselves, and comparative anatomy, as Faber had shown in the 1620s and Severino in the 1640s, yielded other kinds of knowledge about animal and human bodies.

The comparative anatomies of the seventeenth century employed the experimental method Harvey had introduced, and particularly under the influence of mechanical philosophies of nature, developed many different but often complimentary approaches. Influenced by Harvey's discoveries, Descartes learned to dissect in the early 1630s and proposed a number of explanations of bodily functions based on his particular brand of mechanism. He in turn influenced certain Dutch practitioners such as Regius in the late 1630s. As we saw in chapter 2, Descartes witnessed the dissections of François dele Boë, known as Sylvius, who had demonstrated the circulation in Leiden in the late 1630s. In his subsequent career in Amsterdam and Leiden in the 1650s and 1660s, Sylvius continued to dissect, combining anatomy with a mechanistic chemical philosophy derived from Jan Baptista van Helmont (1580–1644). Among his students were Swammerdam, Stensen, and de Graaf, and the latter's study of pancreatic fluid showed clearly the influence of his mentor.[9]

The eclecticism of mechanical anatomy manifested itself across Europe. In Naples in the 1640s, Severino had employed an atomistic mechanical philosophy based, he said, on Democritus. A decade later, Marcello Malpighi's brand of comparative anatomy drew, like Pecquet's, from several sources. Both followed an atomistic matter theory inspired by Galileo and Gassendi, but in Malpighi we can find traces of Cartesian ideas in his notion of individual micromechanisms, which can also be seen in the work of his mentor Giovanni Alfonso Borelli. Malpighi's career took him from Messina to Pisa to Bologna.[10] Stensen's mechanism was equally eclectic, showing the influence of Borelli as well as Descartes. In Britain, the corpuscular philosophy of Robert Boyle provided a basis for the anatomical work at the Royal Society by Richard Lower, Robert Hooke, and others. These "Oxford physiologists," although they traced an intellectual genealogy directly to Harvey, nonetheless displayed a variety of approaches, some but not all filtered through Boyle, including Helmontian chymistry, Gassendi's atomism, and Cartesian mechanism.[11]

Animal mechanism therefore constituted a broad but essential concept for these men, and Le Clerc and Manget's *Bibliotheca anatomica* included several variations on this theme. Tyson summarized mechanistic anatomy thus:

> Natures *Synthetic* Method in the composure and structure of Animal Bodies, is best learn't by this *Analytic*; by taking to pieces this *Automaton*, and viewing asunder the several Parts, Wheels and Springs that give it life and motion.[12]

Harvey's experimental methods and an eclectic mechanical philosophy were the common threads across European anatomical science, and by the 1660s no single theory dominated. The *compagnie* dissected a wide range of common and uncommon animals to replicate and question the work of their immediate predecessors and contemporaries. Their comparative anatomies differed from these contemporaries in two distinctive ways. They were courtly, in the sense of relying on royal patronage both for their supply of animals and for the conduct of their science. Certainly this reliance was not absolute, and other patrons persisted after 1666, as we saw in the case of blood transfusion. But *Histoire des animaux* constituted the public expression of a royal project, and it bore close connections to other royal productions in art, music, architecture, and literature.

The comparative anatomies of the *compagnie* were also distinctive in seeking differences rather than similarities. Mechanical philosophies of nature assumed that nature was uniform at the most profound level, usually the atom or a similar particle. This assumption implied regularity at higher levels of organization as well. But the *Histoire des animaux* project, under Perrault's leadership, declared that each animal was distinctive and unique. This owed both to the nature of the academy's experimental subjects and to philosophical conviction, particularly Claude Perrault's Galenic medical training. This point of view enabled the *compagnie* to rethink topics as diverse as the role of the pineal gland and of the thoracic duct and the lymphatic system, the purpose of the lungs and respiration, and the structure and use of the eye and the tongue, to give only a few examples. The avowedly anti-Cartesian Claude Perrault could describe a *"méchanique des animaux"* while at the same time proposing a theory of animal soul that ascribed volition and memory to animals. Duverney's and Perrault's work on the tortoise and on hearing and the ear revealed a mode of explanation that was uniquely Parisian, at once mechanical and vitalist. Anatomical science therefore defies simple categorizations, and broadens our conceptions of the production of knowledge in the seventeenth century.

3. Natural Histories, Anatomies, and the Crisis in Taxonomy

Harvey's "historical" method took the collection and description characteristic of natural history as its model. Reaching back to Aristotle and Pliny, dissection completed the act of describing the natural history of animals. But by revealing parts that had hitherto been hidden, and particularly by revealing the inner processes of living animals, dissection could also be demonstrative in a way that mere external morphology could not. While many anatomists believed that mechanism provided the cause, and therefore the proof, that Harvey had refused to acknowledge, Perrault and Duverney did not believe that mechanism could fully explain the workings of animate nature. They combined natural history, dissection, and a mechanical philosophy to make claims about the structure and function of the human and animal body that were causal and not merely descriptive. But the cause was not mechanical, but vital; *péristaltique* constituted an innate quality of living matter, and posited an innate soul as the basis of life.

At the same time, the Parisian insistence on individuality complicated the classifying imperative of seventeenth-century naturalists. These naturalists devoted more attention to finding a principle of classification for plants than for animals, in part because, as Aristotle had recognized, classifying animals was just too complex. In *Historia animalium*, he had rejected the notion that any single characteristic could serve as the basis of a comprehensive system of classification, criticizing the Platonic method of division as too rigid. The encyclopedic natural histories of sixteenth-century naturalists such as Gessner and Aldrovandi began from the broadest of classificatory categories: "quadrupeds," "birds," and "serpents." Within these categories, Gessner organized animals alphabetically by their Latin names, with those of similar structure treated together: so various cattle-like animals were grouped together. But Gessner also placed the "camelopard," or giraffe, with the camel, something Perrault explicitly avoided with his careful (if ultimately incomplete) discourse on the name of the camel or dromedary. Aldrovandi organized birds according to whether they nested on the ground or in trees, whether they were domestic or wild, and what they ate. In the 1650s, Johnstone made further attempts to classify, employing categories such as the structure of the feet. In addition, his illustrations presented multiple examples of the same or similar animals, constituting in themselves a kind of classification based on morphological similarities.[13]

Two decades later, Francis Willughby detailed a thorough classification of birds in his 1676 *Ornithologia*, beginning with aquatic versus land-based and employing a varied number of criteria, from size and shape of beaks and feet to broader characterizations of "the thrush kind" and "the poultry kind."[14] John Ray, Willughby's editor, attempted to classify all animals by natural relationships rather than by a single criterion in two volumes of *Synopsis methodica*, published in 1693 and (posthumously) in 1713. Ray had offered a definition of "species" based on the notion of the seed as the repository of identity: one species can never arise from the seed of another. This applied to animals as well.[15] But the classification of animals he offered in *Synopsis methodica* employed multiple criteria.[16] Amid these attempts to find a natural system of taxonomic relationships, Perrault and the *compagnie* had neither the material nor the inclination to engage in such an enterprise, and what I have called Perrault's epistemological modesty led him to avoid classification altogether rather than attempt to build an incomplete system. Seventy years later, Buffon initially agreed with Perrault that only individuals carried ontological meaning and that classification was therefore impossible.[17]

4. Science as Culture

This book provides a different perspective on early modern science than most accounts have offered. The new science I have described had its roots in ancient anatomy and natural history and their humanist revivals. To characterize this science as a triumph of the moderns over the ancients is therefore to overlook this critical role of humanism. The ideas, events, and practices of the seventeenth and eighteenth centuries that took their cue from humanist anatomy and natural history produced new knowledge that was not necessarily the causal knowledge of the exact sciences, but was nonetheless experimental and experiential knowledge and not merely description.

In addition, *The Courtiers' Anatomists* has presented human and animal anatomy in seventeenth-century Paris as part of a much wider cultural milieu that intertwined intellectual and courtly activities across a wide spectrum. Here too ancient and modern ideas coexisted as well as clashed. The *honnêtes hommes* of the academies and salons of Paris freely mingled science, art, music, and literature as interdependent forms of knowledge. Once we jettison our modern disciplinary blinders, we are able to see *Histoire des animaux* as a tour de force of art, anatomy, and mechanical philosophy, the Labyrinth at Versailles as an expression of science as well as art and literature, and Charles Perrault's fairy tales as an example of the values surrounding animals as well as those of the court and the salon. A science based on the work of Marin Cureau de la Chambre and Claude Perrault as well as that of Descartes and Galileo is a very different science than we have up to now acknowledged. *The Courtiers' Anatomists* has, I hope, given readers a new perspective on the messy and idiosyncratic origins of modern science.

Epilogue: The Afterlife of the *Histoire des animaux*

1. The Reprinting Project

Fontenelle, who had assumed the position of "perpetual secretary" of the academy at the time of its reorganization in 1699, signed a brief note, dated 14 February 1728, in the front matter of his history of the academy, published in 1733. The note stated that the *compagnie* decided in August 1727 to undertake the reprinting of its pre-1699 works in a uniform format.[1] The eleven volumes (in thirteen parts) that constituted this project appeared between 1729 and 1734 and included the *Mémoires pour servir à l'histoire naturelle des animaux*. They were not printed in order; Fontenelle's history occupied volume 1, and the *Histoire des animaux*, labeled volume 3, appeared in three parts in 1733 and 1734. The "Avertissement" for volume 1 gave an additional motivation for the project, an unauthorized Dutch edition of the academy's earlier works. This edition, published in The Hague in 1731, was edited by the eminent men of science Willem s'Gravesande and Bernhard Siegfried Albinus.[2] The 1727 decision, however, already had ample political motivation in establishing the priority of France in the new science of the previous century and in presenting Fontenelle's history of the academy that characterized it as a Baconian, empirical body much like the Royal Society.

The Dutch volumes also indicated the importance of the *Histoire des animaux* relative to the academy's other publications. The first two volumes of the Dutch edition consisted of the *Histoire des animaux* (1676 edition), with new engravings based on the originals. Fontenelle's 1733 "Avertissement" in turn took particular notice of the *Histoire des animaux*, claiming that the new edition that was about to appear was "quite different" from the Dutch one, containing many corrections, new plates, and additional descriptions.[3] Fontenelle did not mention the excerpts from the *Histoire des animaux* that had appeared in 1720 (with very inferior copies of the original engravings) in Michael Bernhard Valentin's *Amphitheatrum zootomicum*.

Fontenelle and Bignon seem to have persuaded the elderly Duverney in the late 1720s to release the manuscripts he held for the republication project, but that had not yet happened when he died in September 1730.[4] By November the academy leapt into action, appointing a committee to look at the papers Duverney had left and compare them to the existing editions. The committee consisted of three anatomists: two physicians, Jacques-Bénigne Winsløw (1669–1760) and François Petit (1664–1741), and the surgeon Sauveur-François Morand (1697–1773). Winsløw, originally Danish, had come to France in the 1690s and studied anatomy with Duverney at the garden. He came under the influence of Bossuet and converted to Catholicism, changing his name from Jacob to Jacques-Bénigne after Bossuet. Petit also studied with Duverney after receiving a medical degree from Montpellier, and succeeded Duverney following his retirement in 1725 in the position of *pensonnier-anatomiste* at the academy.[5] The editing came to be done by Winsløw and Petit.

Winsløw and Petit explained this background of the manuscript (although they did not mention the 1695 drawings) in their "Avertissement" to the 1733–34 reprinting of the *Histoire des animaux*.[6] The new edition had several changes. Apart from the new text that Perrault had left at his death in 1688, and possible subsequent additions by Duverney, Winsløw and Petit also added descriptions contemporary with the *Histoire des animaux* that had not been part of its project. These included a 1669 study of a viper by Moises Charas as well as the descriptions of several exotic animals that had been published in the 1688 *Observations physiques et mathématiques pour servir à l'histoire naturelle* by the Jesuit Thomas Gouye, which included additional observations by Duverney. Winsløw and Petit also subtly updated the original text, not only adding revisions from the manuscripts Duverney had held, but judiciously modernizing text that had not previously been reedited. So the description of the bear, for example (originally written in 1668), omitted a passage that cited Galen on the resemblance of the bear to humans, and numerous other additions, omissions, and rewritings can be found throughout the text. But they are only evident if the editions are compared side by side.

Such updating did not occur with the other texts republished between 1729 and 1734, the purpose of which was to provide a record of the academy's early scientific activity. The purpose of the *Histoire des animaux*, then, was rather different. Although Winsløw and Petit criticized the haphazard organization of the volumes, which they believed was dictated by the order in which animals came to the academy to be dissected, their careful editing indicated that the information contained in them was nonetheless still valuable. They rearranged the order of some of the animals, but did not offer any justification for this.[7] Winsløw and Petit had two additional reasons to reedit *Histoire des animaux* apart from commemoration. By incorporating Perrault's additions and corrections as well as new descriptions, including the one of the elephant, they completed the work of one of the most illustrious early academicians. With their subtle updating and reordering, Winsløw and Petit demonstrated the continued value of the work that the Dutch edition also acknowledged while retaining the academy's priority and prescience in publishing the work a half century earlier.

Winsløw and Petit added material to Perrault's preface that underlined his policy of particularity and lack of generalization: "If it sometimes happens that we establish general maxims . . . one must understand that we reserve to ourselves to add to them exceptions and restrictions that new observations can provide when they present themselves, so that they can serve those who would make up the General History, for which these Memoires are drawn up."[8] The "General History," in other words, had not yet been written.

2. Buffon and the *Histoire des animaux*

Around the time these words were written in the early 1730s, a young man from Burgundy came to Paris and began to frequent some of its many salons devoted to natural philosophy. Two institutions tie Buffon and the *Histoire des animaux* together. One was the Paris Academy. The twenty-six-year-old Georges-Louis Leclerc de Buffon (1707–88) of Montbard, near Dijon, whose family had purchased the village of Buffon a little more than a decade earlier, was appointed *adjoint-mécanicien* to the academy in January 1734. Like Perrault, Buffon was a bourgeois whose family had worked assiduously on its advancement. At this time, the academy still kept its division between what had been known as *mathématique*, meaning mechanics and the physical sciences, and *physique*, encompassing the nonmathematical sciences, including chemistry and the sciences of life. Buffon worked on both topics during the 1730s, but by the end of the decade the balance tipped toward the life sciences and he transferred to the *physique* side as *adjoint-botaniste* in the spring of 1739. That summer he was appointed *intendant* of the Jardin du roi.[9]

The garden provided the second institutional tie to the *Histoire des animaux*. The royal physician Gui-Crescent Fagon had assumed the position of *surintendant* of the garden in 1699, and he was succeeded in 1718 by Pierre Chirac, academician and physician to the regent, the Duc d'Orléans. But when he died in 1730, his successor was not a physician but a fellow academician, Charles DuFay, whose interests were mainly in physics. DuFay's untimely death in 1739 provided Buffon with an opening, and, pulling in numerous favors, he managed to be named the new *surintendant*. The garden had been gradually losing its medical identity: the Cabinet of Drugs became the Cabinet of Natural History in 1729, and many of its animal specimens, ranging from pieces of skins to full skeletons, dated from the *Histoire des animaux* project. The well-attended lectures on human and animal anatomy that Duverney had given for forty years continued, first with his son Emmanuel-Maurice and then with François-Joseph Hunauld, but with Buffon, the transition from a physic garden to a natural history museum moved toward completion.

The republication of the *Histoire des animaux* occurred just as Buffon entered the academy. Around this time, the young Louis-Jean-Marie Daubenton (1716–1800), also from Montbard, began to attend Hunauld's anatomy lectures at the garden rather than theology lectures at the Sorbonne. He also attended the surgical demonstrations of Christofle Duverney, who had been appointed in 1727 to a new position at the garden as "démonstrateur en anatomie et chirurgie."[10] Daubenton completed his medical training in 1741, and Buffon soon summoned him from Montbard to work with him at the garden. By 1745 he had an official title as *garde et demonstrateur* of the natural history collections and was a member of the Paris Academy. The instigation for the *Histoire naturelle* was the desire of the government minister and Buffon's patron Jean-Frédéric Phélypaux de Maurepas for an inventory of the Royal Cabinet, which occupied several rooms at the garden. Daubenton dissected and provided the anatomical descriptions of the animals in volumes 3–15, which covered quadrupeds. Although he was not the anatomy professor at the garden—Buffon appointed Winsløw to that post after Hunauld's death in 1742—he later became a *pensionnaire anatomiste* of the Paris Academy. There are parallels between the respective roles of Buffon and Daubenton in comparison to Perrault and Duverney, although Perrault also dissected, and Buffon, as far as I know, did not.[11]

3. The *Histoire naturelle* and the *Histoire des animaux*

Was Buffon's *Histoire naturelle* the "General History" that Winsløw and Petit called for? Buffon seemed to think so, titling his work *Histoire naturelle, générale et particulière*. Unlike Perrault and the *compagnie*, he intended to write an encyclopedic natural history of everything in the world, modeling himself on Pliny rather than Aristotle. But he borrowed freely from the earlier work. Like the *Histoire des animaux*, Buffon's title page named no author, filling that space with a large royal symbol. His introduction to the first volume was "De la manière d'étudier & de traiter l'Histoire naturelle" ("How to study and treat natural history"). Jacques Roger compared it to Descartes's *Discourse on Method* as a manifesto of a new method.[12] But it resembles even more Claude Perrault's preface to the 1671 *Histoire des animaux*. Buffon omitted the homage (which rapidly turned into a critique) of the ancients with which Perrault, still fighting the battles of the new science, had begun. But they shared the same preoccupations: How can we know nature? How do we get from particulars to generalizations? What are the relative roles of reason and observation? Both emphasized the role of observation: "One ought to begin by seeing much and re-seeing often," wrote Buffon.[13] Both urged consultation of other authors on the subject, but only after making one's own detailed observations. And both warned against finding an order in nature that might be more in the mind of the human observer than in nature itself.

Perrault commented, "Our principal design is to report." From the point of view of the early academy, there was so much that was unknown, particularly about the internal parts of animals, and so much myth to disprove. He and the academy admitted no fact they had not observed themselves, and this empirical, particularist approach made generalization difficult if not impossible.

Buffon expressed a similar epistemological modesty: "All that is possible for us, is to perceive some particular effects, to compare them, to combine them, and at last, to recognize an order relative to our own nature." "One sees clearly," he added, "that it is impossible to give a general system."[14] Like Perrault, Buffon claimed that the ultimate object of his study was the human, and they frequently compared animal and human structures.[15] But both believed that the natural history of animals had value in itself, and they compared animals to each other more than to the human. In volume 4, Buffon introduced his volumes on individual animals with a lengthy "Discourse on the Nature of Animals" that included a section on their description, written by Daubenton. Much like Perrault, Buffon and Daubenton emphasized the value of observation and description as the basis of natural history—uncontroversial statements, to be sure—but Daubenton also asserted that "there are neither principles nor rules to guide the observer."[16] On the other hand, there should be clear principles guiding the description of animals, and primary among these must be rules of naming.

Historians have characterized Buffon's diatribe against arbitrary systems of naming as his response to Linnaeus.[17] But Buffon and Daubenton also reached back to Perrault's statements on this topic. Working within a humanist tradition of natural history, Perrault complained of the difficulty of ascertaining whether animals named and described by the ancients were the same animals with those names in his time. He concluded that in many cases the ancient descriptions were too inexact to make such a judgment, so that his goal therefore was simply to provide a very exact description. Daubenton wrote on the same topic, "It is thus absolutely necessary to establish the principles and rules that are exactly followed in all the descriptions, and to propose a method of description in place of methods of nomenclature, which has occupied up to now the majority of naturalists."[18]

Many of the descriptions in the *Histoire des animaux* began with a discussion of the various names of the creature. As we have seen, Perrault spent some time discussing whether an animal in question was a camel or a dromedary, citing numerous observations of other authors as well as his own interview of an Arab ambassador. On the basis of information he gained from the latter, he decided it was a dromedary, defining the dromedary as the variety with one hump, although there continued to be much confusion about this.[19] By 1733, one of each variety had been described, but Winsløw and Petit's chapter title confusingly referred to both as "camels" and only clarified this further in the text, referring to the one-hump variety as a camel "because camel is the name that one ordinarily gives to that with only one hump."[20] Buffon and Daubenton spent even more time on the proper name, going back to Perrault's definition of one-hump dromedary and two-hump camel. They went on to distinguish several varieties of each according to geographic area. The dromedary skeleton that resided in the Cabinet du roi was, he thought, the specimen described by Perrault. But Buffon referred to the 1733 description of it as a camel rather than the original designation of dromedary, indicating that he worked from this edition instead of the earlier ones.[21]

Buffon claimed in the early volumes of the *Histoire naturelle* that species were artificial constructs and that we could have knowledge only of individuals, not of any larger classificatory entity.[22] But by the early 1750s, when the volumes on quadrupeds began to appear, this stance was already changing. The "Discours sur la nature des animaux" and Daubenton's essays on "Description" and "Distribution," all in volume 4 (published in 1753), set forth a system of organization for the animal volumes, something Perrault and the *compagnie* had never done. They divided quadrupeds first into domesticated and wild, then into local and exotic, with further divisions according to teeth and other criteria. So volume 4 began with the horse, the ass, and the cow, and volume 5 covered sheep, goats, pigs, and dogs. Volumes 6–8 described wild animals native to France, including carnivores such as the wolf and the badger and others including bats, hares, and hedgehogs. Of these, only the hedgehog had appeared in the *Histoire des animaux*, as a foil to the porcupine. Only at the end of volume 8 (1760) did exotic animals begin to appear. Buffon's natural history was therefore more general (and more systematic) than Perrault's in two ways: in the number and variety of animals described (not dependent on chance arrivals from the royal menageries) and in establishing a system of organization that the academy's work entirely lacked.

Buffon, like Perrault, often began his accounts with a consideration of the animal's name. Among the first of his exotic animals was the coati and the coatimundi, and Buffon's discussion of the animal's name drew heavily from Perrault's, which had first appeared at the end of the 1671 volume. Both cited Georg Marcgraf; Buffon cited André Thevet, another sixteenth-century traveler, whose journey to Brazil antedated that of Jean de Léry, whom Perrault had cited. On the difference between the coati and the coatimundi, Buffon cited Marcgraf via Perrault. However, Buffon omitted Perrault's discussion of the size and shape of the coati's muzzle. While Perrault and the *compagnie* had diligently measured every animal that they examined, Buffon and Daubenton went much further, measuring every possible animal part, internal and external.[23]

Buffon's and Daubenton's borrowings and references to the *Histoire des animaux* occur throughout the *Histoire naturelle*; on the elephant, for example, Buffon noted, "I have taken from that work [the *Histoire des animaux*] the facts that can enter into my plan of description," and Buffon and Daubenton's description of the elephant constantly engaged with Perrault's work.[24] As in the *Histoire des animaux*, the article on the elephant in the *Histoire naturelle* is among the longest; but Buffon spent considerably more pages on the animal's native habitat and natural behavior, citing a large number of travelers' accounts.[25] Several of the specimens and skeletons from the *Histoire des animaux* project were still at the garden in the *salle des squelettes* and were depicted in the *Histoire naturelle*, including the skeletons of the dromedary and the elephant (see fig. 6.2).

The reediting of the *Histoire des animaux* by Winsløw and Petit acknowledged the continuing value of the work. No other work of its scale or depth had appeared since the 1670s. It continued to be reprinted, with another issue of the Dutch edition in 1736, and a further reprint of the 1733 Paris edition in Leipzig in 1758. As Jeff Loveland suggests, Buffon's decision to include anatomy in the *Histoire naturelle* and therefore to bring Daubenton into the project was certainly inspired and informed by the *Histoire des animaux.*[26]

The "General History" of animals of Buffon and Daubenton may be seen as an extension and even a completion of the *Histoire des animaux* project. The legacy of Perrault's era and the *Histoire des animaux* permeates the *Histoire naturelle* to an extent that has not hitherto been recognized, highlighting the historical place of the *Histoire des animaux* between the Renaissance and the Enlightenment. Perrault and the academy believed they looked forward and not back, and writing seventy years after their last volume came from the press, Buffon apparently agreed with them.

This book took a long time to write, and I have incurred more than the usual number of debts along the way. Karen Merikangas Darling at the University of Chicago Press deserves a medal for her astute editing and constant support. The first inklings of this project were planted during a sabbatical year in 1999–2000 at the Centre Alexandre Koyré in Paris, thanks to an exchange program between the University of California, Santa Barbara (UCSB), and France's Centre national de la recherche scientifique. I also had support that year from a grant from the National Science Foundation, which is to be heartily thanked for its continued support of the history of science. Subsequently I benefited from grants from the Academic Senate and the Interdisciplinary Humanities Center at UCSB; a University of California President's Fellowship in the Humanities; and a Franklin Grant from the American Philosophical Society. Since 2008 my research has been generously supported by the Horning Endowment in the Humanities at Oregon State University. I am grateful every day to the foresight and wisdom of Dr. Benjamin Horning for his gift of the endowment.

Second, I thank the libraries, librarians, and archivists who made my research both possible and pleasurable. First on this list must be Mme. Florence Greffe and her wonderful staff at the Archives of the Paris Academy of Sciences. In Paris, I am also grateful to Guy Cobolet, Estelle Lambert, and Stéphanie Charreaux at the Bibliothèque interuniversitaire de santé; Maxime Préaud at the Département des étampes at the Bibliothèque nationale–Richelieu; the staffs at salle Y at the Bibliothèque nationale–Mitterand and at the Arsenal; and those at the CARAN of the Archives nationales, and the bibliothèque of the Paris Muséum national d'histoire naturelle. In London, the staff of Rare Books and Music at the British Library let me look at five different versions of the *Histoire des animaux* at the same time, sprawled across two desks. The staffs of Manuscripts at the BL and at the Wellcome Library (particularly Richard Palmer) also deserve thanks. David Weston and his staff at Special Collections at the University of Glasgow were extraordinarily helpful, and the late Helen Brock's checklist of the Douglas manuscripts at Glasgow made my task very much easier. In the United States, Jeffrey Rankin of Special Collections at the Charles Young Library at UCLA facilitated my work in scattered visits over several months, as did the staff at the Bancroft Library at UC-Berkeley and at the Huntington Library. Visits to and correspondence with librarians at the Library of Congress and with Stephen Greenberg at the History of Medicine Division, National Library of Medicine, helped on a number of issues. I kept the interlibrary loan departments at UCSB and OSU busy for several years. I spent many happy hours in the lovely reading room of Special Collections, Valley Library, OSU and owe thanks to Cliff Mead and Anne Bahde. For assistance with images, I am particularly grateful to Nancy Green of the Linda Hall Library and Robbi Siegel of Art Resource.

Robert Nye, Michael Osborne, and April Shelford heroically read the entire manuscript and offered numerous helpful suggestions. Domenico Bertoloni Meli and Penelope Gouk have read portions. Three referees for the University of Chicago Press also offered astute criticisms—many thanks to all of you. Copyeditor Susan Cohan made sense of my messy endnotes. For conversations, references, and hand-holding, I thank Katherine Acheson, Bisi Agboola, Nicole Archambeau, Deb Bahn, Nico Bertoloni Meli, Charlotte Bigg, Juliette Cherbouliez, Sarah Cline, Greg Clingham, Sarah Cohen, Jenny Dugan and Dave Hubbard, Paul Farber, Dena Goodman, Penny Gouk, Julien Guillaumot, Sue Haig, Florence Hsia, Linda Hulten, Rina Knoeff, Sachiko Kusukawa, John W. I. Lee, Paula Young Lee, James Lennox, Pamela Long, Rafael Mandressi, Patrick McCray, Bob and Mary Jo Nye, Oded Rabinovitch, Kapil Raj, Robin Rider, Harriet Ritvo, Antonella Romano, Anna Marie Roos, Peter Sahlins, Lisa Sarasohn, Paul Sonnino, Alice Stroup, and Charles Wolfe. Anyone I have forgotten, please accept my gratitude. I thank Veit Erlmann, Oded Rabinovitch, Evan Ragland, and Peter Sahlins for sharing unpublished work. Jole Shackelford and Peter Sobol answered late-night e-mails on Latin translation. The late Josiah Quincy Bennett of the Lilly Library, Indiana University, taught me most of what I know about rare books, and Peter Krivatsy at the History of Medicine Division, National Library of Medicine, taught me the rest. My research assistants, Tina Schweickert, Barbara Canavan, and Tamara Caulkins, performed various tasks with good cheer. Bob Peckyno brilliantly Photoshopped the Paris map in figure 1.1.

None of these people, of course, are responsible for any errors or strange interpretations herein.

Various incarnations of portions of this book have been presented to audiences in Santa Barbara; Los Angeles; San Marino; Berkeley; Corvallis; Bloomington, Indiana; Cambridge, Massachusetts; Minneapolis and Northfield, Minnesota; Pittsburgh; Madison, Wisconsin; Louisville; Tampa; Bethesda; Copenhagen; Leiden; London; Paris; Pisa; Vancouver; and at numerous meetings of the American Society for Eighteenth-Century Studies and the History of Science Society, as well as the Society for Literature, Science, and the Arts, the American Historical Association, and the Western Society for French History. Bits and pieces have appeared at various times in various publications over the past decade. These are indicated in the notes.

Since this is a book about animals, I would be remiss if I did not mention my constant companions during the writing of this book, our cats Isadora (who died in 2012 after a long life) and Milou, who constantly reminded me of the ineffable mystery of animals as well as the propensity of cats to sit on keyboards.

Above all, I thank my husband, Michael Osborne, who for over thirty years has been my best friend, my constant cheerleader, and the best colleague I could ever hope for. He, with our wonderful sons, Paul and Henry Osborne, makes my life worth living. I dedicate this book to the memory of my parents, who always loved me.

NOTES

INTRODUCTION

1. I am unapologetic of my use of the term "science"; see on this point the introduction to Long, *Artisan/Practitioners*, 1–3.
2. Caroline Walker Bynum, "Miracles, Marvels, Magic," *London Review of Books* 31, no. 13 (9 July 2009): 32–33, at 33.
3. Shapin, *Scientific Revolution*, 1.
4. I am of course not alone in seeking a reassessment of standard historiographical categories of this period: notable recent examples include Smith, *Body of the Artisan*; Pomata and Siraisi, *Historia*; Mandressi, *Regard de l'anatomiste*; Cook, *Matters of Exchange*; and Bertoloni Meli, *Mechanism, Experiment, Disease*.
5. The locus classicus on this topic is Elias, *Court Society*.
6. On the problematic nature of "ancients versus moderns" as a characterization of this era, see esp. Shelford, *Transforming the Republic of Letters*.
7. This is not the place for a survey of animal studies. Prominent practitioners, mainly from literary and cultural studies, who refer to early modern evidence include Erica Fudge and Matthew Senior, among others. These practitioners take an implicitly activist stance on current animal uses. Explicitly activist is the sub-subdiscipline known as "critical animal studies." There are many historians whose work on animals does not fall under these characterizations; see, for example, the essays in Brantz, *Beastly Natures*.
8. Stensen, *Discours de M. Stenon*, 54.
9. Platter, *Beloved Son Felix*, 64, 85, 99, 126–28. A summary of various accounts of the public execution of Damiens appears in Foucault, *Discipline and Punish*, 3–6.
10. A textbook example of this argument appears in Dear, *Revolutionizing the Sciences*.
11. Bertoloni Meli, *Mechanism, Experiment, Disease*, 2–3, offers a succinct characterization of the value, and some of the limitations, of *Bibliotheca anatomica* as a snapshot of the state of anatomy in the late seventeenth century.
12. On the early modern distinction between "indigenous" and "exotic," see Cooper, *Inventing the Indigenous*. However, she does not discuss animals.
13. The literature on reading in this period is large and growing; particularly relevant here are Johns, *Nature of the Book*, 380–443; and Chartier, *Order of Books*.
14. On these categories, see Daston, "Description by Omission."

15. *"Physique"* included anatomy, botany, and chemistry. The academy's other section of *"mathématique"* included such exact sciences as astronomy and mechanics.
16. Frank, *Harvey and the Oxford Physiologists.*

CHAPTER ONE

1. J. B. D. M. Saunders and C. D. O'Malley, "Preparation of the Human Skeleton," 450.
2. See Klestinec, *Theaters of Anatomy;* Carlino, *Books of the Body;* French, *Dissection and Vivisection;* and Cunningham, *Anatomist Anatomis'd.*
3. Habicot, *La semaine ou pratique anatomique* (1660), 4–8.
4. On religious and other restrictions, see esp. Mandressi, *Regard de l'anatomiste,* chap. 1; Mandressi, "Dissections et anatomie," 311–13; and Guerrini, "Inside the Charnel House." Carlino, *Books of the Body,* emphasizes the transgressive nature of dissection in disturbing the social fabric.
5. Le Maire, *Paris ancien et nouveau,* 3:359–62; Mercier, *Tableau de Paris,* 5:17–18; Hillairet, *Connaissance du vieux Paris,* pt. 1, 83–86; and Fierro and Sarrazin, *Paris des Lumières,* 43–45.
6. Darnton, *Great Cat Massacre,* 102.
7. Alfred Franklin, *Vie privée d'autrefois,* 2:203–8, 2:96–97, 2:159.
8. Lister, *Journey to Paris,* 194–95.
9. Alfred Franklin, *Vie privée d'autrefois,* 2:144–45; Delort, *Animaux ont une histoire,* 461–70; Robbins, *Elephant Slaves and Pampered Parrots,* 80–85; and Isherwood, *Farce and Fantasy,* 209–11.
10. Farge, *Vivre dans la rue à Paris.*
11. Chardans, *Châtelet;* a good description, though from a later period, is in Schwartz, *Spectacular Realities,* 50–51. The very poor of Paris were referred to as "animaux urbains": Farge, *Vivre dans la rue à Paris,* 9.
12. Chardans, *Châtelet,* 35.
13. Lister, *Journey to Paris,* 69.
14. Guivarc'h, "Lieux des dissections," 434.
15. Mercier, *Tableau de Paris,* 3:232.
16. Fosseyeux, *Hôtel-Dieu de Paris;* Fosseyeux, "Prix des cadavres à Paris"; and Harding, *Dead and the Living,* 113–16.
17. Harding, *Dead and the Living,* 115–16.
18. Fosseyeux, *Hôtel-Dieu de Paris,* 336; and Cordier, *Paris et les anatomistes,* 11–12.
19. Brockliss and Jones, *Medical World,* 221–23.
20. Fosseyeux, "Prix des cadavres à Paris."
21. LeMaguet, *Monde médical parisien,* 256.
22. Fosseyeux, *Hôtel-Dieu de Paris,* 337; and Fosseyeux, "Prix des cadavres à Paris," 53–54.
23. See chap. 5.
24. Harding, *Dead and the Living,* 113.
25. Ibid., 115.
26. Brièle, *Déliberations de l'ancien bureau de l'Hôtel-Dieu,* 1:221, 1:275.
27. Bauhin, *Theatrum Anatomicum.*
28. Portal, *Histoire,* 2:177; and Pineau, *Opusculum physiologicum,* 191–95.
29. Brockliss and Jones, *Medical World,* 103.
30. Lonie, "'Paris Hippocratics.'"
31. Much confusion surrounds the dating of these events—1494, 1498, 1506. See Cor-

dier, *Paris et les anatomistes*, 10–11; Fosseyeux, "Prix des cadavres à Paris"; and Corlieu, *Ancienne faculté*.

32. C. D. O'Malley, "Vesalius, Andreas," in *DSB*; Huard and Grmek, "Oeuvre de Charles Estienne"; and Eloy.

33. Guivarc'h, "Lieux des dissections," 433; and LeMaguet, *Monde médical parisien*.

34. LeMaguet, *Monde médical parisien*, 25–26; and Corlieu, *Ancienne faculté*.

35. la Ramée, "Advertissements," 149–51.

36. Corlieu, *Ancienne faculté*; and M. D. Grmek, "Estienne, Charles," in *DSB*.

37. Corlieu, *Ancienne faculté*, 125.

38. Brockliss and Jones, *Medical World*, 94; and Mandressi, *Regard de l'anatomiste*, 179.

39. Corlieu, *Ancienne faculté*, 8.

40. Riolan, *Opuscula anatomica*, 257; and Stensen, *Discours de M. Stenon*, 34–36.

41. Guivarc'h, "Lieux des dissections"; and Corlieu, *Ancienne faculté*, 132.

42. Germain Courtin, *Leçons anatomiques et chirurgicales*; on Binet and Courtin, see Eloy. According to Eloy, Riolan claimed that this edition had many errors.

43. Riolan is recorded as witnessing the disputation of one of Perrault's theses in 1639: Claude Perrault, *An ut corporis sic animae senectus?*, thèse quodlibetaire, 1639, in *Recueil des thèses de médecine*, fol. SA 940, vol. 2, no. 128, Bibliothèque de l'Arsenal. For the titles of Perrault's theses, see Baron, *Scholis facultatis*, 39–40.

44. Hallays, *Les Perrault*, 14–15; *Commentarium 1636–1652*, BIUS, MS 13, fols. 122r, 133r; and Picon, *Claude Perrault*, 33, 37. We have no evidence about the content of his lectures, but espousing the circulation would have caused comment.

45. Corlieu, *Jacques Mentel*, 6.

46. For intellectual changes in the Paris Faculty over the course of the century, see Brockliss, "Medical Teaching."

47. Quesnay, *Recherches critiques et historiques*, 99.

48. Brockliss and Jones, *Medical World*, 219–20.

49. Gelfand, *Professionalizing Modern Medicine*, 21, citing Quesnay.

50. Brockliss and Jones, *Medical World*, 173–75; and Gelfand, *Professionalizing Modern Medicine*, 21–22.

51. Gelfand, *Professionalizing Modern Medicine*.

52. The church was at the corner of what is now rue Racine and rue de l'École de Médecine in the sixth arrondissement: Guivarc'h, "Lieux des dissections," 435.

53. "Open doors and without reading"—that is, not following the usual scholastic practice of reading from a text. Brockliss and Jones, *Medical World*, 222; and Eloy.

54. Ruhlmann, *Chirurgiens*, 36; and Eloy.

55. Pecquet, preface to *Experimenta nova anatomica* (1651).

56. *Commentarium 1636–1652*, BIUS, MS 13, fol. 418r; and Pecquet, *Experimenta nova anatomica* (1651), 19.

57. Fosseyeux, "Prix des cadavres à Paris," quoting Paré.

58. Pecker, "Chirurgie au xviiᵉ siècle," 196, 193. See also McHugh, "Establishing Medical Men."

59. Pecker, "Chirurgie au xviiᵉ siècle," 192–94; and Fosseyeux, *Hôtel-Dieu de Paris*.

60. Brockliss, *French Higher Education*, 13n.

61. Collège de France, *Liste des professeurs*; LeMaguet, *Monde médical parisien*, 49; Brockliss and Jones, *Medical World*, 208–9; and Sturdy, *Richelieu and Mazarin*, 9–10.

62. "Faire la demonstration oculaire et manuelle de toutes et chacunes des opérations de chirurgie, de quelque nature qu'elles puissant estre": Hamy, "Origines de

l'enseignement," 5; and Laissus, "Jardin du roi," 313. On the founding of the garden, see Lunel, *Maison médicale du roi*, 161–84; Howard, "Guy de la Brosse"; and Howard, "Medical Politics." Marin Cureau de la Chambre's family name was "Cureau"; the "de la Chambre" referred to a very recently acquired family residence.

63. This designation was less exclusive than it might seem: although there was one *médecin ordinaire* and six to eight *médecins par quartier*, there were sixty-six *médecins consultants du roi*; these latter positions, like other royal appointments, could be bought and sold. LeMaguet, *Monde médical parisien*, 197–99; and Lunel, *Maison médicale du roi*.

64. Laissus, "Jardin du roi," 287–89.

65. Corlieu, *Ancienne faculté*, 135; and *Commentarium facultatio medicinae Parisiensis ab anno domini 1636 ad annum 1652*, BIUS, MS 13, fol. 18r (January 1637).

66. On the importance of Latin poetry, see Shelford, *Transforming the Republic of Letters*, esp. chap. 2.

67. Cureau de la Chambre, *Nouvelles conjectures*.

68. Brockliss, "French Galenists," 82–83.

69. Hamy, "Origines de l'enseignement," 6–9; Harrison, "Virtues of Animals," 473–76; Darmon, *Corps immatériels*; Diamond, "Marin Cureau de la Chambre," exaggerates his modernity in my view. On physiognomy and chiromancy, see Courtine, "Miroir de l'âme."

70. Mersenne to Peiresc, 2 août 1634, Mersenne, *Corres.*, 4:301; and Descartes to Mersenne, end of December 1637, Mersenne, *Corres.*, 6:347.

71. Peter N. Miller, *Peiresc's Europe*; France, *Politeness and Its Discontents*; and Fumaroli, *Âge de l'éloquence*. See Huppert, *Style of Paris*, for similar debates in the sixteenth century.

72. "Tout est donc dans la manière et non dans la matière," Fumaroli, *Âge de l'éloquence*, 647–49, at 648; and Pintard, *Libertinage érudit*. A contemporary claim for the French language is Le Laboureur, *Avantages de la langue Françoise*. For the roles of rhetoric and language, see chap. 6, below.

73. Sturdy, *Richelieu and Mazarin*, 93; see chap. 3, below.

74. Hamy, "Origines de l'enseignement," and Laissus, "Jardin du roi," skirt around this, but they both say there is no evidence Cureau de la Chambre lectured, and significantly there was no *sous-demonstrateur* in anatomy at that time, unlike in chemistry and botany. See esp. Legée, "Bio-bibliographie de Marin Cureau de la Chambre," 21–32, at 25–26, who concludes, "Il n'exerça guère ses fonctions de demonstrateur au Jardin Royal." P. A. Cap asserted that François Vautier, later Louis XIV's first physician, introduced anatomical teaching to the *jardin* in 1643, but I have found no corroboration for this assertion: Cap, *Muséum d'histoire naturelle*, 6–7. Cunningham, *Anatomist Anatomis'd*, 99–100, claims that Cureau de la Chambre lectured but cites no new evidence, relying on Hamy.

75. Fontenelle, *Éloges*, 2:438.

76. See chap. 2.

77. Pomata, "*Praxis historialis*"; Mandressi, *Regard de l'anatomiste*, 118–19; and Kusukawa, "Sources of Gessner's Pictures."

78. Furetière, *Dictionnaire*, s.v. "*histoire*"; cf. Mandressi, *Regard de l'anatomiste*, 118.

79. Siraisi, *Medieval and Early Renaissance Medicine*, 96; and Du Laurens, *Historia anatomica*. A cursory look at the National Library of Medicine (NLM) catalog shows over sixty texts.

80. On natural philosophy textbooks, see Reif, "Textbook Tradition."

81. Estienne and la Rivière, *Dissection des parties du corps humain.* The plates were drawn by a number of artists and engraved mainly by Geoffroy Tory. For Estienne see Huard and Grmek, "Oeuvre de Charles Estienne"; and Carlino and Cazès, "Plaisir de l'anatomie."

82. On the model of dissection, see Mandressi, "Dissections et anatomie," 321–25; on technique, see Klestinec, "Civility, Comportment, and the Anatomy Theater."

83. Du Laurens, *Historia anatomica*, 1–6.

84. Gweneth Whitteridge, "Bauhin, Gaspard," in *DSB.*

85. Ibid.

86. Platter, *Beloved Son Felix*, 88–90, 92–93.

87. Whitteridge, "Bauhin, Gaspard," in *DSB.*

88. Riolan, *Oeuvres anatomiques*, 91: "Le subiect de l'Anatomie est le corps de l'homme, la composition & conformation de qui elle nous represente par vne dissection fort exacte."

89. Ibid., 67.

90. Ibid., 95–96.

91. Du Laurens, *Historia anatomica*, 14–15; and Riolan, *Oeuvres anatomiques*, 93–94.

92. Riolan, *Oeuvres anatomiques*, 121, 162. He referred to the "seeing hand" (*oculatis manibus*) in *Encheiridium*, "Praemonitio ad lectorem & auditorem," n.p.

93. Gelfand, *Professionalizing Modern Medicine*, 23, 25; and Ruhlmann, *Chirurgiens*, 36. I have found editions in 1610, 1631, 1650, and 1660.

94. Habicot, *Semaine*, 15–16; subtitle: "Œuvre utile & necessaire à ceux qui desirent parvenir à la parfaite connoissance d'eux mesmes: & specialement à celuy qui veut faire profession de la Medecine & Chirurgie."

95. Lyser, *Culter anatomicus.* On Bartholin, see chap. 2 below.

96. Descartes to Mersenne, 18 décembre 1629, Mersenne, *Corres.*, 2:352.

97. Baillet, *Vie de Monsieur Descartes*, 196–97; Rodis-Lewis, *Descartes*, 85; Descartes, *Anathomia*, in AT, 11:546–634; 588n on Bauhin. Most of the notes date from the 1630s; there is one short excerpt dated 1648.

98. Descartes, *Anathomia*; Cook, *Matters of Exchange*, 231–33; and Descartes, *Dioptrique*, in AT, 6:81–147. Vision and the eye will be further discussed in chap. 5.

99. Sorbière, *Lettres et discours*, 689–90.

100. Brockliss, "Harvey, Torricelli, and the Institutionalization of New Ideas," 119.

101. Wellman, *Making Science Social.*

102. Biagioli, *Galileo's Instruments of Credit*; and Kettering, *Patrons, Brokers, and Clients.*

103. Goldgar, *Impolite Learning*, esp. 1–6; and Shelford, *Transforming the Republic of Letters*, 1–6.

104. Shelford, *Transforming the Republic of Letters*, 5.

105. These various terms are to a degree interchangeable. Their meaning will be further refined in the course of this narrative.

106. Yates, *French Academies*; and Freedberg, *Eye of the Lynx.*

107. Pintard, *Libertinage érudit*, 92–98; Fumaroli, "Fouquet," 248–49; and Fumaroli, *Âge de l'éloquence*, 581–83 and 589–90.

108. Sarasohn, "Peiresc." On correspondence in general, see Chartier, *Correspondance.*

109. Armogathe, "Groupe de Mersenne"; Pearl, "Role of Personal Correspondence"; and Ultee, "Republic of Letters."

110. Peiresc, *Corres.*; and Pearl, "Role of Personal Correspondence."

111. Sarasohn, "Peiresc"; Kettering, *Patrons, Brokers, and Clients*; and Petitfils, *Fouquet*, 98–99. On Renaudot, see Wellman, *Making Science Social.*

112. Fumaroli, "Peiresc," 22–26; and Mandrou, *From Humanism to Science*, 184; cf. Pintard, *Libertinage érudit*.
113. On Colbert, see Soll, "Jean-Baptiste Colbert's Republic of Letters."
114. Peter N. Miller, *Peiresc's Europe*, 10.
115. Brown, *Scientific Organizations*, 8–9; Pearl, "Role of Personal Correspondence," 111; Peiresc to Jacques Dupuy, 25 juillet 1634, Mersenne, *Corres.*, 4:251–52; Gassendi to Elia Diodati, 29 aôut 1634, Mersenne, *Corres.*, 4:338; and Peiresc to Gassendi, 2 février 1634, Peiresc, *Corres.*, 4:433–39.
116. Peter N. Miller, *Peiresc's Europe*, 69–71; and Revel, "Uses of Civility," 192–94. On these questions, see Goldgar, *Impolite Learning*, 235–38; Matthew L. Jones, *Good Life*, 146–49; DeJean, *Ancients against Moderns*, 128–130; Huppert, *Style of Paris*, 99–100; Fumaroli, *Âge de l'éloquence*; and Pintard, *Libertinage érudit*.
117. France, *Politeness and Its Discontents*, 68, who cites D'Alembert's *Essai sur la société des gens de lettres avec les grands*; see also his *Rhetoric and Truth in France*, chap. 3. On the salon as an egalitarian space, see Gordon, "'Public Opinion' and the Civilizing Process in France"; and Goodman, *Republic of Letters*.
118. Faret, *Honneste-homme*, dedication (n.p.), 49.
119. Mitton, "Pensees sur l'honnêteté," quoted in Matthew L. Jones, *Good Life*, 147. See also Pintard, *Libertinage érudit*, 350–51.
120. Fumaroli, "Peiresc," 47–49; and Fumaroli, *Âge d'eloquence*, 657–59.
121. Revel, "Uses of Civility," 195. On Montmor and civility, see Adkins, "Montmor."
122. For an earlier period, see Ogilvie, *Science of Describing*, esp. 74–85.
123. Taton, *Origines*, 18; Sarasohn, "Peiresc," 78–81; and Descartes to Mersenne, 27 February 1637, end of May 1637: AT, 1:348–51, 1:366–68, translated in *PWD*, 3:52–54, 3:56–57.
124. Dear, *Mersenne and the Learning of the Schools*; and Mersenne to Peiresc, 23 mai 1635, Mersenne, *Corres.*, 5:209.
125. Armogathe, "Groupe de Mersenne"; and Gauja, "Origines," 1:22–33.
126. Eloy, s.v. "Bourdelot, Edmé" and "Michon, Pierre." Eloy gives the death date for Edmé Bourdelot as 1620. See also Lévy-Valensi, *Médecine et les médecins*, 579–611.
127. Le Maire, *Paris ancien et nouveau*, 3:443. Pintard, *Libertinage érudit*, 352–55, gives a much less flattering portrait.
128. Villiers to Mersenne, septembre 1633, Mersenne, *Corres.*, 3:482–83; Gabriel Naudé to Mersenne, 12 novembre 1633, Mersenne, *Corres.*, 3:535; Villiers to Mersenne, novembre 1633, Mersenne, *Corres.*, 3:550; Peiresc to Jacques Dupuy, 25 juillet 1634, Mersenne, *Corres.*, 4:251–52; and Gassendi, *Mirrour of True Nobility and Gentility*, 45. Villiers, like Bourdelot, was from Sens.
129. Peiresc to Cassiano dal Pozzo, 17 March 1634, Peiresc, *Lettres à Cassiano dal Pozzo*, no. 41, p. 131; Mersenne, *Corres.*, 4:368; on Cassiano and his collection, see Freedberg, *Eye of the Lynx*, esp. chap. 1. The influence of Cassiano's *museo cartaceo* on Perrault's *Histoire des animaux* will be discussed in chap. 4. On Barberini and Peiresc, see Fumaroli, *Âge de l'éloquence*, 557n.
130. Bouchard to Mersenne, 14 janvier 1634, Mersenne, *Corres.*, 4:3.
131. Pintard, "Académie Bourdelot"; René Taton, "Bourdelot, Pierre Michon," in *DSB*; and Eloy, s.v. "Michon, Pierre."
132. Pascal, *De l'esprit géométrique et l'art de persuader*, 103, 132–33, 146–48; Fumaroli, "Préface," in Pascal, *De l'esprit géométrique*, 12, 41–48; Wellman, *Making Science Social*, 4–5; and Le Maire, *Paris ancien et nouveau*, 3:443–44.

133. Solomon, *Public Welfare*, 55–56, 64; Denonain, "Problèmes de l'honnête homme"; Bigourdan, "Première sociétés" (1916); and Wellman, *Making Science Social*, chaps. 1 and 2.
134. Gauja, "Origines," 1:13; and Solomon, *Public Welfare*, 73–74.
135. Sturdy, *Richelieu and Mazarin*.
136. Taton, *Origines*, 16–17; Le Gallois, *Conversations*, preface to vol. 1; and Le Maire, *Paris ancien et nouveau*, 3:236, 3:442.
137. Pintard, "Académie Bourdelot," 80.
138. Bourdelot to Cassiano dal Pozzo, 1 janvier 1649, 20 juillet 1649, 25 octobre 1649, Mersenne, *Corres.*, 16:530, 16:533, 16:534. Pierre spelled his family name "Mercenne."
139. Brockliss, *French Higher Education*, 20–22; and Garber and Ayers, *Cambridge History of Seventeenth-Century Philosophy*, 18–19.
140. Biographical accounts of Pecquet are both vague and contradictory about his early life. I have pieced together an account that seems to make sense, but I make no claims for absolute truth. See Gilis, "Jean Pecquet"; Pierre Huard and Marie-José Imbault-Huart, "Pecquet, Jean," in *DSB*; and Eloy. On life at the Collège de Clermont, see Motley, *Becoming a French Aristocrat*, 108–22.
141. Eloy; and Le Gallois, *Conversations*, preface to vol. 1.
142. Pecquet, "Observations." Emmerez would be one of the main practitioners of blood transfusion in the 1660s. (See below, chap. 3.)
143. Petitfils, *Fouquet*, 38. Pecquet to Mersenne, 3 août 1648, Mersenne, *Corres.*, 16:466–67. This letter is signed "J. Pecquet s.j.," and on the basis of this suffix, Armand Beaulieu (the editor of Mersenne's correspondence) attributed this letter to an unknown Jesuit Pecquet. From the context, it is certainly the anatomist Pecquet; it is too much of a coincidence for two J. Pecquets to be in Agde with François Fouquet. Without seeing the original, I cannot explain the "s.j.," although it could be a misreading.
144. Pecquet to Mersenne, 3 août 1648, Mersenne, *Corres.*, 16:466–67.
145. Pecquet, *Experimenta nova anatomica* (1651). Dedication, n.p.: "domi tuae natus."
146. Briggs, *Early Modern France*, 126–34.
147. Lewis, "Jean Pecquet," 89–90, plausibly surmises that Bourdelot may have taken a copy with him; see Rudbeck, "Translation," 306–7, 321. On Bourdelot, see Pintard, *Libertinage érudit*, 350–62, 379, 389–403.
148. Gauja, "Origines," 1:24, 1:36.
149. Packard, *Guy Patin*, 167–68.
150. Pecquet does not appear in the faculty's minutes. Picon, *Claude Perrault*, 103–4; Charles Perrault, Claude Perrault, and Pierre Perrault, *Murs de Troye*; and Bonnefon, "Poème inédit de Claude Perrault."
151. Gilis, "Jean Pecquet," 222; and Lévy-Valensi, *Médecine et les médecins*, 280–89.
152. Quoted by LeFanu, "Jean Martet," 1:34.
153. Pecquet, *Experimenta nova anatomica* (1651), 134–35; cf. Gilis, "Jean Pecquet," 223. Roger French (*Harvey's Natural Philosophy*, 162n, 264) claims that Rivière supported the circulation and was threatened with expulsion from his professorship because of it, but I have found no primary evidence to support this claim, which seems to rely on a chain of secondary assertions dating back to Robert Willis in the nineteenth century. Weil, "Echo," 172, notes Willis's claim and states that Rivière's supposed book does not exist.
154. LeFanu, "Jean Martet," 34–41; and Martet, *Abbrégé*. Martet does not appear on Weil's

list of works on the circulation up to 1657. Pecquet, *Experimenta nova anatomica* (1654), 134–35; and Gilis, "Jean Pecquet," 223.

155. Martet, *Abbrégé*, 11–25, 38–43; and Lewis, "Jean Pecquet," 111.

CHAPTER TWO

1. Le Clerc and Manget, *Bibliotheca anatomica* (1711), 1:12, and *Bibliotheca anatomica* (1699), preface to vol. 1, n.p.; and Stensen, *Discours de M. Stenon*, 56.

2. Bacon, *New Organon*, Aphorisms, bk. 1, CXXIV.

3. Dear, "Method and the Study of Nature," 147–77; and Dear, *Discipline and Experience*. See also Garber, "Experiment, Community."

4. Cole gave a genealogy of the term and the practice of comparative anatomy in the introduction to *History of Comparative Anatomy*, 3–23.

5. Salomon-Bayet, *Institution de la science*; and Pomata, "*Praxis historialis*," 115.

6. Ogilvie, *Science of Describing*, 17–23; Dear, "Method and the Study of Nature"; Garber, "Experiment, Community"; and Bertoloni Meli, *Thinking with Objects*, 10–12.

7. Bacon, *New Organon*; Schmitt, "Experience and Experiment"; and Ragland, "Making Trials."

8. Koyré, *Metaphysics and Measurement*, 90–91; Naylor, "Galileo: Real Experiment"; Naylor, "Galileo's Method of Analysis"; and Bertoloni Meli, *Thinking with Objects*, 12.

9. Bertoloni Meli, *Thinking with Objects*, 12.

10. Desmond Clarke, *Descartes' Philosophy of Science*, 19; Garber, "Descartes and Experiment"; and Cook, *Matters of Exchange*, 226–37.

11. Royal Society, Boyle Papers 35: fol. 202r, quoted in Westfall, "Unpublished Boyle Papers," 117.

12. On *historia* and *scientia*, see esp. Pomata, "*Praxis historialis*." The Greek word that du Laurens and others use is "αυτοψία." Du Laurens, *Historia anatomica*, 14. See also Mandressi, *Regard de l'anatomiste*, 118.

13. Bacon, *New Atlantis*, 31.

14. Bernard Lamy, *Entretiens*, 296–97.

15. Dear, *Discipline and Experience*, 129–32, 54–57. For a critique of aspects of this explanation, see Ragland, "Making Trials."

16. On plants, see Ogilvie, *Science of Describing*, 21; Stroup, *Company of Scientists*; and Dagognet, *Catalogue de la vie*.

17. Harvey, *De generatione animalium*, dedication, n.p.

18. Daston, "Marvelous Facts."

19. *Anatomia cophonis*, in Grant, *Source Book*.

20. Dear, "Method and the Study of Nature," 148: "The humanist tradition established a model of knowledge as an interconnected whole, which method might map out."

21. Shakespeare, *Tragedie of Hamlet* (1623); and Bauhin, *Theatrum anatomicum*, 24.

22. Guillaume Lamy, *Discours anatomiques*, 18.

23. Guerrini, *Experimenting with Humans and Animals*, 21–22.

24. Perfetti, *Aristotle's Zoology*; and Pinon, *Livres de zoologie*.

25. Harvey, *De generatione animalium*, dedication, n.p.

26. Ogilvie, *Science of Describing*; and Pickstone, *Ways of Knowing*.

27. Perfetti, *Aristotle's Zoology*; and Beullens and Gotthelf, "Gaza's Translation of Aristotle's De Animalibus."

28. Blair, *Too Much to Know*.

29. Kusukawa, "Sources of Gessner's Pictures"; and Kusukawa, *Picturing the Book of Nature*.

30. Aelian, *Opera*.
31. Kusukawa, "Sources of Gessner's Pictures"; Asúa and French, *New World of Animals*, 108–9, 190–96; Ashworth, "Emblematic Natural History"; on Gessner's style, see West, *Theatres and Encyclopedias*, 106–7.
32. Asúa and French, *New World of Animals*, 193; on Johnstone, see chap. 4, below.
33. Guerrini, *Natural History and the New World*.
34. Gessner, *Historiae animalium*, 1:981–82 ("simivulpa"); George, "Sources and Background," 81; and Asúa and French, *New World of Animals*, 193–96.
35. L'Écluse, *Exoticorum libri decem*; Harvey, *De generatione animalium*, dedication, n.p.; and Harvey, *Anatomical Exercitations*, epistle dedicatory, n.p.
36. Rondelet, *Libri de piscibus marinis*, 459–73; Belon, *Nature et diversité des poisons*; Glardon, *Histoire naturelle*; and Cole, *History of Comparative Anatomy*, 60–72.
37. Belon, *Histoire de la nature des oyseaux*, 75. On the Aristotelian notion of animal, see Cunningham, *Anatomical Renaissance*, chap. 6.
38. Kusukawa, "Sources of Gessner's Pictures"; Asúa and French, *New World of Animals*, 100–114; du Laurens, *Historia anatomica*; see chap. 1, above.
39. Glardon, *Histoire naturelle*, 380.
40. The works of Belon and Rondelet on marine animals were reprinted several times with Gessner's volume on the same topic, e.g., *Conradi Gesneri, . . . Historiae animalium liber IIII., qui est de piscium et aquatilium animantium natura . . . Continentur in hoc volumine Gulielmi Rondeletii, . . . et Petri Bellonii, . . . de aquatilium singulis scripta* (1620). This undercuts Ashworth's claim that Belon and Rondelet "faded" from view in the seventeenth century: Ashworth, "Emblematic Natural History," 30. All of Aldrovandi's work on animals appeared posthumously, the last in 1640 (see below). Only Coiter's work remained in a single printing, although his work on the fetal skeleton was reprinted in 1659 and by Le Clerc and Manget in 1699.
41. Ruini, *Anatomia del cavallo*; and Aldrovandi, *De quadrupedibus*, 13–14.
42. Héroard, *Hippostologie*. It consists of twenty-four pages and a lengthy dedication to the king.
43. Cunningham, *Anatomical Renaissance*, chap. 6; and French, *Dissection and Vivisection*, 231–32.
44. Cunningham, *Anatomical Renaissance*, 176–77; and Pomata, "*Praxis historialis*," 116–18.
45. Fabricius, *De visione, voce, auditu*.
46. Casserio, *De vocis auditusque organis historia anatomica*.
47. Le Clerc and Manget on Harvey, *Bibliotheca anatomica*, 839.
48. Bylebyl, "Nutrition, Quantification, and Circulation," 369. See also Bylebyl, "Medical Side of Harvey's Discovery."
49. Cole, "Harvey's Animals." It is not clear whether Harvey experimented on all these animals or whether he was citing others' work.
50. Riolan, *Oeuvres anatomiques*, 980–90; and du Laurens, *Historia anatomica*.
51. French, *Harvey's Natural Philosophy*; and Wear, "Harvey and the 'Way of the Anatomists.'"
52. This is the argument of Whitteridge, *William Harvey and the Circulation of the Blood*.
53. Harvey, *De motu cordis*, 5–9; and Harvey, *Circulation of the Blood*, 4–6.
54. Harvey, *De motu cordis*, 23; and Harvey, *Circulation of the Blood*, 20, my emphasis. Harvey, *De motu cordis*, 10; and Harvey, *Circulation of the Blood*, 7.
55. Harvey, *De motu cordis*, chap. 4; Riolan, *Oeuvres anatomiques*, 984–86; and Bauhin, *Theatrum anatomicum*, 410–11, 418–19. Bauhin used "dogs and monkeys" (p. 410).
56. Aselli, *De lactibus*, 20.

57. Ibid., 56–64. The modern definition of chyle is a mixture of lymph and intestinal fats. Eales, "History of the Lymphatic System," 280–81; and Ambrose, "First Priority Dispute," 1–3.

58. Pomata, *"Praxis historialis,"* 119–21.

59. Aselli, *De lactibus*, 21: "Ita nihilominus, idque pro certo statuo, quae in tot brutis visa mihi sunt, iis fieri nullo modo posse, unus & solo homo, ut deficiatur."

60. Bylebyl, "Cesalpino and Harvey"; and Pagel, *New Light on William Harvey*, 128–31.

61. Gassendi to Mersenne, 4 février 1629, Mersenne, *Corres.*, 2:182, 2:189n.

62. Gassendi to Peiresc, 28 août 1629, Mersenne, *Corres.*, 2:268, says "M. du Puy" will send a copy of Harvey's book to Peiresc. "Son opinion de la continuelle circulation du sang par les arteres et veines est fort vraysemblable et establie."

63. Peiresc to Dupuy, 25 juillet 1634, Mersenne, *Corres.*, 4:251–52. He described it to Bourdelot a month later: Peiresc to Bourdelot, 6 septembre 1634, Peiresc, *Corres.*, 7:712.

64. Gassendi to Diodati, 29 août 1634, Mersenne, *Corres.*, 4:335–341, at 4:339–40.

65. Descartes to Mersenne, November or December 1632, Mersenne, *Corres.*, 3:346: "J'ay veu le livre . . . et me suis trouvé un peu différent de son opinion, quoyque je ne l'aye vu qu'après avoit achevé d'écrire de cette matière."

66. Descartes, *Discours de la méthode*, in AT, 6:50; and PWD, 1:136. See also Gorham, "Mind-Body Dualism."

67. Grmek, *Première revolution biologique*, 130–31; cf. Cook, "Victories for Empiricism," 23–26.

68. Grmek, *Première revolution biologique*, chap. 5, esp. pp. 129–36; French, *Harvey's Natural Philosophy*; Cook, *Matters of Exchange*; and Descartes, *Anathomia*, in AT, vol. 11. On Cartesian natural philosophy more generally, see Desmond Clarke, *Occult Powers and Hypotheses*.

69. French, *Dissection and Vivisection*, 210, on Colombo; survey of *Index*.

70. Cook, *Matters of Exchange*, 148–49, 235; Pagel, *New Light on William Harvey*, 114; and Schouten, "Johannes Walaeus," 260.

71. Schouten, "Johannes Walaeus"; *Johannis Walaei epistolae duae: De motu chyli, et sanguinis. Ad Thomam Bartholinum*, in Bartholin, *Institutiones anatomicae*, 443–88; the letters are dated October and December 1640. The Countway Library at Harvard has a copy of the pamphlet.

72. Pagel, *New Light on William Harvey*, 113–35; and Schouten, "Johannes Walaeus."

73. Baron, *Scholis facultatis*, 41. Riolan was not present at this defense.

74. Ibid., 42.

75. Riolan, *Opuscula anatomica*, 3.

76. Ibid., 20–145.

77. This discredits the common notion that the Paris Faculty was monolithically Galenic. It held several members who supported the circulation, including Mentel, who served as professor of surgery from 1648 to 1650. Corlieu, *Jacques Mentel.*

78. Riolan, *Opuscula anatomica*, 10.

79. On this debate, see Guerrini, "Experiments, Causation."

80. Severino's editor, Johann Georg Volkamer, mentions Harvey with approval in his preface, n.p.

81. Severino, *Zootomia democritaea*, 36; Schmitt and Webster, "Harvey and M. A. Severino," 57n33.

82. Severino, *Zootomia democritaea*, 44, 49.

83. On Du Prat, see below; Bartholin, *Institutions anatomiques*, 242–49, 589–616.

84. Bartholin, *Institutiones anatomicae*, 448–49. The text (191) asserts the existence of the pores, but Thomas Bartholin added passages in brackets that refuted this; this is one of many examples of Thomas arguing with his late father in the text, and was not resolved until Thomas undertook a full revision in the 1650s.

85. Gassendi, *Opera omnia*, 6:203b–204a.

86. Sorbière, *Discours sceptique*, 3. Sorbière and Du Prat were at this time both Protestants.

87. Ibid., 82–84, 40–41.

88. Ibid., 56, 68–69.

89. Ibid., 54–55.

90. Galen, *On the Usefulness of the Parts*, 49; and Sorbière, *Discours sceptique*, 58, 55.

91. On Descartes, see Nonnoi, "Against Emptiness." Mersenne, *Corres.*, 13:234–248. See also 13:177–89, 13:391–92 on Torricelli. See also Zouckermann, "Air Weight and Atmospheric Pressure."

92. Mersenne, *Corres.*, 14:648; see Matthew L. Jones, "Writing and *Sentiment*," on Pascal.

93. See Mersenne, *Corres.*, vols. 13 and 14, for relevant letters. Pecquet, *Experimenta nova anatomica* (1651), 50–54; Bertoloni Meli, "Collaboration between Anatomists and Mathematicians," 670–77; Matthew L. Jones, "Writing and *Sentiment*," 139–43; on Bourdelot, see Peumery, "Conversations médico-scientifiques," 129.

94. Pecquet, *Experimenta nova anatomica* (1651), 4; and Pecquet, *New Anatomical Experiments*, 7.

95. Pecquet, *Experimenta nova anatomica* (1651), 3; Pecquet, *New Anatomical Experiments*, 4; and Riolan, *Opuscula anatomica*, 109.

96. Riolan, *Oeuvres anatomiques*, 164. On Fabrici, see Klestinec, "Civility, Comportment, and the Anatomy Theater."

97. Pecquet, *New Anatomical Experiments*, 7–8; and Pecquet, *Experimenta nova anatomica* (1651), 4–5.

98. Pecquet, *Experimenta nova anatomica* (1651), 31–32; and Pecquet, *New Anatomical Experiments*, 54–55, 57. The anonymous translator of *New Anatomical Experiments* translated "αυτοψία" as "ocular inspection."

99. Pecquet, *Experimenta nova anatomica* (1651), 25.

100. Pecquet, *Experimenta nova anatomica* (1651), 49; and Pecquet, *New Anatomical Experiments*, 90.

101. Pecquet, *Experimenta nova anatomica* (1651), 86; and Pecquet, *New Anatomical Experiments*, 165. On the implications for physics of this work, see Bertoloni Meli, *Thinking with Objects*, 225.

102. Bertoloni Meli, *Mechanism, Experiment, Disease*, 37.

103. Bartholin, *De lacteis thoracicis*; and Bartholin, *Vasa lymphatica*, 59.

104. Sorbière, "Praefatio," n.p., in Gassendi, *Opera omnia*, vol. 1; and Brown, *Scientific Organizations*, 69–70.

105. Molière, *Précieuses ridicules*, scene 10.

106. Gauja, "Origines," 1:37–44; Taton, *Origines*, 21–27; Bigourdan, "Premières sociétés" (1917), 131–34; and Brown, *Scientific Organizations*, 64–72. Brown cites a source that claims Montmor's income was 100,000 livres a year, an enormous sum. Pintard, *Libertinage érudit*, 403–4.

107. Brown, *Scientific Organizations*, 71–72.

108. Trevor McClaughlin has written extensively on Rohault; see in particular "Concept de science chez Jacques Rohault" and "Descartes, Experiments." See also John Shuster, "Rohault, Jacques," in *DSB*.

109. Lewis, "Jean Pecquet," 155.
110. Petitfils, *Fouquet*, 166–71; and Brice, *Description nouvelle*, 1:209–10. Fouquet's house was adjacent to the Vincennes menagerie.
111. E. S. Saunders, "Politics and Scholarship," 5–9; and Petitfils, *Fouquet*, 201–15, 258–79. Moreau (d. 1656) was a faculty physician, professor of medicine at the Collège Royal, and part of Peiresc's circle. Fouquet's library included twenty-four works on anatomy when it was inventoried in 1665.
112. Petitfils, *Fouquet*, 183–88. Pecquet must have gone with him.
113. Brown, *Scientific Organizations*, 75; and Bigourdan, "Premières sociétés" (1917), 132–33. On the history of meetings, see Van Vree, *Meetings, Manners and Civilization*.
114. Gauja, "Origines," 1:37–44; Brown, *Scientific Organizations*, 87; and Bigourdan, "Premières sociétés" (1917), 159–62, 216–20.
115. Sorbière, *Lettres et discours*, 23 (Sorbière to Mazarin, 10 février 1659). He published a second volume later that year that included an *éloge* for Du Prat, who had recently died: *Relations, lettres, et discours*, 301–20. He dedicated both volumes to Mazarin. See Perkins, "Uses of Science."
116. Oldenburg to Saporta, 28 June 1659, Oldenburg, *Corres.*, 1:259–65; Oldenburg to Hartlib, 6 September 1659, ibid., 1:306–10; and Brown, *Scientific Organizations*, 92–107.
117. Oldenburg to Hartlib, 29 November 1659, Oldenburg, *Corres.*, 1:331–33.
118. Huygens, *Oeuvres*, 22:536, 22:550.
119. Brown, *Scientific Organizations*, 111–12; and Huygens, *Oeuvres*, 22:539, 22:544.
120. Huygens, *Oeuvres*, 22:546, 22:554.
121. Ibid., 22:560.
122. Sorbière, *Lettres et discours*, 22–59; and Perkins, "Samuel Sorbière," 217–23.
123. Sorbière, *Lettres et discours*, 24, 57, 58; and Perkins, "Samuel Sorbière," 218–19. Perkins suggests that Mazarin was previously unaware of Pecquet's discovery, which seems plausible.
124. Petitfils, *Fouquet*, 338–455.
125. Gilis, "Jean Pecquet," 222–23; and Mme. de Sévigné to Pomponne, 26 décembre 1664, Sévigné, *Correspondance*, 1:81, 1:918n.
126. Rabinovitch, "Versailles as a Family Enterprise." See also Hallays, *Les Perrault*; and Soriano, *Dossier Perrault*. On the importance of "horizontal" relationships, see Cavallo, *Artisans of the Body*.
127. Portal, *Histoire*, 3:2.
128. Brown, *Scientific Organizations*, 132–33; and Huygens, *Oeuvres*, 4:481–82.
129. Huygens, *Oeuvres*, 5:70; and Brown, *Scientific Organizations*, 133.
130. Brown, *Scientific Organizations*, 135–37; Gauja, "Origines," 1:44–46; McClaughlin, "Rapports"; Nordström, "Swammerdamiana"; and Modlin, "Regnier de Graaf."
131. Poulsen and Snorrason, *Nicolaus Steno*, 15–17; and Stensen, *De musculis*.
132. Unsigned review of *De musculis*, *JdS* (March 1665): 139–42.
133. Nordström, "Swammerdamiana," 29–38.
134. Du Laurens, *Discours . . . de la veuë*, 1–12; Riolan, *Oeuvres anatomiques*, 563; Stensen, *Discours de M. Stenon*, 3 and "Epistre," n.p. This is the first printing of Stensen's work; a Latin edition appeared in 1671: Poulsen and Snorrason, *Nicolaus Steno*, 215.
135. Stensen, *Discours de M. Stenon*, 11–13, 31. The *Traité de l'homme*, along with the rest of *Le Monde*, had been published in Latin in 1662 and in French two years later. On the horse, see Stensen to Thomas Bartholin, 5 March 1663, quoted in Grell, "Between Anatomy and Religion," 210.
136. *JdS* (10 février 1670): 7–9.

137. Brown, "Jean Denis," 15n; and Denis, *Recueil des mémoires*.
138. Gabbey, "Bourdelot Academy," 96; and Molière, *Femmes sçavantes*, act 3, scene 2. On his earlier meetings, see chap. 1 above.
139. Brown, *Scientific Organizations*, 161–84; Brown, "Cosmopolite du Grand Siècle"; and *ODNB*, s.v. "Justel, Henri."
140. Brown, *Scientific Organizations*, 223; and Huygens, *Oeuvres*, 5:61.
141. Modlin, "Regnier de Graaf"; and de Graaf, *Suc pancréatique*. De Graaf credits Bourdelot in the preface to the French translation of his work on reproduction: de Graaf, *Histoire anatomique*, n.p.
142. De Graaf, *Opera omnia, dedicatio*, n.p.
143. Burke, *Fabrication of Louis XIV*, 49–50.
144. See Hahn, *Anatomy*; Gauja, "Origines"; Taton, *Origines*; Fauré-Fremiet, "Origines"; and Brown, *Scientific Organizations*.
145. Charles Perrault and Claude Perrault, *Mémoires de ma vie*, 35–38; and Berger, *Garden of the Sun King*, 7–8, 14–15.
146. "Note de Charles Perrault à Colbert pour l'établissement d'une Académie générale," in Colbert, *Lettres*, 5:512–13; and Préaud, "'Académie des sciences et des beaux-arts,'" 73–81.
147. Lux, "Colbert's Plan for the *Grande Académie*," 177–88; Tits-Dieuaide, "Colbert et l'Académie royale des sciences," 214–20; and Hahn, *Anatomy*, 4–5.
148. Tits-Dieuaide, "Colbert et l'Académie royale des sciences," 214.
149. Huygens, *Oeuvres*, 4:325–29. See Sturdy, *Science and Social Status*, 70–72; Salomon-Bayet, *Institution de la science*, 37–43; and Hahn, *Anatomy*, 8.
150. Huygens, *Oeuvres*, 6:95–96. See also Salomon-Bayet, *Institution de la science*, 37–43.
151. Sturdy, *Science and Social Status*, 77–79.
152. Salomon-Bayet, *Institution de la science*, 51; Sturdy, *Science and Social Status*; Justel to Oldenburg, 13 October 1666, Oldenburg, *Corres.*, 3:240. On the Chapelain comment, see Roger, *Life Sciences*, 593n40. Roger noted that Jean-Baptiste Du Hamel was also classified as an anatomist, "but this title was purely fictive."
153. Taton, "Mariotte et l'Académie royale des sciences," 14. He argues for a later date of entry, perhaps June 1668.
154. Roger, *Life Sciences*, 139; and Hahn, *Anatomy*, 14.

CHAPTER THREE

1. This division lasted until 1684, after which the two sections met together. The meeting days of the two sections had reversed by 1675: Tits-Dieuaide, "Institution sans statuts," 4.
2. Meynell, "Académie des sciences"; and Balayé, *Bibliothèque nationale*, 71–82. Balayé's account differs in some details from Meynell's; in particular, she states that the library and academy were at numbers 6 and 8, not 8 and 10 (80–82). On the earliest meetings, see Tits-Dieuaide, "Institution sans statuts," 3.
3. Brice, *Description nouvelle*, 1:54; and Brice, *New Description of Paris*, 1:29.
4. "Pourquoy et comment l'Observatoire a esté basti," août 1667, in Colbert, *Lettres*, 5:515; cf. Balayé, *Bibliothèque nationale*, 75.
5. E. S. Saunders, "Politics and Scholarship"; Petitfils, *Fouquet*, 169; and Balayé, *Bibliothèque nationale*, 77–78.
6. Balayé, *Bibliothèque nationale*, 82–83; see also Grafton, "Libraries and Lecture Halls." However, he concludes that "humanist schools and massive libraries were not the central stages on which the dramas of the new natural philosophy were acted" (250).

7. *HARS*, 1:16. Fontenelle here is talking about 1699, but it is reasonable to assume the same room was used since 1666; see Meynell, "Académie des sciences," 29.

8. Meynell, "Académie des sciences," offers a detailed account of the buildings. Huygens described his rooms to his family in December 1666: Christiaan Huygens to Lodewijk Huygens, 3 December 1666, Huygens, *Oeuvres*, 10:727.

9. Tits-Dieuaide, "Institution sans statuts," 4–6.

10. *CdB*, 1:299–300; and René Taton, "Cassini, Gian Domenico," in *DSB*. For an example of collective authorship elsewhere, see Bertoloni Meli, "Authorship and Teamwork."

11. Stroup, *Company of Scientists*, 3–10, 13–14. It should be noted, however, that a "*compagnie*" also had a less formal meaning of simply a gathering: *Dictionnaire de l'Académie française*, 1694, *Dictionnaires d'autrefois*, ARTFL.

12. Demeulenaere-Douyère, "Esquisse d'une histoire des archives de l'Académie des sciences," 45–46; and Tits-Dieuaide, "Institution sans statuts." For a different interpretation of the academy's policies of secrecy and collaboration, see Biagioli, "Etiquette, Interdependence, and Sociability," 216–25.

13. Duclos had outlined a program in chemistry at the end of December 1666; but both Du Hamel and Fontenelle place Perrault's address first in their accounts: PV 1 (1667), fols. 9–10.

14. Picon, *Claude Perrault*, 14.

15. Picon, *Claude Perrault*, 13–19; and Rabinovitch, "Versailles as a Family Enterprise." On Claude's architectural career, see also Herrmann, *Theory of Claude Perrault*.

16. *HARS*, 1:18: "les deux parties les plus utiles & les plus curieuses de la Philosophie naturelle." Accounts of Perrault's address to the academy are in Du Hamel, *Historia*; *HARS*, vol. 1; PV 1 (1667); and PdS (1667). None of these accounts are exactly alike. The version in the Pochette de séance, which consists of thirteen closely written sides, is much more of a manifesto.

17. Demeulenaere-Douyère, "Sources documentaires," 61–67.

18. *HARS*, vol. 1: "Avertissement," n.p.

19. Du Hamel, *Historia*; and *HARS*, vol. 1. On Fontenelle's role in shaping the academy's persona, see esp. Mazauric, *Fontenelle*. See also Gaukroger, "Académie des Sciences and the Republic of Letters"; and Shank, *Newton Wars*.

20. PV 1 (1667), fol. 22, "Proiet pour les Experiences et observations Anatomiques."

21. PdS (1667).

22. PdS (1667); this list appears in a different form in *HARS*, 1:18–19. PV 1 (1667), fol. 23.

23. *HARS*, 1:18.

24. PV 1 (1667), fol. 24.

25. PV 1 (1667), fols. 25–26.

26. *HARS*, 1:18–19. On Perrault's epistemology, see Wright, "Embodied Soul"; Azouvi, "Entre Descartes et Leibniz"; and below, chap. 5.

27. Salomon-Bayet, *Institution de la science*, 68.

28. PV 1 (1667), fols. 29–30. On Colbert and his cats, see Paradis de Moncrif, *Chats*.

29. PdS (1667); PV 1 (1667), fol. 202; and Du Hamel, *Historia*, 19–21.

30. This number includes, besides those already named in chap. 2, five "élèves" and a secretary, Jean-Baptiste Du Hamel. A second secretary, Abbé Jean Gallois, was added in 1668: Sturdy, *Science and Social Status*, 77–79.

31. *HARS*, 1:7. The 1671 image, which appeared as a frontispiece to all of the 1670s volumes on animals and plants, has been analyzed by Stroup (*Company of Scientists*,

6–8) and also by Watson ("Early Days")." As Stroup points out, it is physically impossible to see the *observatoire* from either the Bibliothèque du roi or the Jardin du roi. The cartoon by Henri Testelin, based on a painting by Charles Le Brun, has been dated 1667, but its depiction of the observatory and of Cassini dates it to the early 1670s. Cassini moved to Paris in 1669, and the observatory was completed in 1672.

32. Peumery, *Origines de la transfusion sanguine*, 5. "An account of the rise and attempts of a way to conveigh [sic] liquors immediately into the mass of blood," *Phil. Trans.* 1 (1665–66): 128–30, states this occurred "at least six years since." Timothy Clark, writing in 1668, claimed this occurred in 1656: Clark, "A letter, written to the Publisher by the learned and experienced Dr Timothy Clarck [sic] one of his Majesties Physitians in Ordinary, Concerning some anatomical inventions and observations, particularly the origin of the injection into veins, the transfusion of blood, and the parts of generation," *Phil. Trans.* 3 (1668): 172 [sic]–682, at 678.

33. Denis, *Lettre écrite à M. Sorbière*; and Rodis-Lewis, "Écrit de Desgabets."

34. Lower, *Tractatus de corde*, 174. Lower states that this event took place in February 1665; however, the new year in Britain began March 25, and since the *Philosophical Transactions* article cited below states that the experiment took place a few months earlier, I assume the date is February 1666.

35. "The Success of the Experiment of Transfusing the Bloud of One Animal into Another," *Phil. Trans.* 1, no. 19 (1666): 352 (issue dated 19 November 1666); "The Method Observed in Transfusing the Bloud out of One Animal into Another," *Phil. Trans.* 1, no. 20 (1666): 353–58 (issue dated 17 December 1666), translated in "Extrait dv iournal d'Angleterre, Contenant le maniere de faire passer le sang d'vn animal dans vn autre," *JdS* (31 janvier 1667): 31–36.

36. There is a large secondary literature on the transfusions of the 1660s; most relevant to the discussion to follow is Delaporte, "Animal Blood"; Peumery, *Origines de la transfusion sanguine*; Schiller, "Transfusion sanguine"; Brown, "Jean Denis"; Farr, "First Human Blood Transfusion"; Guerrini, "Ethics of Animal Experimentation"; and Hoff and Guillemin, "First Experiments," which translated into English the account in the PV. Recent popular accounts include Tucker, *Blood Work*; and Moore, *Blood and Justice*. Schiller gives by far the best account of the academy's work; Delaporte gives a judicious survey of the whole.

37. *HARS*, 1:37.

38. PV 1 (1667), fol. 202. Auzout had assisted Pecquet in the 1640s. Schiller, "Transfusion sanguine," 38, compares the PV and Perrault's account.

39. Claude Perrault, *Essais*, 4:409–10.

40. Ibid., 4:406, 4:417, 4:421–22.

41. Brown, "Jean Denis"; and Peumery, *Origines de la transfusion sanguine*. Denis at some point was named one of the sixty-six royal physicians-in-ordinary; it is not clear whether he was at this time.

42. Farr, "First Human Blood Transfusion"; and Brown, "Jean Denis."

43. On Descartes, see Desmond Clarke, "Descartes's Philosophy of Science and the Scientific Revolution," 272. Thomas Hobbes expressed a similar view; see Shapin and Schaffer, *Leviathan and the Air-Pump*.

44. *HARS*, 1:2: "Le regne des mots & des termes est passé; on veut des choses." The word "*terme*" was most often defined as "end" or "goal" in this era (*Dictionnaire de l'Académie francaise*, 1694, *Dictionnaires d'autrefois*, ARTFL). Fontenelle may refer here to teleological philosophies or to the terms of a syllogism.

45. *HdA* (1671), *préface*, n.p.

46. Schiller, "Transfusion sanguine," 36–37; and Claude Perrault, *Essais*, 4:426–27.
47. Claude Perrault, *Essais*, 4:426, 4:417–18. The question of the fetal circulation would again arise at the academy in the 1690s: see chap. 6.
48. Harvey, *De generatione animalium*, 162; and Harvey, *Anatomical Exercitations*, 296.
49. Kuriyama, "Interpreting the History of Bloodletting"; and Brain, *Galen on Bloodletting*.
50. Tardy, *Traitté de la monarchie du cœur*; Tardy, *Cours de médecine*; and Tardy, *Traitté de l'ecoulement du sang*.
51. Claude Perrault, "Rélation du voyage fait en 1669," in Charles Perrault and Claude Perrault, *Mémoires de ma vie*, 139–218. Jean Perrault became ill on 1 October 1669 and died on the 30th; he was also subjected to daily enemas and numerous purgatives and emetics.
52. La Martinière, *A Monseigneur Colbert*, 1.
53. PV 5, fols. 181r–181v (23 novembre 1669). Pecquet emphasized the therapeutic possibilities of transfusion, "la chirurgie transfusoire," which had been successful in Germany.
54. On the notion of credit see Biagioli, *Galileo's Instruments of Credit*.
55. Kronick, "*Devant le deluge*," 96.
56. Tits-Dieuaide, "Institution sans statuts," 8; PV 1 (1667): fol. 200v; and PV 2 (1667–68), fols. 159–60.
57. Claude Perrault, *Essais*, 4:405–6; and Schiller, "Transfusion sanguine," 34.
58. [Henry Oldenburg], "An Account of More Tryals of *Transfusion*, accompanied with some *Considerations* thereon, chiefly in reference to its circumspect Practice on *Man*, together with a farther *Vindication* of this Invention from Usurpers," *Phil. Trans.* 2, no. 28 (21 October 1667): 517–25. Oldenburg described the experiment of weighing the dogs that Gayant had performed for the academy without mentioning the academy.
59. Kronick, "*Devant le deluge*," 99; and Oldenburg, *Corres.*, 5:xxv.
60. On Renaudot, see Wellman, *Making Science Social*, 52–55; on the *Journal des sçavans*, see Camusat, *Histoire critique des journaux*, 1; Birn, "Journal des savants"; and Brown, "History and the Learned Journal." Sturdy, *Science and Social Status*, 88, plausibly claims that "contemporaries believed that the *Journal* could be to Colbert what Renaudot's *Gazette* had been to Richelieu," but gives no reference.
61. "Imprimeur au lecteur," *JdS* (5 January 1665), n.p.
62. On Chapelain, see Sturdy, *Science and Social Status*, 64–65, 70–72; and Pintard, *Libertinage érudit*, 78, 91, and passim. Chapelain had been a patron of de Graaf during his 1666 stay in Paris (see above, chap. 2); on Gallois, see Sturdy, *Science and Social Status*, 87–89; he had served as *précepteur* in Colbert's household.
63. *JdS* (5 janvier 1665): 10–11; *JdS* (23 mars 1665): 139–43; Brown, *Scientific Organizations*, 193–98; and Sturdy, *Science and Social Status*, 88.
64. *JdS* (30 mars 1665): 156; *JdS* (4 janvier 1666), "Au roy," n.p.; "Imprimeur au lecteur," *JdS* (4 janvier 1666), n.p.; Birn, "Journal des savants," 20–21; and Camusat, *Histoire critique des journaux*, 1:79–81.
65. *JdS* (20 décembre 1666): 491–501; PV 4 (1668), fol. 1r; Birn, "Journal des savants," 22; and Sturdy, *Science and Social Status*, 88–89. As Sturdy points out, the minutes under Gallois are much better organized than under Du Hamel.
66. PV 2, fols. 159r, 159v (23 mars 1667).
67. *HARS*, 1:37; and *JdS* (4 avril 1667): 84.
68. *JdS* (4 avril 1667): 81–84.
69. "Communicatio canalis thoracici cum emulgente, inventa a D. Gayant, praesente

Dn. Pecqueto & Dn. Perraulto," 5–6; "Chirugia infusoria variis hominibus in diversis morbis cum successu adhibit," 7–8; and "Clysmatica nova," 9–11—all in *Celeberrimorum anatomicorum . . . varia opuscula anatomica.*

70. *JdS* (14 mars 1667): 69–72; and Denis, *Recueil des mémoires*. He billed them as continuations to the *JdS*, but the *JdS* was still printing in 1672. Kronick, "*Devant le déluge*," 4.

71. On Cusson, see Renouard, *Répertoire*, 109.

72. On Duverney, see Fontenelle, *Éloges*, and chap. 5, below.

73. Justel to Oldenburg, 5 octobre 1667, Oldenburg, *Corres.*, 3:484. On the academy's ostriches, see chap. 5.

74. Fontenelle, *Éloges*; and Roger, *Life Sciences*, 140.

75. Denis, *Recueil des mémoires*. This includes the 1674 compilation of the *conférences*.

76. Ibid., 39–43.

77. Ibid., 328. Duverney regularly dissected for Denis even after he entered the academy in 1674.

78. PV 1 (1667–68); PV 4 (1668); on Huygens's air pump, see PV 4, fols. 6v–7r (14 avril 1668).

79. Claude Perrault, dossier, AdS. The pages in the dossier are not numbered; this description consists of two manuscript pages in Perrault's hand, headed "Observations anatomiques faites le premier 2 3 4 et 5e jour de février 1667." This dissection is also mentioned in PV 1, 52–57, with considerably less detail.

80. Christiaan Huygens to Lodewijk Huygens, 4 February 1667, Huygens, *Oeuvres*, 6:104. Huygens used the term "*céans*," which, according to the *Dictionnaire de l'Académie française* of 1694, "ne se dit que des maisons" (*Dictionnaires d'autrefois*, ARTFL).

81. Perrault, dossier, AdS.

82. *CdB*, vol. 1, cols. 299–300; for a previous year, see cols. 161–62.

83. PV 1 (1667), fol. 204.

84. PV 4, fol. 7v (14 avril 1668).

85. For Duverney's purchases, see chap. 6, below.

86. Cordey, "Colbert, Le Vau"; and Iriye, "Le Vau's Menagerie," 23–30. *Dictionnaire de l'Académie française* (1694), s.v. "mesnagerie," *Dictionnaires d'autrefois*, ARTFL.

87. Colbert to Mazarin, 31 May 1658, in Colbert, *Lettres*, 1:294–95.

88. Brice, *Description nouvelle*, 209–10; and Brice, *New Description of Paris*, 131.

89. Cordey, "Colbert, LeVau," 283–84; and Hamy, "Royal Menagerie of France," 511.

90. George, "Sources and Background"; and Baratay and Hardouin-Fugier, *Zoo*, 17–28.

91. *Mercure galant*, août 1682, 184–87; cf. Loisel, *Ménageries*, 2:99.

92. Mabille, "Ménagerie de Versailles."

93. Scudéry, *Promenade de Versailles*, 93–94, 96–97; La Fontaine, preface to *Les amours de psyche et de cupidon* (1669), in *Œuvres diverses*, 168–69; and Iriye, "Le Vau's Menagerie," 39–41. Claire Goldstein compares the descriptions of Vaux-le-Vicomte and Versailles of La Fontaine and Scudéry in *Vaux and Versailles*, 132–53.

94. Loisel, *Ménageries*, 2:103 and chap. 7; Baratay and Hardouin-Fugier, *Zoo*, 35–36, 39; and Gaillard, "Bestiaire réel, bestiaire enchanté," 186.

95. On food animals, see Flandrin, "Dietary Choices and Culinary Technique, 1500–1800," 405. He argues that birds such as herons and peacocks had disappeared from the table by 1650, but they lingered for at least few years more: La Varenne, *Cuisinier François*, n.p.

96. Lacroix, "Approvisionnement des ménageries." On the gift economy and early modern science, see Findlen, "Economy of Scientific Exchange," and Eamon, "Court,

Academy, and Printing House." Both these studies refer to sixteenth-century Italy, whose circumstances do not directly map onto seventeenth-century France; see also Mousnier, *Age d'or du mécénat*.

97. Loisel, *Ménageries*, 2:112–15; "Observations astronomiques et physiques faites en l'isle de Caïenne," in *MARS*, vol. 7, pt. 1, 325–26.

98. Loisel, *Ménageries*, 2:112–15; and Colbert to Chevalier d'Hailly, 3 juin 1671, Colbert, *Lettres*, 5:311.

99. PV 1 (1667), fol. 204. Claude Perrault mentioned "le grand hyver de l'année 1670," in *Essais*, 3:30, noting that a *chat-pard* (serval) froze to death.

100. "Liste des animaux décrits dans l'Académie, & dont les Descriptions publiées se trouvent dans les Mémoires. 1666–1698," in Académie royale des sciences de Paris, *Table alphabétique*, 1, 17–20.

101. Loisel, *Ménageries*, 2:98–99; see below, chap. 6.

102. PV 4, fols. 227r–233v (20 septembre 1668); and PV 5, fols. 176r–179v (13 octobre 1669).

103. PV 5, fols. 58r–59r (9 mars 1669). Peristaltic motion was a central concept for Perrault, who believed it was innate to the organism; see below, chap. 5.

104. On the geography of Paris, see above, chap. 1; Fierro and Sarazin, *Paris des Lumières*; Ranum, *Paris*; and Hazan, *Invention of Paris*.

105. Stroup, *Company of Scientists*, 211; Charles Perrault, *Mémoires de ma vie*, 46–47; and Sturdy, *Science and Social Status*, 133–37. On "invisible technicians," see Shapin, *Social History of Truth*, chap. 8 (355–407).

106. Tits-Dieuaide, "Institution sans statuts," 5.

107. *CdB*, vol. 1, cols. 448, 460, 648; and PV 4, fol. 45r (2 juin 1668). The badger was dissected the following week: PV 4, fols. 45r–48r (9 juin 1668). Loisel, *Ménageries*, 2:114n; for example, in 1688, "Monsieur Couplet a fait son rapport dun Chameau qu'il a veû à Versailles, qui est fort galeux": PV 12, fol. 98v (18 août 1688).

108. The historiography on Cureau de la Chambre's and, particularly, Perrault's scientific work is thin. On Cureau de la Chambre, see Darmon, *Corps immatériels*; Legée, "Bio-bibliographie de Marin Cureau de la Chambre"; Diamond, "Marin Cureau de la Chambre"; and Wright, "Embodied Soul," which treats both Cureau de la Chambre and Perrault. On Perrault, see Picon, *Claude Perrault*, chaps. 4–5; Azouvi, "Entre Descartes et Leibniz"; Tenenti, "Claude Perrault"; Baratay, "Naturalistes dans les cages"; and Des Chene, "Mechanisms of Life." Herrmann, *Theory of Claude Perrault* is, like most work on Perrault, about architecture. See chap. 5, below.

109. Bayle, *Nouvelles de la république des lettres*, mars 1684, 1:19–20.

110. The publication history of this work is extraordinarily tangled because reprints of the earlier volumes, sometimes with different titles, overlapped with the issue of later volumes; however, this also indicates how popular this work was.

111. On the six nonnaturals, see Rather, "'Six Things Non-natural.'" Cureau de la Chambre, *Charactères des passions*, vol. 1: "Advis necessaire au Lecteur," n.p.

112. Cureau de la Chambre, *Système de l'âme*, 217, quoted and translated in Wright, "Embodied Soul," 25.

113. Cureau de la Chambre, "Quelles est la Connoissance des Bestes, et iusques où elle peut aller," in *Charactères des passions*, 2:2 (separately paginated). *Dictionnaire de l'Académie française*, 1694, *Dictionnaires d'autrefois*, ARTFL.

114. Cureau de la Chambre, "Connoissance des Bestes"; and Thomas, *Man and the Natural World*, 30–41.

115. Cureau de la Chambre, "Connoissance des Bestes," 19; and Harrison, "Virtues of Animals," 472–76. Harrison here follows Boas, *Happy Beast*.

116. Chanet, *Considerations de la sagesse de Charon [sic]*, 57. The identity of Pierre Chanet is obscure; it would be nice, but it is unlikely, that he was the same person as Pierre Chanut (1601–62), friend of Descartes and Montmor and French ambassador to Sweden in the late 1640s.

117. Chanet, *De l'instinct et de la connoissance des animaux*, 2, 4, 120, 161.

118. Rorario, *Quod animalia bruta ratione*; and Des Chene, "*Animal* as Category: Bayle's 'Rorarius.'"

119. Guillaume Lamy, *Discours anatomiques*, "Seconde lettre," n.p.

120. Cureau de la Chambre, *Traité de la connoissance des animaux*, dedication (n.p.), 3–4. This work was translated into English in 1657: *Discourse of the Knowledg [sic] of Beasts*. See also Boas, *Happy Beast*, 74–82; and Harrison, "Virtues of Animals," 475–79.

121. Cureau de la Chambre, *Traité de la connoissance des animaux*, 324, 324–76.

122. Descartes, *Passions de l'âme*, in AT, 11:387, 11:397–98; Sorell, *Descartes Reinvented*, 89–92; and Harrison, "Virtues of Animals."

123. Garber, "Descartes and Occasionalism."

124. On Cordemoy, see Ablondi, "Géraud de Cordemoy"; and Rosenfield, *From Beast-Machine to Man-Machine*, 38–41.

125. Malebranche, *Search after Truth*, 6.2.7; see also the discussion of Malebranche in Radner and Radner, *Animal Consciousness*, 70–91.

126. See particularly Rosenfield, *From Beast-Machine to Man-Machine*.

127. Sedgwick, *Jansenism*; Mandrou, *From Humanism to Science*, 230–32, 48; Schmaltz, "What Has Cartesianism"; McClaughlin, "Censorship and Defenders"; and Ariew, "Damned If You Do."

128. Mme. de Sévigné to Mme. Grignan, 16 septembre 1676, Sévigné, *Correspondance*, 2:398.

129. Schmaltz, "What Has Cartesianism"; McClaughlin, "Censorship and Defenders"; and Ariew, "Damned If You Do."

130. Roger, *Life Sciences*, 184–85: "The science of the Academy was a Christian science."

131. Ariew, "Damned If You Do"; and McClaughlin, "Censorship and Defenders."

132. On this point cf. Roger, *Life Sciences*, 167: "Everyone of any importance was a mechanist." He briefly considers the *Essais de physique* on p. 177. On the varieties of mechanism in the physical sciences, see Bertoloni Meli, *Thinking with Objects*; and Westfall, *Force in Newton's Physics*.

133. On these points see esp. Bertoloni Meli, *Mechanism, Experiment, Disease*, 12–16; and Guerrini, "Varieties of Mechanical Medicine."

134. Claude Perrault, *Essais*, 3:1, "Avertissement."

135. Claude Perrault, *Essais*, 3:3–4.

136. Ibid., 2:217–18, 2:260. See chap. 5, below.

137. Claude Perrault, *Essais*, 3:23, 3:57, 3:59, 3:68. On voice, see *Essais*, 2:330–31.

138. Ibid., 2:308, 2:330–31; and *HdA* (1676), 126.

139. Claude Perrault, *Essais*, 2:264, 2:275; on memory traces, see John Sutton, *Philosophy and Memory Traces*.

140. Claude Perrault, *Essais*, 2:276–78.

141. Ibid., 2:283; and Azouvi, "Entre Descartes et Leibniz," 11.

142. Pardies, *Discours de la connoissance des bestes*, 15; and Rosenfield, *From Beast-Machine to Man-Machine*, 80–86. Gallois favorably reviewed this work in *JdS* (25 July 1672): 125–29.

143. Pardies, *Discours de la connoissance des bestes*, 148.
144. Du Hamel, *De corpore animato*, pt. 3, chap. 1 (301–26). See Rosenfield, *From Beast-Machine to Man-Machine*, 85, 221.
145. Wright, "Embodied Soul," 34, quotes Leibniz (via Wolfgang Hermann), who noted around 1676 "the opinion of Mr Perrault . . . that the soul is spread throughout the body."
146. Derrida, *Animal That Therefore I Am*, 9.

CHAPTER FOUR

1. *Extrait d'une lettre*, 1–10. Gayant and Pecquet are named as the dissectors in *HdA* (1676), "Avertissement," n.p.
2. Loisel, *Ménageries*, 2:96–97; and *Extrait d'une lettre*, 10.
3. *Extrait d'une lettre*, 13–27.
4. Malpighi and Fracassati, *Tetras anatomicarum epistolarum*; Malpighi, "Account of Some Discoveries"; Bellini, *Gustus organum*; and *Extrait d'une lettre*, 4, 14.
5. Although Duverney's dissection of a badger was included in a folder of drawings compiled in 1695 for the revision of the *Mémoires pour servir à l'histoire naturelle des animaux*, it was not included in the final volume: MS 220, folder 3, fols. 55–60, MNHN.
6. Lokhorst, "Descartes and the Pineal Gland."
7. Stensen, *Discours de M. Stenon*; see chap. 2, above.
8. On Johnstone, see *ODNB*; J. K. Crellin, "Jonston, John," in *DSB*; and esp. Gordon L. Miller, "Beasts of the New Jerusalem." His surname is variously spelled as "Jonston," "Johnston," "Johnstone," "Jonstonus" and his first name as "Jon," "John," "Jan," "Johannes." *ODNB* uses "John Johnstone." The natural history volumes, originally published by the Merian brothers in Frankfurt am Main, were reprinted in Amsterdam in 1657.
9. Johnstone, *Quadrupetibus*, 1.
10. On commonplace books, see Blair, *Too Much to Know*; Johnstone, *Quadrupetibus*, 31–34; and Johnstone, *Four-Footed Beasts*, 19–21.
11. The Amsterdam edition of 1657 was reprinted in 1665–67, and the first volume on quadrupeds was translated into Dutch (1660) and English (1678). A French translation of some of the work on birds appeared in 1772–73. In 1718 a collected edition, edited by Frederik Ruysch, appeared: *Theatrum universale omnium animalium*. Further reprints of this collected edition appeared throughout the eighteenth century.
12. Di Renzi, "Writing and Talking of Exotic Animals"; Hernández, *Thesaurus*, 475; and Freedberg, *Eye of the Lynx*, 277.
13. Freedberg, *Eye of the Lynx*, chap. 9 (245–74). A few of these copies were dated 1649.
14. Hernández, *Thesaurus*, 538–81.
15. *Extrait d'une lettre*.
16. Huygens, *Relation d'une observation faite a la Bibliotheque du roy*.
17. I have looked at five different copies of the pamphlet, and the engravings are in different places in each one.
18. "Observations faites sur une grande poisson dissequé à la Bibliothèque du Roy, le 24 juin 1667," *JdS* (28 novembre 1667): 157–59; "Extraits d'une lettre écrite a Monsieur de la Chambre, qui contient les Observations qui ont été faites sur un Lion dissequé à la Bibliothèque du Roy le 28. Juin 1667," *JdS* (5 décembre 1667): 171–74; and "Observations faites sur un GRAND POISON [*sic*] & un LION," *Phil. Trans.* 2 (1666–67): 535–37.

19. It is not, however, clear when the *pochettes* were compiled. See Demeulenaere-Douyère, "Sources documentaires," 65–68.

20. "Extrait d'une lettre de M. Petit Intend. Des Fortific. &c. au R. P. de Billy de la Comp. des Jesus, Touchant une nouvelle machine pour mesurer exactement les Diamètres des astres," *JdS* (16 mai 1667): 102–8; "Observations faites sur une grande poisson disséqué à la Bibliothèque du Roy, le 24 juin 1667," *JdS* (28 novembre 1667): 157–59; and "Extraits d'une lettre écrite à Monsieur de la Chambre, qui contient les Observations qui ont été faites sur un Lion disséqué à la Bibliothèque du Roy le 28. Juin 1667," *JdS* (5 décembre 1667): 171–74.

21. On Denis, see *JdS* (14 mars 1667): 69–72; (25 avril 1667): 96; and (28 juin 1667): 133–34. On Tardy, see *JdS* (16 mai 1667): 117–18. "Diverses pièces touchant la Transfusion du Sang," *JdS* (6 février 1668): 13–24.

22. "*La presence de Jesus dans le Saint Sacrement*" (review), *JdS* (28 juin 1667): 121–33; and "Lettre de M. Denis Professeur de Philosophie et de Mathématique, à M. de Montmor premier Maistre des Requestes: touchant deux Expériences de la Transfusion faites sur des hommes," ibid., 134–36.

23. *JdS* (30 juillet 1668): 66–68 (worms), 69–72.

24. Birn, "Journal des savants," 23–24.

25. Denis, *Recueil des mémoires*. See Kronick, *History of Scientific and Technical Periodicals*, 86–88.

26. PV 8, fol. 1r (9 janvier 1675). Gallois must have relinquished his duties as editor at about this time.

27. Shapin, "Pump and Circumstance," 483.

28. Magalotti, *Saggi di naturali esperienze*. A copy was presented to the Royal Society (see "An Account of Two Books," *Phil. Trans.* 3 [1668]: 640), but I have found no evidence that the academy received one, and I have not found a notice in the *JdS*.

29. On the making of the image of Louis, see Burke, *Fabrication of Louis XIV*; Apostolidès, *Roi-machine*; and Sabatier, "Gloire du roi." On the particular role of the menageries, see Senior, "Ménagerie and the Labyrinth"; Sahlins, "Royal Menageries of Louis XIV"; and below. On the gift economy, see Findlen, "Economy of Scientific Exchange."

30. *Description anatomique*.

31. *Lettres écrites sur le sujet d'une nouvelle découverte touchant la veü*. This pamphlet was reprinted with the 1682 reprint of the 1669 *Description anatomique* and also appears bound together with the 1667 and 1669 works; the National Library of Medicine and the British Library have such copies. An expanded version, including a letter by Claude Perrault, appeared in *Recueil de plusieurs traitez*, a collection of treatises. The letters on vision are discussed in chap. 5.

32. This impatience was clear at the meetings, when members elected to perform dissections first so that they were not superseded by the interminable readings on chemistry of Duclos. See PV, vol. 4.

33. *Description anatomique*, 3.

34. Aristotle, *History of Animals*, 503b; Ashworth, "Marcus Gheeraerts"; and Hernández, *Thesaurus*, 721–43.

35. Scudéry, "Histoire de deux caméléons," 534–38. The chameleons of the academy and Mlle. de Scudéry have been exhaustively examined in several recent works: Geoffrey V. Sutton, *Science for a Polite Society*, 123–26 (who incorrectly asserts that the academy's chameleon was examined and later dissected at Mlle. de Scudéry's salon); Harth, *Cartesian Women*, 98–106; Grande, "Vedette des salons"; Rabinovitch, "Chameleons between Science and Literature"; and Sahlins, "Story of Three Chameleons."

36. PV 4, fol. 227r.

37. PV 4, fols. 227r–233v; and *Description anatomique*, 3–48.

38. PV 4, fols. 231r–233v.

39. PV 4, fols. 260r–294v (10 novembre 1668).

40. PV 4, fols. 236r–238r (13 octobre 1668). In fact, they did not dissect additional lizards until 1683: "De quatre lezards, Duverney-Perrault 28 août 1683," Duverney dossier, AdS.

41. *Description anatomique*, 3–4.

42. Ibid., 3–21.

43. Ibid., 22–23. This material does not appear in the PV.

44. Ibid., 25–26, 28–30. On this issue before the seventeenth century, see Lindberg, *Theories of Vision*, 36–37, 190–93; in the seventeenth century, see Wade, *Natural History of Vision*, 100; and Descartes, *Traité de l'homme*, in AT, 11:175–79.

45. *Description anatomique*, 40–41. Perrault used the term "*aduste*" to refer to the dark and bitter humor of the blood, which "only refers to humors of the human body": *Dictionnaire de l'Académie française*, 1694, s.v. "*aduste*" (*Dictionnaires d'autrefois*, ARTFL).

46. *Description anatomique*, 50. On the iconography of the chameleon, see Ashworth, "Marcus Gheeraerts"; and Guerrini, "King's Animals and the King's Books."

47. Gessner, *Historia animalium*, 1:336–44; Topsell, *Foure-Footed Beastes*, 44–50; Johnstone, *Four-Footed Beasts*, 79–81. The European beaver had gone extinct in Britain in the sixteenth century, but was still extant in France; see Martin, *Castorologia*, 29. On the iconography of the beaver, see Acheson, "Gesner, Topsell, and the Purposes of Pictures."

48. *HARS*, 1:52. The skeleton was mounted by the botanist Nicolas Marchant. Fontenelle claimed that the beaver was the first animal to be dissected. The account of the beaver in PV 4, fols. 175v–196v, credited to Gallois, appears in *Description anatomique*, 51–72.

49. *Description anatomique*, 51. The European and North American beavers are distinct species. The beaver apparently died of natural causes: ibid., 67.

50. *Description anatomique*, 51–52. On the French beaver trade, see Le Blant, "Commerce compliqué des fourrures Canadiennes"; and Innis, *Fur Trade in Canada*.

51. Hernández, *Thesaurus*, 473; Harvey, *De motu cordis*, 28, 64; and Severino, *Zootomia democritaea*, 400. He wrote the term in Greek.

52. See Bertoloni Meli, *Mechanism, Experiment, Disease*, 40–41.

53. *Description anatomique*, 52. On the uses of beaver fur, see Grant, "Revenge of the Paris Hat."

54. *Description anatomique*, 54, 58–62.

55. Polecat ("putois"), PV 4, fols. 61r–71r; civette, Du Hamel, *Historia*, 69; and *HARS*, 1:82–83.

56. Topsell, *Foure-Footed Beastes*, 35–43; and Johnstone, *Four-Footed Beasts*, 68–70 and plate LV. The symbolism of the bear is examined in Pastoureau, *Bear*; performing bears are mentioned in Fournel, *Tableau du vieux Paris*, 147, 149, 155, 158, 208. See also Delort, *Animaux ont une histoire*, 137–40. Brunner, *Bears*, 190, shows the "Grand Carrousel." "Les plaisirs de l'isle enchantée" is noted by Gaillard, "Bestiaire réel, bestiaire enchanté," 186.

57. *Description anatomique*, 85–104, at 88–89. Sørensen, "Animaux du roi," 182, gives the date of dissection as 27 February 1668 but does not provide a reference.

58. Pastoureau, *Bear*.

59. Stroup, *Company of Scientists*, 29, 293n, citing the notebooks of the chemist Claude Bourdelin, Bibliothèque nationale, MS n.a.r. 5147, fol. 21v.

60. *Description anatomique*, 94; Malpighi at about the same time thought they resembled a bunch of cherries: Bertoloni Meli, *Mechanism, Experiment, Disease*, 121.
61. *Description anatomique*, 94–95; Highmore, *Corporis humani*; and Bellini, *De structura et usu renum*.
62. *Description anatomique*, 95; and Highmore, *Corporis humani*, 75.
63. *Description anatomique*, 71–84; and Johnstone, *Four-Footed Beasts*, 54–56 and image XLI–XLIV; the horse-faced camel is on the latter page on the bottom. See Margócsy, "Camel's Head."
64. *Description anatomique*, 103–21.
65. "Description anatomique," *Phil. Trans.* 4 (1669): 991–96; "Description anatomique," *JdS* (16 décembre 1669): 37–42.
66. "Description anatomique," *JdS* (16 décembre 1669): 41–42.
67. *Description anatomique*, "L'imprimevr av lectevr," n.p.
68. I have based this count on an examination of the PV, the PdS, and the histories of Fontenelle and Du Hamel.
69. Bernard, *Histoire de l'Imprimerie royale*, "Catalogue chronologique," 140–43.
70. Bernard, *Histoire de l'Imprimerie royale*; and Balayé, *Bibliothèque nationale*, 80–81.
71. Grivel, "Cabinet du roi," 36; Grivel and Fumaroli, "Genèse d'un manuscrit," 108. See chap. 2, above, 88.
72. Quoted in Grivel, "Cabinet du roi," 38. On the status of the *graveurs du roi*, see Savill, "Triple Portrait."
73. The *mémoire* is "Mémoire que Monseigneur a dressé touchant la publication des ouvrages où Il y a des Planches gravées," n.p., AN, MS O/1/1964, Cotte 2, dated 22 février 1670.
74. "Mémoire" (1670). As we have seen, these were not precisely in quarto format, but they were smaller than a folio. The manuscript attributes these to Bailly. Of the others, Bosse did the chameleon and Leclerc the rest.
75. Ibid.
76. *HdA* (1671), "Preface," n.p.
77. Several of the copies I have seen of the two editions have similar binding. Grivel, "Cabinet du roi," 40–42, gives details of the paper and binding.
78. "Mémoire" (1670). On the techniques employed, see Préaud, *Leclerc*.
79. Two hundred copies were ordered of the 1676 edition: "Mémoire de toutes les Planches qui ont esté gravées pour le Roy, depuis l'année 1670 jusqu'au 15 may 1675," AN, MS O/1/1964/2, Cotte 2. Pitfeild, "Publisher to the Reader," n.p.
80. Some copies exist of *Suite des memoires pour servir à l'histoire naturelle des animaux*, which did not list Perrault on the title page but is paginated continuously with the 1671 edition; those who already had the 1671 edition could complete it with this one. Hans Sloane's copy in the British Library is an example of the two bound together. However, the description of the coati, which was incomplete in the 1671 edition, does not appear in the *Suite* but does appear, with two coatis, in the full 1676 edition.
81. On "paper museums," see Burke, "Images as Evidence"; and Carlino, *Paper Bodies*. On Cassiano's *museo cartaceo*, see Freedberg, *Eye of the Lynx*, esp. chap. 1.
82. The literature on *spectacles* is huge. Useful (although both focus more on the eighteenth century) are Campardon, *Spectacles de la foire*; and Isherwood, *Farce and Fantasy*; see also Fournel, *Tableau du vieux Paris*.
83. Aldrovandi, *De quadrupedibus*, 221–22. He also had a category of "usus in triumphis."
84. Gaillard, "Bestiaire réel, bestiaire enchanté."

85. Although the preface is not signed, it bears all the marks of Perrault's composition. It is likely that the academy as a whole read and approved it, but I have found no mention of this.

86. Burke, *Fabrication of Louis XIV*, 3, 28, 35; Sabatier, "Gloire du roi"; Posner, "Lebrun's *Triumphs of Alexander*"; and Keeble, "Sincerest Form of Flattery."

87. See Grivel and Fumaroli, "Genèse d'un manuscrit," 21; and La Fontaine, *Fables*, no. 152.

88. *Physiologus*, 3–4.

89. See chap. 2, above, and Pomata, "*Praxis Historialis.*"

90. Hahn, *Anatomy*, 15.

91. *HdA* (1671), preface, n.p., my translation. Pitfeild's, in this case, is imprecise: Pitfeild, preface, n.p.

92. Salomon-Bayet, *Institution de la science*, 72.

93. *HdA* (1671), preface; *Dictionnaire de l'Académie francaise* (1694), 110:1. "Naturel, sans fard, sans artifice," 2. "Il signifie aussi, Qui represente bien la verité, qui imite bien la nature." *Dictionnaires d'autrefois*, ARTFL.

94. Descartes, *Discours de la méthode*, in AT, 6:7–9; and Descartes, *Meditationes*, in AT, 7:349–50.

95. *HdA* (1671), preface; Goldberg, *Mirror and Man*, 147; and Melchior-Bonnet, *Mirror*, 136. On the physical characteristics of early modern mirrors, see Dupré, *Optics, Instruments, and Painting*.

96. Melchior-Bonnet, *Mirror*, 35–46.

97. Ibid., 27 (Fouquet); 46–48 (Versailles); and Goldberg, *Mirror and Man*, 163–68.

98. Dupré, *Optics, Instruments, and Painting*.

99. On the definition of comparative anatomy, see Cole, *History of Comparative Anatomy*.

100. *HdA* (1671), preface.

101. There are several different states of the two editions and several different versions of some of the engravings. For the engravings, see Préaud, *Leclerc*, 235–55. On the subsequent complex publication history of the *Histoire des animaux*, see chap. 6, below.

102. *HdA* (1676), 121–22.

103. Johnstone, *Four-Footed Beasts*, 91 and table LXVIII; and *HdA* (1676), 112–13, 131.

104. This argument is made in particular by Sahlins, "Royal Menageries of Louis XIV," following the work of Norbert Elias.

105. *HdA* (1671), preface. This complicates the arguments made in Daston and Galison, *Objectivity*, chap. 2. Daston made a similar argument in Daston and Park, *Wonders and the Order of Nature*, chap. 9. See below.

106. *HdA* (1671), preface; PV 5, fols. 180v–181r (23 novembre 1669); and *HdA* (1676), "Avertissement," n.p.

107. The 1694 *Dictionnaire de l'Académie française* defined "*espèce*" as "Terme de Logique qui est sous le genre, & qui contient sous soy plusieurs individus," using as one example "*l'espèce la plus parfaite des animaux c'est l'homme*": *Dictionnaires d'autrefois*, ARTFL. On the pre-Linnaean definition of species, see Atran, *Cognitive Foundations of Natural History*; Atran et al., *Histoire du concept d'espèce*, 1–36; and Aquinas, *Summa theologica*, pt. III, question 74, article 3, objection 2: "Figura est signum speciei in rebus naturalibus." See also Marcialis, "Species."

108. *HdA* (1676), 112–13; and *HdA* (1671), 83–84.

109. *HdA* (1671), 49; and Aristotle, *History of Animals*, 606b, 15–25.

110. *HdA* (1671), 49–52.

111. Pitfeild, 115; and *HdA* (1671), 89.

112. *HdA* (1671), 89.
113. The 1676 edition included all of the descriptions from 1671 plus several new ones. Some of the additional descriptions will be discussed in chap. 5.
114. [Claude Perrault], *Dix livres d'architecture de Vitruve*, 12–13. On the trompe l'oeil technique, see Kaufmann, *Mastery of Nature*, chap. 1 (11–48).
115. Ogilvie, *Science of Describing*, 175; and Siraisi, "Vesalius and Human Diversity," 60n3.
116. Pitfeild. The engravings were copied by Richard Waller. This edition was a folio, but not an elephant folio.
117. An inventory of Duverney's cabinet is contained in the PdS (1756), Archives, AdS. On multiple specimens, see Findlen, *Possessing Nature*, 208–20; and Guerrini, "Duverney's Skeletons."
118. West, *Theatres and Encyclopedias*, 102–7; on the originality of Gessner's illustrations, see Kusukawa, "Sources of Gessner's Pictures."
119. AN, MS O/1/1964/7, fols. 25–27, lists a series of seventy-five animal designs for tapestries by "Boels" that includes most of the animals depicted in *HdA*. Sørensen, "Animaux du roi"; Foucart-Walter, *Pieter Boel*; on La Hire, see Sturdy, *Science and Social Status*, 195–205. La Hire was a protégé of Abraham Bosse, who, like him, was interested in mathematics.
120. *HdA* (1671), preface.
121. [Charles and Claude Perrault], *Cabinet des beaux-arts*, 1–4. The preface is signed by Charles Perrault, but the text appears to have been written by Claude: see Hallays, *Les Perrault*, 198–99.
122. *HdA* (1671), 88.
123. Ashworth, "Marcus Gheeraerts," 133–37. Ashworth does not mention the *Histoire des animaux*.
124. For Leclerc's biography, see Préaud, *Leclerc*. Grivel and Fumaroli, *Devises*.
125. On the relationship between artistic genre and scientific illustration, see Elkins, "Art History."
126. Senior, "Ménagerie and the Labyrinthe," 218. Préaud, *Leclerc*, 247, says only that it depicts "une terrasse dans le jardin d'un château." The labyrinth at Versailles was constructed between 1674 and 1677; see chap. 5. See also Friedman, "Evolution of the Parterre d'eau"; and Friedman, "What John Locke Saw." For the orange trees, see Sweetser, *La Fontaine*, 78.
127. Campardon, *Spectacles de la foire*, 311; D'Assoucy, *Combat de Cyrano de Bergerac*; Molière, *Tartuffe*, 32 (act 2, scene 3); and La Fontaine, *Fables choisies*, pt. 3, 37–40. His scene also included a performing bear.
128. *HdA* (1676), 120. Galen, *On the Usefulness of the Parts*, 1:69. Galen devoted the entire first book of this work to the hand.
129. Joint authorship is implied in PV 8, fol. 15v (27 février 1675).
130. *HdA* (1676), 126. The *"philosophes"* included Anaxagoras, Aristotle, and Galen but not Descartes.
131. It therefore might be classified as a "boundary object": Star and Griesemer, "Institutional Ecology"; and Chartier, *Cultural Uses of Print*.
132. Colbert's gift, which also included a silver medal for each member, is described in PV 7, fol. 245r (3 juin 1679). "Account of Some Books," *Phil. Trans.* 11 (1676): 591–98, at 591–96; *HdA* (1733), pt. 1, "Avertissement des libraires," n.p.; and *HdA* (1733), pt. 3, "Avertissement," n.p. This is further discussed in the epilogue.
133. "Account of Some Books," 591.
134. Daston and Galison, *Objectivity*, 63.

135. Ibid., 67.
136. Ibid., 91. The image, from the Archives of the Paris Academy of Sciences, is not from Cartonnier 13 (Dessins Textes non datés pour Hist. nat. des animaux par PERRAULT), as claimed, but from the dossier for the artists Louis and Philippe Simonneau, and is one of many illustrations Philippe Simonneau did for Jean Méry in the 1690s during his debate with Duverney over the *trou ovale*. See chap. 6, below.
137. Two versions of the two-headed calf may be found in Claude Perrault, Dossier, Archives, Paris Academy of Sciences.

CHAPTER FIVE

1. On D'Artagnan's role in Fouquet's arrest, see Marie de Sévigné to Pomponne, 22 December 1664, Sévigné, *Correspondance*, 1:79.
2. Biographical accounts of Gayant are sketchy at best. See Sturdy, *Science and Social Status*, 115–19; and Eloy. His dossier in Archives, Paris Academy of Sciences, adds little.
3. The inventory of his goods after death noted 4.5 *muids* of Burgundy wine, or over 1,200 liters, in his *cave*: Sturdy, *Science and Social Status*, 122. Du Hamel, *Historia*, 137 notes the deaths of the three anatomists: Cureau de la Chambre, Gayant, and Pecquet. He erroneously dates Cureau de la Chambre's death to 1671 rather than 1669.
4. Mariotte to Jean Lantin, March 1674, in Taton, "Mariotte et l'Académie Royale des Sciences," 31.
5. *HdA* (1671), preface, n.p. A different interpretation of this image appears in Watson, "Early Days"; see also Stroup, *Company of Scientists*. Jombert, *Catalogue*, 1:152, identifies the animal as a fox. See also Préaud, *Leclerc*, 241.
6. PdS (1672); and PdS (1673). The dissections of twelve animals are recorded in the PdS for 1673. All but one of the accounts are in Perrault's hand—the other is by Pecquet—but it is not clear who dissected. Because of the absence of the PV from 1670 to 1674, it is difficult to ascertain what exactly happened in these years. Only three animal dissections appear in the PdS for 1674.
7. Roger, *Life Sciences*, 138–40; and Licoppe, *Formation*, 46. According to L. F. Peltier, the letter of recommendation was from a certain Guisoni; however, Peltier gives no reference for this assertion: "Joseph Guichard Duverney," 308. There are few other biographical accounts of Duverney; the main ones include: Fontenelle, *Éloges*, 2:436–51; Eloy; Renaud, *Communautés de maîtres chirurgiens*, 109–15; Wesley C. Williams, "Duverney, Joseph-Guichard," *DSB*; and Sturdy, *Science and Social Status*, 188–92.
8. Brockliss and Jones, *Medical World*, 199, 486–87; and Marchand, *L'Université d'Avignon*, 117–18.
9. See chap. 3, above. Thomas Willis had published his *Anatomy of the Brain* in 1664.
10. Fontenelle, *Éloges*, 2:437; and Roger, *Life Sciences*, 140.
11. Fontenelle, *Éloges*, 2:437.
12. Ibid., 2:440.
13. *DSB*; and Blegny, *Livre commode*, 1:165–66n. Geoffroy, the father of the chemist Étienne Geoffroy, was an apothecary. See Stroup, *Company of Scientists*, 196–97. On Duverney's title, see Laissus, "Jardin du roi," 329; the title of professor of anatomy was first held by his son Emmanuel-Maurice.
14. *HdA* (1676), "Avertissement," n.p. Pitfeild did not include this in his translation of the 1676 edition.
15. Du Hamel, *Historia*, 140, notes Duverney dissecting in 1674. *CdB*, vol. 1, col. 783, lists a payment to Duverney for 750 livres (half his eventual pension at the academy)

on 31 May 1675; however, it appears under the accounts for 1674, indicating that he was working there during the last six months of that year. See Meynell, "Surgical Teaching at the Jardin des plantes." He first appears in the PV sharing a presentation with Perrault: PV 8, fol. 15v (27 février 1675). Employment before appointment was not uncommon; Mariotte presented work to the academy for at least a year before his appointment: Taton, "Mariotte et l'Académie Royale des Sciences," 14.

16. On Hooke, see Shapin, *Social History of Truth*; and Revel, "Uses of Civility," 192, 196.
17. Sturdy, *Science and Social Status*, 188–92.
18. Brièle, *Déliberations de l'ancien bureau de l'Hôtel Dieu*, 1:206 (8 janvier 1677).
19. Perrault read his report in May 1668: PV 4, fols. 17v–19r.
20. Claude Perrault, "Nouvelle insertion du canal thoracique," in *Essais*, 1:305–17. This includes Pecquet's letter to the *JdS* from 1667 and Perrault's "Decouverte d'une communication du Canal Thoracique avec la veine-cave inferieure," which had originally been published in the *JdS* in 1672 (8 February 1672, 45–48). He expressed skepticism in *Essais*, 3:223–24. On the composition of bones, see, for example, Duverney's lectures on osteology in BIUS, MS 2067, "Lectiones anatomicae Domini Duverney amphitheatro chirirgico."
21. On this topic see Stroup, *Company of Scientists*, chap. 10 (131–44), and below. Perrault read a treatise on this topic in July 1668: PV 4, fols. 93r–98v.
22. Francis Vernon to Henry Oldenburg, 12 June 1669, Oldenburg, *Corres.*, 6:5–7, at 6:6; and PV 5, fols. 98r–99v (8 juin 1669).
23. PV 4, fol. 146r (18 août 1668); PV 5, fols. 58r–59r (9 mars 1669); and Du Hamel, *Historia*, 69.
24. PV 8, fols. 15v–37r (27 février 1675). At the same meeting, Perrault described a sapajou (capuchin monkey), which was then turned over to Duverney to dissect, and it was suggested that another animal be dissected (or possibly vivisected; the text is unclear) to see the connections between the "ventricule" (stomach) and the veins.
25. Claude Perrault, "De la pesanteur des corps, de leur ressort et de leur dureté," in *Essais*, 1:1–130, at 1:2. Le Clerc and Manget, preface to *Bibliotheca anatomica* (1685).
26. Des Chene, "Mechanisms of Life." The discussion that follows owes much to this article.
27. On the distinction between mechanism as method and as ontology, see Des Chene, "Mechanisms of Life," 249–50.
28. Claude Perrault, *Essais*, 1:2. On his influences, see Picon, *Claude Perrault*, 89; and Des Chene, "Mechanisms of Life."
29. Claude Perrault, *Essais*, 1:4–5. AT, 8.1: 105 (bk. 3, sec. 52), "Tria esse hujus mundi aspectabilis elementa"; and *PWD*, 1:258.
30. Claude Perrault, *Essais*, 1:4–10; and Christiaan Huygens, "Extrait d'une lettre de M. Hugens [*sic*] de l'Académie Royale des Sciences à l'auteur de ce Journal, touchant les phenomènes de l'eau purgée de l'air," *JdS* (25 July 1672): 133–40.
31. PV 8, fol. 90v (29 juillet 1676).
32. Claude Perrault, "Du mouvement péristaltique," in *Essais*, 1:131–72, at 1:131–38.
33. Claude Perrault, *Essais*, 4:27.
34. Fontenelle particularly noted this conclusion in his account of Perrault's treatise on peristaltic motion: *HARS*, 1:195.
35. Claude Perrault, "De la circulation de la sève des plantes," in *Essais*, 1:173–304; and Stroup, *Company of Scientists*, 133–44.
36. Claude Perrault, *Essais*, 3:1. See above, chap. 3.
37. Claude Perrault, *Essais*, 3:2–4.

38. Ibid., 3:5–8. Peter Harrison has argued, in contrast, that the mechanical philosophy was a necessary prerequisite for animal experimentation: "Reading Vital Signs," 195. See chap. 2, above.

39. Aristotle, *Parts of Animals*, 697b; Loisel, *Ménageries*, 2:275; Henri Justel to Henry Oldenburg, 5 octobre 1667, Oldenburg, *Corres.*, 3:484; and PdS (1671).

40. Loisel, *Ménageries*, 2:397 (citing AN MS O/1/1805, folder 3), 115; and Robbins, *Elephant Slaves and Pampered Parrots*, 20, 43. There are many references to Monier's purchases in *CdB*.

41. *HdA* (1676), 180.

42. Loisel, *Ménageries*, 2:108–111; and Scudéry, *Promenade de Versailles*.

43. Iriye, "Le Vau's Menagerie"; Gaillard, "Bestiaire réel, bestiaire enchanté"; and Sahlins, "Royal Menageries of Louis XIV."

44. *HdA* (1733) pt. 3, 161; PV 10, fol. 86r (6–8 décembre 1681); and Scudéry, *Promenade de Versailles*, 95–98: "vn certain animal appellé chapas, plus beau & mieux marqueté qu'vn tigre, doux & flateur comme vn chien," at 98. *HdA* (1671), 13.

45. *HdA* (1733), pt. 3, 91–156; and PV 10, fols. 58r–59r (22, 26, 27 janvier 1681).

46. Berger, *Garden of the Sun King*, 14–15, 30; quote from André Félibien (1676) in Maisonnier and Maral, *Labyrinthe de Versailles*, 9.

47. Maisonnier and Maral, *Labyrinthe de Versailles*, 18–19.

48. Sweetser, *La Fontaine*, 36–37, 44–46; and Sévigné, *Correspondance*, vol. 1, passim.

49. Maisonnier and Maral, *Labyrinthe de Versailles*, 48.

50. Gaillard, "Bestiaire réel, bestiaire enchanté," 184–86; Maisonnier and Maral, *Labyrinthe de Versailles*, 49–50; Baridon, *Gardens of Versailles*, 184–90; Berger, *Garden of the Sun King*, 30–32; and Charles Perrault, *Recueil de diverses ouvrages*, 226.

51. Maisonnier and Maral, *Labyrinthe de Versailles*, 30–41, 102–85. Charles Perrault, *Recueil de diverses ouvrages*, 225–68; Charles Perrault, *Labyrinte de Versailles*; Charles Perrault, *Contes*, 225; and Berger, *Garden of the Sun King*, 35. Perrault's name did not appear on the title page of the *Recueil*, but he was identified as the author in the dedication (to the prince of Conti) by Louis Le Laboureur.

52. *HdA* (1676), 185.

53. Gaillard, "Bestiaire réel, bestiaire enchanté"; Scudéry, *Promenade de Versailles*, 95; Iriye, "Le Vau's Menagerie," 36–40; and Sørensen, "Animaux du roi."

54. *HdA* (1676), 107.

55. *HdA* (1733), pt. 3; and PV 10, fol. 87r (7 janvier 1682).

56. *HdA* (1676), 164–83; and PV 8, fol. 63v (20–27 novembre 1675).

57. Pitfeild, 248; *HdA* (1676), 190; and Claude Perrault, *Essais*, 1:151.

58. Pitfeild, 225; and *HdA* (1676), 173. "*Ventricule*" referred to the "petit ventre," or stomach; the term "*ventre*" applied to the entire area below the thorax. See Claude Perrault, *Essais*, "Table pour l'explication des termes de science," s.v. "ventricule," 3:359.

59. Pitfeild, 225–26; and *HdA* (1676), 173.

60. PV 8, fol. 145v (8 février 1678); and PV 8, fol. 149v (2 mars 1678). Duverney, "Observations sur les parties qui servent à la nourriture des oiseaux," *Œuvres*, 2:446–52, talks mainly about the gizzard and does not progress to the stomach. Claude Perrault, *Essais*, 3:223–24.

61. Pitfeild, 234–35; and *HdA* (1676), 180–81.

62. "Grande tortue des Indes," *HdA* (1676), 192–205.

63. Weber, "L'Inde française de la compagnie de Colbert."

64. Although the minutes state at the time of dissection that Duverney "will make his

report" ("dont il fera son rapporte," PV 8, fol. 63v [27 novembre 1675]), Claude Perrault read the account several months later, "with his reflections" (PV 8, fol. 78r [11 mars 1676]). Perrault read the introduction to the *Essais de physique* on 2 December 1676, and the academy resolved to hear parts of this work every Wednesday thereafter (PV 8, fol. 100v). As we have seen, the essay on peristaltic motion had already been presented.

65. PV 8, fols. 64r, 64v (11 décembre 1675); Pitfeild, 257; and *HdA* (1676), 197.

66. Pitfeild, 260–61; and *HdA* (1676), 200. From the context, it appears that "La méchanique des animaux" was already written at this point. Aristotle, "On Respiration," in *On the Soul, Parva naturalia, On Breath*, 475b, 20–30. I am grateful to James Lennox for this reference.

67. Claude Perrault, *Essais*, 3:266.

68. Ibid.; *Description anatomique*, 13–14, 26–27; Pitfeild, 261; and *HdA* (1676), 200. Duverney and Perrault did not distinguish between tortoises and turtles, and the experiments he describes were most likely performed on domestic turtles rather than tortoises. I shall use the term "tortoise" as they did.

69. Robert Hooke, "An Account of an Experiment Made by Mr. Hook, of Preserving Animals Alive by Blowing through Their Lungs with Bellows, *Phil. Trans.* 2 (1666–67): 539–40; Pitfeild, 262; and *HdA* (1676), 200.

70. Pitfeild, 262; *HdA* (1676), 201; and Hooke, "Account of an Experiment." On the open thorax, see Frank, *Harvey and the Oxford Physiologists*, 158–60, 200–201.

71. Pitfeild, 259–62; *HdA* (1676), 199–201; and Claude Perrault, *Essais*, 3:267. Duverney had a lengthy dispute in the 1690s with Jean Méry on the topic of the fetal circulation; see chap. 6, below.

72. Claude Perrault, *Essais*, 3:266–68; Pitfeild, 262–63; and *HdA* (1676), 201–2.

73. Magalotti, *Saggi di natvrali esperienze*, 118–21.

74. Pitfeild, 263–64; and *HdA* (1676), 202.

75. See chap. 6, below.

76. Claude Perrault, *Essais*, 3:264–67. On Mayow and the aerial nitre, see Frank, *Harvey and the Oxford Physiologists*, chap. 10 (248–74).

77. Claude Perrault, *Essais*, 3:268–71.

78. Pitfeild, 266; *HdA* (1676), 204.

79. Claude Perrault, *Essais*, 4:165.

80. Ibid., 3:30–45, 4:168–228. The historiography of the anatomy and physiology of the eye in this period (as opposed to work on optics) is remarkably thin and is largely focused on Descartes. See Mancosu, "Acoustics and Optics"; Wade, *Natural History of Vision*; Simon, "Theory of Visual Perception"; and Hyman, "Cartesian Theory of Vision." For the earlier period, see Lindberg, *Theories of Vision*.

81. Like many of the early works of the academy, the publication history of the work of Mariotte and his critics on the seat of vision is complex. The first letter of Mariotte with a reply by Pecquet is *Nouvelle découverte touchant la veüe* in 1668. This was partially translated into English in Henri Justel, "A New Discovery Touching Vision," *Phil. Trans.* 3 (1668): 668–71. Mariotte replied to Pecquet in 1669 in *Seconde lettre de M. Mariotte à M. Pecquet pour montrer que la choroïde est le principal organe de la veüe*, translated in "The Answer of Monsieur Mariotte to Monsieur Pecquet," *Phil. Trans.* 5 (1670): 1023–42 (the pages are misnumbered; 1039 follows 1030). Claude Perrault's contribution to the debate appeared in the elephant folio *Recueil de plusieurs traitez* of 1676. This also included Mariotte's final reply to Pecquet and Perrault. The letters of Mariotte and Perrault were reprinted in *Lettres écrites sur le sujet d'une nouvelle*

découverte touchant la vue. An account of this pamphlet appeared in *Phil. Trans.* 13 (1683): 265–67. All five letters appeared in the two editions of Mariotte's *Œuvres.* Robert Root-Bernstein translated the five letters: "New Discoveries Touching Vision, 1668–1676," http://www.princeton.edu/~hos/mike/texts/mariotte/mariotte.html. M. D. Grmek examined the controversy in "Le débat sur le siège de la perception visuelle," in *Première révolution biologique,* 189–229. See also Salomon-Bayet, *Institution de la science,* chap. 3 (67–105); and Licoppe, "Crystallization of a New Narrative Form," 221–24. The latter essay is reprinted (in French) in Licoppe, *Formation,* chap. 2 (53–87).

82. "Proposition d'optique de M. Mariotte," PV 2, fols. 75r–80v, misnumbered as 81 (21 juin 1667). Cf. Grmek, "Débat," 201–2.

83. Lindberg, *Theories of Vision,* chap. 9; and Descartes, *Dioptrique,* in AT, 6:115–17.

84. Mariotte and Pecquet, *Nouvelle découverte touchant la veüe*; and PV 2, fol. 163r (4 juin 1667): "Le 4ᵉ de juin M. Mariotte a donné un Escrit ou il monstre que la Choroide est le principal organe de la veue et non pas la retine." The date of Mariotte's entry into the academy is debated; some sources say May 1667, but René Taton argues it was not until June 1668: Taton, "Mariotte et l'Académie Royale des Sciences," 14. Biographical material includes Sturdy, *Science and Social Status,* 110–12; and Michael Mahoney, "Mariotte, Edmé," in *DSB.*

85. Mariotte and Pecquet, *Nouvelle découverte.* The pamphlet with Mariotte's and Pecquet's letters appeared in May 1668, published, like the 1667 dissection pamphlets, by royal printer Frédéric Léonard. I believe this indicates the academy approved the publication, but Licoppe, "Crystallization of a New Narrative Form," 221, asserts that Mariotte took the initiative. See Grmek, "Débat," 193; and Henri Justel to Henry Oldenburg, 9 May 1668, Oldenburg, *Corres.,* 4:348.

86. Review of *Nouvelle découverte touchant la veüe,* followed by a review of Pecquet's letter on Mariotte's theory, *JdS* (17 septembre 1668): 79–82 and 82–84.

87. Mariotte and Pecquet, *Nouvelle découverte,* 10, 17.

88. Ibid., 8–10; and Picon, *Claude Perrault,* 82.

89. Huygens, *Oeuvres,* 13:787–90; and Grmek, "Débat," 197.

90. PV 5, fol. 165r (17 août 1669); and PV 5, fols. 166r–172r (24 août 1669).

91. *Lettres écrites par MM. Mariotte, Pecquet et Perrault, sur le sujet d'une nouvelle découverte touchant la veü / Faite par M. Mariotte,* 17–18, in *Recueil de plusieurs traitez.* Each section of this volume is paginated separately and has separate publication information in a colophon. Picon, *Claude Perrault,* 82.

92. Grmek, "Débat," 227–28.

93. *Lettres écrites sur le sujet d'une nouvelle découverte touchant la vue,* 22; and PV 11, fol. 68v (24 mai 1684; Mariotte died on the twelfth).

94. Wright, "Embodied Soul"; and Azouvi, "Entre Descartes et Leibniz."

95. Claude Perrault, *Essais,* 4:177.

96. Ibid., 4:166.

97. Ibid., 4:205. Descartes, *Traité de l'homme,* in AT, 11:175–78, 11:184–85; and *PWD,* 106–7. Wade, *Natural History of Vision,* 100, notes that while the accompanying illustration in *Traité de l'homme* demonstrated binocular vision, the corresponding illustration in the Latin version, *De homine,* did not.

98. Claude Perrault, *Essais,* 4:199–200.

99. Ibid., 4:208–9, 4:211.

100. Ibid., 3:24.

101. Ibid., 3:3.

102. Ibid., 3:50.
103. Claude Perrault, *Du bruit,* in *Essais,* vol. 2. The discussion that follows is indebted to Erlmann, *Reason and Resonance,* 69–109, and his "Physiologist at the Opera." For an overview of music and science in this era, see Gouk, *Music, Science, and Natural Magic.*
104. *Dictionnaire de l'Académie française, Dictionnaires d'autrefois,* ARTFL.
105. Claude Perrault, *Essais,* 2:213.
106. Ibid., 3:48, 4:43–44.
107. Ibid., 2:186–213, 3:60–61.
108. PV 8, fol. 117r (2 juin 1677) and fol. 146r (16 février 1678).
109. Du Hamel drew up a memo in April 1675: PV 8, fols. 42r–45r. Because the minutes are missing from 1670 to 1674, it is not possible to know if there were earlier reports. The next one in the minutes is in 1678.
110. PV 8, fol. 161v; also included was his investigation of honey and manna.
111. PV 8, fol. 101r (9 décembre 1676): "Il a esté resolu que l'on proposeroit à Monseigneur Colbert s'il jugea propre que l'on fasse imprimer les ouvrages des particuliers en un volume in folio médiocre." Both the *Histoire des animaux* and the *Recueil* were elephant folios.
112. Duverney, *Traité de l'organe de l'ouïe,* "Avertissement," n.p.
113. PV 10, fol. 61v (5 mars 1681).
114. Duverney, *Traité de l'organe de l'ouïe,* "Avertissement." Duverney and rhetoric are further examined in chap. 6.
115. Duverney, *Traité de l'organe de l'ouïe,* 67–68.
116. Fontenelle detailed the differences between Perrault's and Duverney's theories in *HARS,* 1:395–99.
117. Duverney, *Traité de l'organe de l'ouïe,* 79; and Carl Goldstein, *Print Culture,* 96–98.
118. Duverney, *Traité de l'organe de l'ouïe,* 84–85; and Galileo, *Two New Sciences,* 100.
119. Duverney, *Traité de l'organe de l'ouïe,* 96–98, at 98; and Erlmann, *Reason and Resonance,* 74–75. As Erlmann points out, Duverney's understanding was exactly reversed from the modern view, which holds that the thicker end hears higher tones.
120. Claude Perrault, *Essais,* vol. 2, chap. 4; and Duverney, *Traité de l'organe de l'ouïe,* 195, 206. Fontenelle held to the explanation of tinnitus as a vibration of air in the ear: *HARS,* 1:399. Erlmann, "Physiologist at the Opera."
121. Claude Perrault, *Essais,* 4:260–334.
122. Ibid., 3:60–61; on the chameleon, see *Description anatomique,* 41. Mersenne, *Harmonie universelle,* 2:79; and Couvreur and Favier, introduction to Favier and Couvreur, *Plaisir musical,* 13–15.
123. Claude Perrault, *Essais,* 3:61–62.
124. Ibid., vol. 2, bk. 2, chaps. 10–12, pp. 113–85.
125. Descartes, *Compendium musicae,* in AT, vol. 10; Galileo, *Two New Sciences,* 107; Mersenne, *Harmonie universelle;* Dear, "Marin Mersenne"; and Erlmann, *Reason and Resonance,* 66–68, 98–101.
126. Descartes to Mersenne, 18 March 1630, AT, 1:132–33; *PWD,* 3:19–20; and van Wymeersch, "Descartes."
127. The relationship between rhetoric, subjectivity, and scientific discourse will be further discussed in chap. 6.
128. Claude Perrault, *Ordonnance des cinq espèces de colonnes,* iii–iv; Claude Perrault, *Treatise of the Five Orders of Columns,* iii–iv; and Dear, "Marin Mersenne."
129. Claude Perrault, *Essais,* 2:336.

130. Claude Perrault, "Préface," Bibliothèque nationale, MS Fr. 25,350, in Gillot, *Querelle des anciens et des modernes*, 576–91, reproduced in Claude Perrault, *"Du bruit"* et *"De la musique des anciens."*

131. Claude Perrault, "Préface"; and Bonnefon, "Poème inédit de Claude Perrault."

132. Claude Perrault, "Préface," 587: "il entra mesme en quelque façon dans ses sentimens," *"sentimens"* meaning "sense experience" (*Dictionnaire de l'Académie française*, 1694, s.v. *"sentiment"*).

133. Claude Perrault, "Préface," 587.

134. Erlmann, "Physiologist at the Opera."

135. Claude Perrault to Colbert, 27 janvier 1674, BN, Mélanges Colbert, MS 167, fol. 245, quoted in Quinault, introduction to *Alceste*, x–xi. The opera in question was Lully's *Alceste*.

136. Cowart, *Triumph of Pleasure*, 150–57.

137. Quinault, introduction to *Alceste*; Charles Perrault, *Recueil de diverses ouvrages*, 269–310; Cowart, introduction to *Triumph of Pleasure*, xx–xxi.

138. Quinault, introduction to *Alceste*; and Bonnefon, "Poème inédit de Claude Perrault."

139. Fumaroli, "Abeilles et les araignées," 167–69; and Quinault, introduction to *Alceste*, ix.

140. Claude Perrault, *De la musique des anciens*, "Avertissement," in *Essais*, 2:335–36, at 2:336.

141. Vitruvius, *Dix livres de architecture*, bk. 5, chap. 4, 150–57; and Picon, *Claude Perrault*, 104, 266n14. Picon points out that Mersenne had made many of the same points in his early *Questions harmoniques*.

142. Claude Perrault, *De la musique des anciens*, in *Essais*, 2:356, 2:359.

143. Ibid., 2:388, 2:394, 2:400, 2:387–90; and Picon, *Claude Perrault*, 104–7.

CHAPTER SIX

1. Colbert to Picard, 21 septembre 1679, in Colbert, *Lettres*, 5:403–4; on the mapping project, see Konvitz, *Cartography in France*, 2–6.

2. Fontenelle, *Éloges*, 2:16–17.

3. Colbert to Picard, 21 septembre 1679, in Colbert, *Lettres*, 5:403–4; Du Hamel, *Historia*, 175–76; and Colbert to La Hire, 10 novembre 1679, in Colbert, *Lettres*, 5:407–8.

4. Loisel, *Ménageries*, 2:112–15; and *CdB*, vols. 2 and 3.

5. PV 10, fol. 6v (10 janvier 1680); and *HARS*, 1:308.

6. PV 10, fol. 57v (15 janvier 1681); Du Hamel, *Historia*, 176, 196; Briggs, "Académie des sciences"; and Stroup, *Company of Scientists*, 32.

7. PV 10, fol. 57v (15 janvier 1681), fol. 58r (22 janvier 1681); and *HARS*, 1:322.

8. *CdB*, vol. 2, col. 101, 22 octobre 1681.

9. *HdA* (1733), 3:3, 3:101, 3:119–20, 3:127; Houel, *Histoire naturelle des deux éléphans*, 8, 10–14; and *Osteologie de M^r Du Vernay 1685*, MNHN, MS 1760, fol. 108.

10. Alfred Franklin, *Vie privée d'autrefois*, 1:100–104; Baratay, "Naturalistes dans les cages"; and Peiresc to Dupuy brothers, 26 décembre 1631, Peiresc, *Corres.*, 2:293–94.

11. Posner, "Lebrun's *Triumphs of Alexander*," 237–48, dates the *Entry into Babylon* around 1665, based on Bernini's witnessing of it in progress in October of that year. However, he admits that exact dating is impossible and the last picture in the series may not have been completed until 1673. They were exhibited as a group in 1682. On engraving, see Keeble, "Sincerest Form of Flattery"; on the tapestry, see *CdB*, vol. 3, col. 435 (1690).

12. *HdA* (1733), 3:3; and PV 10, 57v–59r (15, 22, 26 janvier 1681). Biagioli, "Etiquette,

Interdependence, and Sociability," 217, argues that such events were purely spectacles, with no scientific value.

13. *HdA* (1733), 3:91–156; and *HARS* 1:322–26, at 1:322–23.

14. PV 10, fols. 59r, 59v (26, 29 janvier and 5, 12 février 1681).

15. PV 10, fol. 74v (2 juillet 1681), fols. 82r–82v (annual report, December 1681), and fol. 84r ("Livres de physique Prests à Imprimer," December 1681).

16. PV 10, fol. 103v (17 juin 1682), fol. 143v (28 avril 1683).

17. *HdA* (1733), 3:140–50 (trunk: 3:140–46; skeleton: 3:146–50).

18. PV 10, fol. 74v (9 juillet 1681); and *HdA* (1733), 3:1–15, at 3:4–8.

19. *HdA* (1733), 3:5. Cf. Buffon, *Histoire naturelle*, 9:139.

20. *HdA* (1733), 3:1–15.

21. *HdA* (1733), 2:287–88; and Loisel, *Ménageries*, 2:98–99.

22. PV 10, fol. 86r (6–8 décembre 1681); see chap. 5, above.

23. PV 10, fols. 85r, 85v (5 décembre 1681); and Du Hamel, *Historia*, 196–97.

24. *CdB*, vol. 2, col. 221, under "Jardin du Roi," notes payment "au Sr du Verney, démonstrateur, pour ses gages des années 1680, 1681, et 1682."

25. "Contestation Académie-Chirac, Cabinet des squelettes," PdS, avril 1731, Archives, AdS. See Guerrini, "Duverney's Skeletons."

26. Laissus, "Jardin du roi," 323, 327; Fagon was named "sous-démonstrateur des plantes" in 1671 and "démonstrateur et opérateur pharmaceutique" in the following year. "Lettres patentes concernant la surintendance de Jardin Royale des Plantes, Décembre 1671," in Colbert, *Lettres*, 5:533–35. A sketch of Fagon earlier in his career is in Brockliss, "Literary Image of the Médecins du Roi," 132–33.

27. "Declaration du Roy. Pour faire continuer les Exercices au Jardin du Roi des Plantes . . . 23 mars. 1673," AN, AJ/15/01, no. 17; Hamy, "Origines de l'enseignement," 15; and Deleuze, *Histoire du Muséum*, 15.

28. Sibbald, *Memoirs*, 59. In the winter of 1662, Sibbald wrote, "I studied the plants under Junquet in the King's Garden, and heard the publick lectures of Monsieur de la Chambre the younger, and Monsieur Bazalis." There is some ambiguity here as to whether Cureau de la Chambre's lectures were also at the garden. I have not been able to identify Bazalis. Dionis commented that in 1682, lectures had not been given at the garden for "several years" but gives no further detail: Dionis, preface to *Anatomie*, n.p.

29. François Cureau de la Chambre received the payment of fifteen hundred livres, and paid his surrogates an unspecified amount: *CdB*, vol. 1, col. 602.

30. Elizabeth Maxfield Miller, "Molière"; Baron, *Scholis Facultatis*, 47–48; and Hamy, "Origines de l'enseignement," 14–18. He and François Cureau de la Chambre had been classmates at the Paris Faculty in the 1650s.

31. Laissus, "Jardin du roi"; Lunel, *Maison médicale du roi*, 182–83; and Dionis, preface to *Anatomie*. For Dionis, see Eloy. Senior, "Dionis and Duverney," unfortunately contains many inaccuracies.

32. Dionis, *Anatomie*, 134.

33. Dionis, preface to *Anatomie*; "Declaration du Roy," 23 mars 1673, AN, MS AJ/15/501, no. 17; Hamy, "Origines de l'enseignement," 15; Gannal, "Cours d'anatomie"; and Deleuze, *Histoire du Muséum*, 15.

34. *CdB* 1:601; and Dionis, preface to *Anatomie*.

35. Dionis, preface and "Table des titres et sections," both in *Anatomie*.

36. Gelfand, *Professionalizing Modern Medicine*, 24–27; and Lunel, *Maison médicale du roi*, 62–65.

37. Hamy, "Origines de l'enseignement," 19.
38. Molière, *Malade imaginaire*, act 2, scene 6. See Lunel, *Maison médicale du roi*, 183; Chauvois, "Molière"; and Grmek, "L'émergence de la médecine scientifique sous le règne de Louis XIV," in *Première révolution biologique*, 233–59.
39. "Ce dogme fameux et fumeux n'est bon qu'à perturber le corps sain ou malade, c'est la confusion en toutes choses," quoted in Albou, "*Arrêt burlesque* de Boileau," 27. See also Chauvois, "Molière."
40. Albou, "*Arrêt burlesque* de Boileau," 30–31; on Bernier, see the introduction by France Bhattacharya to Bernier, *Voyage*.
41. Dionis, *Anatomie*, 167–68.
42. Descartes, *Traité de l'homme*, in AT, 11:121.
43. Dionis, *Anatomie*, 184.
44. Gelfand, *Professionalizing Modern Medicine*, 17. The theater formed the frontispiece to the second edition of Dionis's *Anatomie*.
45. Duverney's letter of appointment, dated 23 mars 1682, is in AN AJ/15/509, fol. 205, copy of AN, O/1/26, fol. 71v; *CdB*, vol. 2, col. 221, under "Jardin du Roi," notes payment "au Sr du Verney, démonstrateur, pour ses gages des années 1680, 1681, et 1682." Brièle, *Délibérations de l'ancien bureau de l'Hôtel-Dieu*, 1:206, implies that Duverney dissected at the garden as early as 1677. As noted above, the title of professor of anatomy dates from the 1720s.
46. The numbers are from Eloy's biography of Duverney.
47. Guillaume Lamy, *Discours anatomiques*, 120–22: "plusieurs canaille de la fauxbourg, attirez par une veine curiosité de voir dissequer un corps." Cressé's lecture replied to Lamy's critique of him. See Hamy, "Origines de l'enseignement," 16–19.
48. "Ordonnance portant deffences d'assister au lecons qui se sont au Jardin du Roi, avec Espées ou bastons," AN, MS AJ/15/514, folder A16, no. 670. Additional ordinances and declarations on riots are in the same folder, A16, of this archive. On the neighborhood of the garden, see Colin Jones, *Paris*, 140; Stroup, *Company of Scientists*, 187; Perret, *Jardin des plantes*; and Truc, *Quartier St-Victor*.
49. "Extrait d'une lettre écrite à Monsieur le Président Cousin," *JdS* (23 mai 1689): 219–26; and [Duverney], *Lettre*.
50. Duverney, *Œuvres*, 1:v. Lecture notes include "Lectiones anatomicae Domini Duverney amphitheatro chirugico," BIUS, MS 2067; "Osteologie de Mr. Du Vernay 1685," MNHN, MS 1760; [Patrick Mitchell], "Observations upon osteology, taken out of Mr Du Verney's private course, begunn the 22 of October 1697," Wellcome, MS 5433; Jean-Baptiste Thorière, Lecture notes, Paris, 1715–16, Wellcome, MS 4786; James Douglas, "Osteologia," Glasgow University Library, MS Hunter D311; "Traicté de la gangrene, par monsieur du vernay," Paris, 27 July 1683, BL, Sloane MS 1216, fols. 45r–74v; "Traité des maladies des os," BL, Sloane MS 1499; and "Traitté des maladies de la poitrine," BL, Sloane MS 87. These will hereafter be cited by catalog numbers, so "BIUS 2067."
51. BIUS 2067.
52. Duverney, *Traité des maladies des os*. The volumes were small duodecimos without illustrations.
53. For Mitchell, see Desmond, *Dictionary of British and Irish Botanists*, 492. He was twice president of the Royal College of Physicians of Ireland. For Douglas, see *ODNB*; and Guerrini, "Anatomists and Entrepreneurs."
54. MNHN 1760; and Hunter, *Two Introductory Lectures*. On medical note taking, see Guerrini, "Value of a Dead Body." Little has been written about the manuscript

economy of medical students and its implications for medical knowledge. For a general survey of early modern note taking, see Soll, "From Note-Taking to Data Banks"; see also Blair, *Too Much to Know*.

55. On Charas, see Sturdy, *Science and Social Status*, 254–57; and Stroup, *Company of Scientists*, 20–21, 38, 111. He died in January 1698. Mitchell's notes begin April 23 ("I begun a course of pharmacy under Mr Charras") and end August 3 but do not give a year (Wellcome 5433, fols. 154r–161v).

56. Duverney, *Œuvres*, 1:1–14; and Wellcome 5433, fol. 38r.

57. BIUS 2067, p. 2; Duverney, *Œuvres*, 1:16–17; Glasgow Hunter D311, fol. 26r; and Wellcome 5433, fol. 45r.

58. BIUS 2067, p. 2; and Wellcome 5433, fols. 7r, 8r.

59. Wellcome 5433, fol. 114v; Glasgow Hunter D311, f. 46v; and Wellcome 4786, fol. 445.

60. Wellcome 5433, fol. 10v.

61. Glasgow Hunter D311, fol. 46v.

62. MNHN 1760, p. 137; Wellcome 5433, fol. 19v; and Glasgow Hunter D311, fols. 28v–29r.

63. [Duverney], *Lettre*, 9; and MNHN 1760, pp. 99–100.

64. Wellcome 5433, fol. 9v.

65. For example, MNHN 1760, fol. 106; Wellcome 5433; and BIUS 2067.

66. BIUS 2067, fol. 3.

67. On "faculté formatrice," see MNHN 1760, fols. 7–8; and BIUS 2067, fols. 6–8. On generation, see Glasgow Hunter D311, fols. 56r, 56v.

68. BL Sloane 1216, fol. 48r; and BIUS 2067, fols. 16–17.

69. Glasgow Hunter D311, fol. 41v; and BL, Sloane 1216, fol. 51v.

70. Wellcome 5433, fol. 121r; and BIUS 2067, fol. 20. On Claude Perrault's ideas, see *Essais*, vol. 3; and Roger, *Life Sciences*, 271–75.

71. Glasgow Hunter D311, fols. 42r, 35v–36r, 44v.

72. The British Library notes may fall into this category.

73. AN, ser. O/1/2124, *liasse* 2; the entire manuscript, which concerns the garden, occupies a box of loose sheets and is not paginated or foliated. For an example of Couplet's reporting, see PV 12, fol. 98v (18 août 1688).

74. For the Gérards, see *CdB*, vol. 2, cols. 502, 1010, 1198, and others.

75. *CdB*, vol. 1, col. 631. On "invisible technicians," see Shapin, *Social History of Truth*, chap. 8 (355–407). *CdB*, vol. 1, col. 889 (1676), col. 947 (1677–78).

76. PV 10, fol. 98v. On the construction of skeletons, see Guerrini, "Inside the Charnel House"; and Guerrini, "Duverney's Skeletons." On the elephant's skeleton, see PV 10, fols. 107v–108v. On Marchant, see Sturdy, *Science and Social Status*, 112–15; and *HARS*, 1:52.

77. For Dionis, see *CdB*, vol. 2, col. 781. For Duverney, see *CdB*, vol. 1, col. 1,343; vol. 2, col. 1188; and vol. 3, col. 120.

78. AN, ser. O/1/2124, *liasse* 2; the phrase is "chez l'Execution." This section is partially printed in Schiller, "Laboratoires d'anatomie," app., 114–15. See also Meynell, "Surgical Teaching at the Jardin des plantes," who reproduces another portion of this manuscript.

79. AN, ser. O/1/2124, *liasse* 2; the phrase is "Fourni à l'exécuteur pour plusieurs [sujets?] et autres parties des sujets humains." The next entry is for "un sujet humain executé."

80. AN, ser. O/1/2124, *liasse* 2. In "Laboratoires d'anatomie," 109, Schiller mentions the supply of eau-de-vie; in *Royal Funding*, 48, Stroup notes that the anatomists also drank it.

81. *CdB*, vol. 3, col. 730, and vol. 4, cols. 150, 210; AN, ser. O/1/2124, *liasse* 2, shows a "Histoire de remboursement" of Duverney in 1697 for the expenses of the dissection room that amount to 2,333 livres. See also Laissus, "Jardin du roi," 307; Salomon-Bayet, *Institution de la science*, 385–86, notes that this building was for research rather than teaching.
82. PV 10, fol. 87r (7 janvier 1682).
83. PV 11, fol. 3r (14 juillet 1683), fol. 28v (12 janvier 1684).
84. *CdB*, vol. 3, col. 120.
85. Roule, *Médecins du Jardin du roi*, 46; for the reference to Bossuet, see Cole, *History of Comparative Anatomy*, 403n. Cole quotes the nineteenth-century critic George Henry Lewes, but does not cite the work, and I have been unable to find the original quote.
86. Discussion of the rhetorical dimensions of natural philosophy includes Loveland, *Rhetoric and Natural History*; Walmsley, *Locke's "Essay"*; Dear, *Literary Structure*; and Vickers and Struever, *Rhetoric and the Pursuit of Truth*. An earlier version of what follows is Guerrini, "Eloquence of the Body."
87. Fontenelle, *Éloges*, 2:189; and Renaud, *Communautés de maître chirurgiens*, 112.
88. Kennedy, *Classical Rhetoric*, 74–93; Herrick, *History and Theory of Rhetoric*, chap. 4; and Aristotle, *Rhetoric*, chap. 1.
89. Mason, "Miraculous Birth"; see also his "Hommage à M. Despréaux."
90. Herrick, *History and Theory of Rhetoric*, 102; and Kennedy, *Classical Rhetoric*, 113–14.
91. Kennedy, *Classical Rhetoric*, 115–17; and France, *Rhetoric and Truth in France*, 8–9. On Quintilian, see also Murphy, *Rhetoric in the Middle Ages*, chap. 1.
92. France, *Rhetoric and Truth in France*, 8–9.
93. Murphy, *Rhetoric in the Middle Ages*, 357–60; and Grafton and Jardine, *From Humanism to the Humanities*, 67–68, 77–82.
94. Descartes, *Discours de la méthode*, in AT, 6:7; and *PWD*, 1:114. On changes in classical rhetoric in this era, see Fumaroli, *Âge de l'éloquence*.
95. Pascal, *De l'esprit géométrique et l'art de persuader*, 132–33, 146–48; Fumaroli, "Préface," 12, 41–48; and Wellman, *Making Science Social*, 4–5.
96. France, *Rhetoric and Truth in France*, 56–58; and Fumaroli, "Abeilles et les araignées," 13–17.
97. Mason, "Hommage à M. Despréaux," 59–60; Fumaroli, "Abeilles et les araignées," 15–17; Wellman, *Making Science Social*, 379–81; and Goldgar, *Impolite Learning*.
98. On expertise and credibility, see Shapin, *Social History of Truth*.
99. Quintilian, *Institutio oratoria*, trans. H. E. Butler (London: Loeb Classical Library, 1920), VII.X.7, quoted in France, *Rhetoric and Truth in France*, 10.
100. Burke, "Language of Gesture," 74; and Revel, "Uses of Civility," 195–96.
101. Antoine Courtin, *Nouveau traité de la civilité*, 207–14, and introduction, 11–12.
102. Brièle, *Délibérations de l'ancien bureau de l'Hotel-Dieu*, 1:206.
103. [Duverney], *Lettre*. The pamphlet, like the *JdS* article cited above, is written in the third person.
104. The material on teeth is almost identical to the student notes in MNHN 1760, fols. 98–109.
105. [Duverney], *Lettre*, 15.
106. MNHN 1760, fol. 100.
107. Bossuet, *Connaissance de Dieu*, chap. 2; and Wellcome 5433, fol. 45r.
108. Ferrari, "Public Anatomy Lessons." As we have seen, tickets were required at the garden for surgical students, to allow them entry before the general public.

109. DeJean, *Ancients against Moderns*, 42; and Charles Perrault, *Siècle de Louis le Grand*.

110. *HdA* (1671), preface; and Charles Perrault, *Siècle de Louis le Grand*, 258.

111. Fontenelle, *Éloges*, 2:438.

112. DeJean, *Ancients against Moderns*, 45–46.

113. Nicolas Boileau, *Satire X*, in *Oeuvres*, 1:113, lines 435–37: "Puis une femme morte avec son embryon / Il faut chez Du Verney voir la dissection. / Rien n'echappe aux regards de notre curieuse."

114. Antoine Courtin, *Nouveau traité de civilité*, 77; DeJean, *Ancients against Moderns*, 52–56; France, *Rhetoric and Truth in France*, 160; and Mason, "Hommage à M. Despréaux," 61–63.

115. Zipes, introduction to *Beauties, Beasts, and Enchantment*, 7.

116. On this topic, see Darnton, "Great Cat Massacre"; Castle, *Masquerade and Civilization*; and Ferrari, "Public Anatomy Lessons." On the mythical power of metamorphosis, see Warner, *No Go the Bogeyman*, 263–83.

117. Zipes, introduction to *Beauties, Beasts, and Enchantment*, 7; Charles Perrault, "Le petit chaperon rouge," in Perrault, *Contes*, 254–56; and "Little Red Riding Hood," in Zipes, *Beauties, Beasts, and Enchantment*, 58–60.

118. Duverney, *Œuvres*, 2:2.

119. Briggs, "Académie des sciences," 48–49; Sturdy, *Science and Social Status*, 214–16; and Taton, "Presentation d'ensemble," 14–15. Bessé was married to a niece of Boileau.

120. PV 11, fol. 24r (17 novembre 1683); and *HARS*, 1:386.

121. PV 10, fol. 122r (10 mars 1683), fol. 154r (annual report, 1682–83); and Briggs, "Académie des sciences."

122. H. J. M. Bos, "Huygens, Christiaan," in *DSB*; Briggs, "Académie des sciences," 50; and PV 10, fols. 180r–191v, "Traité de l'Aimant par Monsieur Hugens" (summer 1683).

123. Briggs, "Académie des sciences," 49–50; Stroup, *Royal Funding*; Stroup, *Company of Scientists*, 51–56; Sturdy, *Science and Social Status*, 214–20; Hahn, *Anatomy*, 19–20; and esp. Mallon, "Académie des sciences."

124. Sturdy, *Science and Social Status*, 216. On Thévenot's financial contribution, see, e.g., *CdB*, vol. 3, col. 582.

125. PV 11; *CdB*, vol. 3, cols. 305–6, 438–40; and Stroup, *Company of Scientists*, 35–38. On Bessé's attendance, see La Hire to Huygens, 1 juin 1687, in Huygens, *Oeuvres*, 9:164.

126. Fontenelle, *Éloges*, 2:163–64. For biography, see Wesley C. Williams, "Méry, Jean," in *DSB*; and Eloy.

127. Lister, *Journey to Paris*, 66; Fontenelle, *Éloges*, 2:163–75; and K. J. Franklin, "Jean Méry."

128. Eloy, 3:281; Guillaume Lamy, *Explication méchanique*, includes Méry's "Lettre écrite à M. Lamy" on the structure of the ear.

129. PV 11, fol. 56v (19 avril 1684).

130. *CdB*, vol. 2, cols. 781–83 (1685).

131. PV 11, fol. 57v (26 avril 1684); and "Observation faicte sur la peau de la grenouille le 24ᵉ Avril 1684 par Mʳ Merri," followed by two other observations: PV 11, fols. 61v–66r (mai 1684).

132. PV 12, fols. 19r–19v (16 novembre 1686).

133. The articles from the *Journal des sçavans* between 1666 and 1699 were collected in *Memoires de mathématique et de physique, par messieurs de l'Académie royale des sciences: Extraits des Journaux des Sçavans*, pp. 451–744 in *MARS*, vol. 10 (Paris: par la Compagnie des Libraires, 1730). The first half of this volume consisted of the Académie des sciences de Paris, *Mémoires de mathématique et de physique tirez des registres de l'Académie royale des sciences*, 1692–93.

134. *Description anatomique de divers animaux dissequez dans l'Académie royale des sciences*, 2nd ed. (Paris: Laurent d'Houry, 1682).

135. PV 10, fol. 122v (10 mars 1683), fol. 153r. Stroup, *Company of Scientists*, 80, states that Colbert stopped paying for engravings in 1681, but her citation from PV 10, fol. 80v, does not support this conclusion.

136. PV 11 (novembre 1683); Mallon, "Académie des sciences," 21–24; and *CdB*, vol. 2, cols. 541 (1684), 784–85 (1685), 1015 (1686), 1210 (1687). Châtillon is described as "dessinateur de dissections d'animaux et de plantes" in 1690: *CdB*, vol. 3, col. 440.

137. PV 11, fols. 52r, 52v (8 mars 1684).

138. Hahn, *Anatomy*, 27–28; and PV 12, fols. 98v–99r (18 août 1688).

139. For example, PV 12 (novembre 1686); and PV 12, fols. 45r, 45v (19 juillet 1687). Notes for revisions are in cartonnier 13, Archives, AdS.

140. On Claude Perrault's death, see Charles Perrault, *Hommes illustres*, 189. PV 12, fols. 104r, 104v (10–17 novembre 1688).

141. PV 12, fol. 116r (18 décembre 1688); *CdB*, vol. 3, col. 441 (10 septembre 1690); and Stroup, *Royal Funding*, 50.

142. Bernard, *Imprimérie royale*, 148–49. This volume of the *Histoire des animaux* , dated 1688, is in the library of the Institut de France (Cote M 130 C***). It conforms to Winsløw and Petit's description in *HdA* (1733), "Avertissement." It has no title page and ends with the cocq-indien, as Winsløw and Petit described.

143. A copy exists in the Library of the Institut de France, Cote M 130 C**. It is listed in the catalog of the Library of Congress but seems to be a ghost.

144. Gouye, *Observations*; on the identity of the "tok-kaie," see Bodson, "Living Reptiles." The Jesuits published additional observations in 1691–92, but these were the only animals. On Gouye, who was later named to the academy, see Hsia, *Sojourners*; and Sturdy, *Science and Social Status*, 289, 345.

145. Stroup, *Royal Funding*. The move to the Maison du roi meant that the academy's accounts no longer appeared in the *Comptes des bâtiments*; Stroup discovered them in the 1980s.

146. Sturdy, *Science and Social Status*, 221–26; and Jack A. Clarke, "Bignon."

147. Bessé to Huygens, 9 juin 1687, in Huygens, *Oeuvres*, 9:166; and Briggs, "Académie des sciences," 52.

148. Briggs, "Académie des sciences," 52–53, claims these volumes were based on the backlog of academy projects, but he seems to confuse these with the folio volumes from the same years.

149. Académie des sciences de Paris, *Divers ouvrages*; and Académie des sciences de Paris, *Recueil des observations*.

150. Du Hamel, preface to *Historia*, n.p. This statement is omitted in the 1701 edition.

151. The last volume of this series, for 1790, appeared in 1797. See Mazauric, *Fontenelle*, chap. 4; Halleux et al., *Publications*, vol. 1; and Guénoun, "Publications," 107–27.

152. Hahn, *Anatomy*, 60–61.

153. MNHN, MS 220. Some of the *paquets* made their way to the archives of the academy: *cartonnier* 13, Dessins Textes non datés pour Hist. nat. des animaux par PERRAULT.

154. MNHN, MS 220.

155. Stroup, *Royal Funding*, 138–42, notes several unspecified payments to Duverney from 1694 to 1697, and payment of three hundred livres to Châtillon in 1695 for "drawings and maps."

156. Mabille and Pieragnoli, *Ménagerie de Versailles*, 31–55; Loisel, *Ménageries*, 2:122–32. Monier (or Mosnier) appears in every volume of the *CdB*. Loisel, *Ménageries*,

2:298, asserts that the project disappears from the royal accounts in 1696, but gives no source. The minutes for 1695–96, which are appended to PV 13 (1689–93), are sparse and do not mention the *Histoire des animaux*.

157. Duverney, *Œuvres*, 1:iv ("Avertissement," by Emmanuel-Maurice Duverney); and Duverney dossier, Archives, AdS.

158. Eloy; and Tauvry, *Nouvelle anatomie raisonnée*. This was reprinted several times and translated into Latin and English.

159. Pitfeild, 262; and *HdA* (1676), 201.

160. Frank, *Harvey and the Oxford Physiologists*.

161. PV 11, fols. 144r–144v, 152r–152v (novembre–décembre 1685); and PV 12, fols. 72r, 77v (février–mars 1688).

162. Méry, "De la manière dont la circulation de sang se fait dans le fetus," dated 31 mars 1692, in *MARS*, 10:65–67. The other treatises are "Pourquoi le foetus et la tortue vivent très-long-temps sans respirer?," in *MARS*, 10:271–75, and "Pourquoi la respiration est nécessaire pour entretenir la vie de l'homme depuis qu'il est sorti de sa mère, & même lorsqu'il y est encore enfermé; & qu'au contraire la tortuë peut vivre très-long-temps sans respirer," in *MARS*, 10:386–97. Méry's arguments are summarized in *HARS*, 2:175–78. See K. J. Franklin, "Jean Méry"; and Méry, *Nouveau système*.

163. Carton Louis Simonneau, AdS archives. These did not appear in any of Méry's publications on this topic.

164. *MARS*, 10:386.

165. *MARS*, 10:393–96.

166. K. J. Franklin, "Jean Méry"; and *HARS*, 2:209–11.

167. Méry, *Nouveau système*.

168. Méry, *Nouveau système*, front matter.

169. On Christofle, see Sturdy, *Science and Social Status*, 191; and Laissus, "Jardin du roi." He became an anatomy demonstrator at the garden in the 1720s; see epilogue, below. On Pierre, see Eloy, s.v. "Du Verney, Pierre." Eloy conflates the two brothers.

CONCLUSION

1. Mabille and Pieragnoli, *Ménagerie de Versailles*, 81–82.

2. Ibid., 31–55.

3. Laissus, "Jardin du roi"; and Barritault, *Anatomie en France*.

4. Gelfand, *Professionalizing Modern Medicine*.

5. See epilogue, below.

6. Tyson, *Phocaena*, 3, 10.

7. Pitfeild, "Publisher to the Reader," n.p.

8. Valentin, *Amphitheatrum*; and Cole, *History of Comparative Anatomy*, 174–75. Valentin is known as Valentini in some sources.

9. Cook, *Matters of Exchange*, 295–98; G. A. Lindeboom, "Sylvius, Franciscus dele Boë," in *DSB*; and Ragland, "Experimenting with Chemical Bodies." Sylvius dele Boë should be distinguished from his sixteenth-century predecessor, Jacques Dubois or Jacobus Sylvius.

10. Bertoloni Meli, *Mechanism, Experiment, Disease*, 43–44.

11. Frank, *Harvey and the Oxford Physiologists*.

12. Tyson, *Phocaena*, 2.

13. Atran et al., *Histoire du concept d'espèce*; Aristotle, *History of Animals*; Gessner, *Historia animalium*, vol. 1; Aldrovandi, *Ornithologiae*; and Johnstone, *Quadrupetibus*.

14. Willughby, *Ornithology*, 20–28.

15. Ray, *Historia plantarum*, vol. 1.
16. Ray, *Synopsis methodica*.
17. See epilogue, below.

EPILOGUE

1. An earlier version of this chapter appeared as Guerrini, "Perrault, Buffon." *HARS*, 1:ii. No discussion of this project appears in the PV for August 1727. This project led to a renumbering of the *Histoire et mémoires* series begun in 1699. The eleven pre-1699 volumes were inserted in this series and numbered 1–11; post-1699 volumes begin with number 12. The pre-1699 volumes include Fontenelle's *Histoire* (vols. 1 and 2). The *Mémoires* (vols. 3–11, vol. 3 in three parts) include *Histoire des animaux*, *Histoire des plantes*, the 1692–93 volumes, La Hire's volumes of the works of individual academicians, work from the *Journal des sçavans*, and other works. The post-1699 *Histoire et mémoires* were reprinted with the new numbering at the same time, around 1730. See Mazauric, *Fontenelle*, 77–80.
2. *Mémoires pour servir à l'histoire naturelle des animaux et des plantes per Messieurs de L'Academie Roiale des Sciences*. The Dutch edition was the beginning of a series reprinting the academy's work before 1699. *HARS*, 1: "Avertissement," n.p.
3. *HARS*, 1: "Avertissement," n.p.
4. A memorial service was held for Duverney on 23 November 1730. MNHN, MS 89.
5. Eloy. On Winsløw's Catholicism, see also Grell, "Between Anatomy and Religion."
6. *HdA* (1733), 1, "Avertissement des libraires" (by Winsløw and Petit), n.p.
7. The coatimundi, for example, was shifted.
8. *HdA* (1733), 1, xvi–xvii.
9. Hanks, *Buffon avant l'"Histoire naturelle"*; and Roger, *Buffon*. On Buffon's tenure at the garden, see Spary, *Utopia's Garden*.
10. On Hunauld and Christofle Duverney, see Laissus, "Jardin du roi."
11. On the relative roles of Buffon and Daubenton, see Farber, "Buffon and Daubenton"; and Loveland, "Another Daubenton, Another *Histoire naturelle*."
12. Roger, *Buffon*, 81–91.
13. Buffon, *Histoire naturelle*, 1:6.
14. Ibid., 1:12–13.
15. *HdA* (1671), preface; and Buffon, *Histoire naturelle*, 4:4.
16. Buffon, *Histoire naturelle*, 4:113.
17. Sloan, "Buffon-Linnaeus Controversy."
18. Buffon, *Histoire naturelle*, 4:114.
19. *HdA* (1671), 29.
20. *HdA* (1733), 1, "Description anatomique de deux chameaux," 71.
21. Buffon, *Histoire naturelle*, vol. 11 (1764), 211ff., skeleton 282.
22. Farber, "Buffon and the Concept of Species"; and Sloan, "From Logical Universals to Historical Individuals."
23. *HdA* (1671), 89–91; and Buffon, *Histoire naturelle*, 8:358.
24. Buffon, *Histoire naturelle*, 11:96n; the article on the elephant is on pp. 1–173.
25. Buffon, *Histoire naturelle*, 11:1–50.
26. Loveland, "Another Daubenton, Another *Histoire naturelle*," 460–61.

BIBLIOGRAPHY

MANUSCRIPTS
Paris
ARCHIVES NATIONALES

Minutier central: will of Duverney.
Ser. AJ/15 (Muséum d'histoire naturelle).
Ser. O/1/1964 (Maison du roi: Inventaire d'œuvres d'art).
Ser. O/1/2124 (Maison du roi: Jardin des plantes).

BIBLIOTHÈQUE INTERUNIVERSITAIRE DE SANTÉ

MSS 12, 13, 14. Paris Faculty of Medicine. *Commentarium Facultatio Medicinae Parisiensis.*
MS 2067. "Lectiones anatomicae Domini Duverney amphitheatro chirugico."

INSTITUT DE FRANCE, ACADÉMIE DES SCIENCES, ARCHIVES

Carton Louis Simonneau.
Cartonnier 13, Dessins Textes non datés pour Hist. nat. des animaux par PERRAULT.
Dossiers of academy members.
Pochettes de séance, 1667–1756.
Procès-verbaux, 1667–1734.

INSTITUT DE FRANCE, BIBLIOTHÈQUE DE L'INSTITUT

Cote M 130 C**. Plagiarized edition of *Histoire des animaux*, 1700.
Cote M 130 C***. Unpublished revisions of *Histoire des animaux*, ca. 1688.

MUSÉUM NATIONAL D'HISTOIRE NATURELLE, ARCHIVES

MSS 89. Material relevant to Duverney.
MS 220. Manuscripts concerning *Memoires . . . histoire des animaux.*
MS 1760. "Osteologie de Mr. Du Vernay 1685."

London
BRITISH LIBRARY

Sloane MS 87. "Traitté des maladies de la poitrine."
Sloane MS 1216. "Traicté de la gangrene, par monsieur du vernay." Paris, 27 July 1683, fols. 45r–74v.
Sloane MS 1499. "Traité des maladies des os."

WELLCOME LIBRARY

MS 4786. Jean-Baptiste Thorière. Lecture notes. Paris, 1715–16.

MS 5433. [Patrick Mitchell]. "Observations upon osteology, taken out of Mr Du Verney's private course, begunn the 22 of October 1697."

Glasgow

UNIVERSITY LIBRARY, SPECIAL COLLECTIONS

MS Hunter D311. James Douglas. "Osteologia."

PUBLISHED PRIMARY SOURCES

Académie des sciences de Paris. *Divers ouvrages de mathématique et de physique*. Paris: Imprimerie Royale, 1693.

———. *Mémoires de mathématique et de physique, tirez des registres de l'Académie royale des sciences*. 2 vols. Paris: Imprimerie Royale, 1692–93.

———. *Recueil de plusieurs traitez de mathématique de l'Académie royale des sciences*. Paris: Imprimerie Royale, 1676.

———. *Recueil des observations faites en plusieurs voyages par ordre de Sa Majesté, pour perfectionner l'Astronomie et la Géographie*. Paris: Imprimerie Royale, 1693.

Académie royale des sciences de Paris. *Histoire de l'Académie royale des sciences*. 2 vols. Vol. 1, 1666–86. Vol. 2, 1686–99. Paris: Gabriel Martin, Jean-Baptiste Coignard, Hippolyte-Louis Guérin, 1733.

———. *Mémoires de l'Académie royale des sciences, depuis 1666 jusqu'à 1699*. Vols. 3–11 (continuous with *Histoire*, above). Paris: Par la Compagnie des Libraires, 1729–34.

———. *Table alphabétique des matières contenues dans l'Histoire & les Mémoires de l'Académie Royale des Sciences, publiées par son Ordre*. Vol. 1, 1666–98. Paris: Par la Compagnie des Libraires, 1734.

Aelian. Αἰλιανου τα εὑρισκομενα ἁπαντα. *Claudii Æliani . . . Opera quæ extant omnia, Græce Latineque e regione, cura et opera C. Gesneri*. Zürich: Gesner Brothers, 1556.

Aldrovandi, Ulisse. *De quadrupedibus digitatis viviparis libri tres et de quadrupedibus digitatis oviparis libri duo*. Frankfurt: Joannes Treudel, 1623.

———. *Ornithologiae tomus alter*. Frankfurt: Wolfgang Richter, 1610.

Anatomia cophonis. In *Source Book of Medieval Science*, edited by Edward Grant. Cambridge, MA: Harvard University Press, 1974.

Aquinas, Thomas. *The Summa theologica of St. Thomas Aquinas, Literally Translated by the Fathers of the English Dominican Province*. London: Burns, Oates, and Washburne, 1920.

Aristotle. *History of Animals*. Edited by Allan Gotthelf. Translated by D. M. Balme. Cambridge, MA: Harvard University Press, Loeb Classical Library, 1991.

———. *On the Soul, Parva naturalia, On Breath*. Translated by W. S. Hett. Cambridge, MA: Harvard University Press, Loeb Classical Library, 1957.

———. *Parts of Animals*. Translated by A. L. Peck. *Movement of Animals, Progression of Animals*. Translated by E. S. Forster. Cambridge, MA: Harvard University Press, Loeb Classical Library, 1968.

———. *Rhetoric*. Translated by John Henry Freese. Cambridge, MA: Harvard University Press, Loeb Classical Library, 1967.

Aselli, Gaspar. *De lactibus, sive Lacteis venis*. Milan: Io. Bapt. Bidellum, 1627.

Bacon, Francis. *Francis Bacon: The New Organon*. Edited by Lisa Jardine and Michael Silverthorne. Cambridge: Cambridge University Press, 2000.

———. *New Atlantis*. London: Tho. Newcomb, 1659.

Baillet, Adrien. *La vie de Monsieur Descartes.* Paris: D. Horthemels, 1691.

Baron, Hyacinthe-Théodore. *Scholis Facultatis Medicinae Parisiensis, agitatae sunt & discussae, series chronologica; cum doctorum praesidum, et baccalaureorum propugnantium nominibus.* Paris: Jean-Thomas Herissant, 1752.

Bartholin, Thomas. *De lacteis thoracicis.* Copenhagen: Melchior Martzan, 1652.

———. *Institutiones anatomicae, novis recentiorum opinionibus & observationibus, quarum innumere hactenus editae non sunt, figuris que secundo auctae ab autoris filio Thoma Bartholino movendo.* Leiden, Netherlands: Franciscus Hack, 1645.

———. *Institutions anatomiques de Gasp. Bartholin, . . . augmentées et enrichies . . . par Thomas Bartholin, . . . et traduictes en françois par Abr. Du Prat.* Paris: M. Hénault, 1647.

———. *Vasa lymphatica: Nuper Hafnia in animantibus inventa et hepatis exsequiae.* Leiden, Netherlands: Franciscus Hack, 1653.

Bauhin, Caspar. *Theatrum anatomicum.* Frankfurt am Main: Matthäus Becker, 1605.

Bellini, Lorenzo. *Exercitatio anatomica de structura et usu renum.* Florence: Sub signo Stellae, 1662.

———. *Gustus organum.* Bologna: Typis Pisarrianis, 1665.

Belon, Pierre. *L'histoire de la nature des oyseaux, avec leurs descriptions et naïfs portraicts retirez du naturel, escrite en sept livres.* Paris: Chez Guillaume Cavellat, 1555.

———. *La nature et diversité des poissons.* 1555. In *L'histoire naturelle au XVIe siècle: Introduction, étude et édition critique de La nature et diversité des poissons de Pierre Belon (1555),* edited and with an introduction by Philippe Glardon. Travaux d'humanisme et Renaissance, vol. 483. Geneva: Droz, 2011.

Bernier, François. *Voyage dans les états du Grand Mogol.* Introduction by France Bhattacharya. Paris: Fayard, 1981.

Blegny, Nicolas. *Le livre commode des addresses de Paris pour 1692 par Abraham du Pradel.* Edited by Edouard Fournier. 2 vols. Paris: Paul Daffis, 1878.

Boileau-Despréaux, Nicolas. *Oeuvres complètes.* 7 vols. Paris: Société les Belles Lettres, 1934–60.

Bonnefon, Paul, ed. "Un poème inédit de Claude Perrault." *Revue d'histoire littéraire de la France* 7 (1900): 449–73.

Bossuet, Jacques-Bénigne. *Introduction à la philosophie, ou de la connaissance de Dieu et de soi-même.* Paris: Robert-Marc d'Espilly, 1722.

Brice, Germain. *Description nouvelle de ce qu'il y a de plus remarquable dans la ville de Paris.* 2 vols. Paris: Abraham Arondeus, 1685.

———. *A New Description of Paris.* 2nd ed. 2 vols. London: Henry Bonwicke, 1688.

Brièle, Léon, ed. *Délibérations de l'ancien bureau de l'Hôtel-Dieu.* In *Collection de documents pour servir à l'histoire des hôpitaux de Paris.* Vols. 1–2. Paris: Imprimerie Nationale, 1881–83.

Buffon, Georges-Louis Leclerc. *Histoire naturelle, générale et particulière, avec la description du Cabinet du roy.* Vols. 1–15. Paris: Imprimerie royale, 1749–67.

Camusat, Denis-François. *Histoire critique des journaux.* 2 vols. Amsterdam: J. F. Bernard, 1734.

Casserio, Giulio. *De vocis auditusque organis historia anatomica.* Ferrara, Italy: Vittorio Baldini, 1600.

Celeberrimorum anatomicorum Severini, Castrensis, Jasolini et Cabrolii varia opuscula anatomica praemissae sunt Observationes Chirurgiae Infusoriae Hominibus adhibitae; Dissertation de Generatione Animalium Theodori Aldes, Angli contra Harvejum, & nova ductus Thoracici cum emulgente communio. M. Gayani Parisiensis, Ex Gallico sermone in Latinum versa. Frankfurt: Herman à Sande, 1668.

Chanet, Pierre. *Considerations de la sagesse de Charon [sic]*. Paris: Claude Le Groult and Jean Le Mire, 1643.

———. *De l'instinct et de la connoissance des animaux. Avec l'examen de ce que Monsieur de la Chambre a escrit sur cette matière*. La Rochelle, France: Toussainct de Gouy, 1646.

Colbert, Jean-Baptiste. *Lettres, instructions, et mémoires de Colbert*. Edited by Pierre Clément. 8 vols. in 10. Paris: Imprimerie impériale, 1861–82.

Courtin, Antoine. *Nouveau traité de la civilité* (1671). Edited by Marie-Claire Grassi. Saint-Étienne, France: Publications de l'Université de Saint-Étienne, 1996.

Courtin, Germain. *Leçons anatomiques et chirurgicales de feu M.ᵉ Germain Courtin . . . recueillies, colligées & corrigées par Estienne Binet*. Paris: D. Langlois, 1612.

Cureau de la Chambre, Marin. *Les charactères des passions*. Vol. 1, 1640. Reprint, Amsterdam: Antoine Michel, 1658. Vol. 2, 1645. Reprint, Paris: P. Rocolet, 1660.

———. *A Discourse of the Knowledg [sic] of Beasts*. London: Tho. Newcomb for Humphrey Mosele, 1657.

———. *Nouvelles conjectures sur la digestion*. Paris: Pierre Rocolet, 1636.

———. *Le système de l'âme*. Paris: Jacques D'Allin, 1664.

———. *Traité de la connoissance des animaux*. 1648. Reprint, Paris: Jacques d'Allin, 1664.

D'Assoucy, Charles Coypeau. *Le combat de Cyrano de Bergerac avec le singe de Brioché, au bout du Pont-Neuf*. 1653. Reprint, Paris: M. Rebuffe le jeune, 1704.

De Graaf, Regnier. *Histoire anatomique des parties génitales: De l'homme et de la femme, qui servent à la génération: Avec un Traité du suc pancréatique, des clistères et de l'usage du syphon*. Basel, Switzerland: Emanuel Jean George König, 1699.

———. *Opera omnia*. Leiden, Netherlands: Officina Hackiana, 1677.

———. *Traitté de la nature et de l'usage du suc pancréatique, où plusieurs maladies sont expliquées, principalement les fièvres intermittentes*. Paris: O. de Varennes, 1666.

Denis, Jean-Baptiste. *Lettre écrite à M. Sorbière, . . . par Jean Denis touchant l'origine de la transfusion du sang, et la manière de la pratiquer sur les hommes*. Paris: J. Cusson, [1668].

———. *Recueil des mémoires et conférences qui on esté présentées a Monseigneur Le Dauphin pendent l'année M.DC.LXXII*. Paris: Frédéric Léonard, 1672[–74].

Descartes, René. *Œuvres*. Edited by Charles Adam and Paul Tannery. New ed. 11 vols. Paris: CNRS, 1974–86.

———. *The Philosophical Writings of Descartes*. Edited and translated by John Cottingham, Robert Stoothoff, and Dugald Murdoch. 3 vols. Cambridge: Cambridge University Press, 1985–91.

Description anatomique d'vn caméléon, d'vn castor, d'vn dromadaire, d'vn ours, et d'vn gazelle. Paris: Frédéric Léonard, 1669.

Dionis, Pierre. *L'anatomie de l'homme*. 2nd ed. Paris: Laurent d'Houry, 1694.

Du Hamel, Jean-Baptiste. *De corpore animato*. Paris: Étienne Michallet, 1673.

———. *Regiæ scientiarum academiæ historia*. Paris: J. B. Delespine, 1699.

Du Laurens, André. *Discours de la conservation de la veuë, des maladies mélancoliques, des catarrhes, et de la vieillesse*. 1594. Reprint, Paris: Jamet Mattayer, 1597.

———. *Historia anatomica humani corporis*. Paris: Marc Orry, 1600.

[Duverney, Joseph-Guichard]. *Lettre à Monsieur ***, contenant plusieurs nouvelles observations sur l'ostéologie*. Paris: Laurent d'Houry, 1689.

———. *Œuvres anatomiques de M. Duverney*. 2 vols. Paris: Charles-Antoine Jombert, 1761.

———. *Traité de l'organe de l'ouïe*. Paris: Estienne Michallet, 1683.

———. *Traité des maladies des os*. 2 vols. Paris: Le Bure l'ainé, 1751.

Estienne, Charles, and Estienne de la Rivière. *La dissection des parties du corps humain, divisée en trois livres*. Paris: Simon de Colines, 1546.

Extrait d'une lettre écrite à Monsieur de La Chambre, qui contient les observations qui ont esté faites sur un grand poisson disséqué dans la Bibliothèque du Roy, le 24e juin 1667.—Observations qui ont esté faites sur un lion disséqué dans la Bibliothèque du Roy, le 28e juin 1667, tirées d'une lettre écrite à M. de La Chambre. Paris: Frédéric Léonard, 1667.

Fabricius ab Acquapendente. *De visione, voce, auditu.* Venice: Francesco Bolzetta, 1600.

Faret, Nicolas. *L'honneste-homme, ou, L'art de plaire à la court.* Paris: Toussaincts du Bray, 1630.

Fontenelle, Bernard le Bovier de. *Eloges des académiciens avec l'histoire de l'Académie royale des sciences en MDCXCIX.* 2 vols. The Hague: Isaac van der Kloot, 1740.

Galen. *On the Usefulness of the Parts of the Body.* Translated and edited by Margaret Tallmadge May. 2 vols. Ithaca, NY: Cornell University Press, 1968.

Galilei, Galileo. *Two New Sciences.* Translated by Stillman Drake. Madison: University of Wisconsin Press, 1974.

Gassendi, Pierre. *The Mirrour of True Nobility and Gentility. Being the Life of the Renowned Nicolaus Claudius Fabricius Lord of Pieresk, Senator of the Parlement at Aix.* Translated by William Rand. London: J. Streater for Humphrey Moseley, 1657.

———. *Opera omnia.* 6 vols. Lyon: L. Anisson, 1658.

Gessner, Conrad. *Historiae animalium.* 5 vols. Zurich: Christian Froschauer, 1551–58, 1587.

Gouye, Thomas, ed. *Observations physiques et mathématiques pour servir à l'histoire naturelle, & à la perfection de l'astronomie & de la géographie.* Paris: Veuve Edmé Martin, Jean Boudot, and Estienne Martin, 1688.

Habicot, Nicolas. *La semaine ou pratique anatomique.* 1610. New ed. Paris: Michel Bobin, 1660.

Harvey, William. *Anatomical Exercitations Concerning the Generation of Living Creatures.* London: Octavian Pulleyn, 1653.

———. *The Circulation of the Blood and Other Writings.* Translated by Kenneth J. Franklin. London: Dent, 1990.

———. *De motu cordis.* Frankfurt: Wilhelm Fitzer, 1628.

———. *Exercitationes de generatione animalium.* London: Octavian Pulleyn, 1651.

Hernández, Francisco, et al. *Rerum medicarum novae Hispaniae thesaurus, seu, plantarum animalium mineralium Mexicanorum historia.* Rome: Vitalis Mascardi, 1651.

Héroard, Jean. *Hippostologie, c'est à dire, discovrs des os dv cheval.* Paris: Mamert Patisson, 1599.

Highmore, Nathaniel. *Corporis humani disquisitio anatomica.* The Hague: Samuel Brown, 1651.

Houel, Jean-Pierre-Louis-Laurent. *Histoire naturelle des deux éléphans, mâle et femelle, du Muséum de Paris, venus de Hollande en France en l'an VI.* Paris: Chez l'Autheur [et al.], An XII–1803.

Hunter, William. *Two Introductory Lectures, Delivered by Dr. William Hunter, to His Last Course of Anatomical Lectures, at His Theatre in Windmill-Street.* London: J. Johnson, 1784.

Huygens, Christiaan. *Œuvres complètes.* 22 vols. The Hague: Martinus Nijhoff, Société hollandaise des sciences, 1888–1950.

———. *Relation d'une observation faite à la Bibliothèque du roy, à Paris, le 12. may 1667. sur les neuf heures du matin, d'un halo ou couronne à l'entour du Soleil; avec un discours de la cause de ces météores, & de celle des parelies.* Paris: Frédéric Léonard, 1667.

Johnstone, John. *A Description of the Nature of Four-Footed Beasts: With Their Figures Engraven in Brass.* Translated by J.P. London: Moses Pitt, 1678.

———. *Historiae naturalis de quadrupetibus [sic] libri: Cum aeneis figuris. [de avibus; de piscibus et cetis; de exanguibus aquaticis; de insectis . . . de serpentibus et draconibus].* 4 vols. Frankfurt am Main: heirs of Matthäus Merian, [1650–53].

————. *Theatrum universale omnium animalium, piscium, avium, quadrupedum, exanguium aquaticorum, insectorum et angium, CCLX tabulis ornatum, ex scriptoribus tam antiquis quam recentioribus*. Edited by Frederik Ruysch. 2 vols. Amsterdam: R. and G. Wetsteen, 1718.

Jombert, Charles-Antoine. *Catalogue raisonné de l'œuvre de Sébastien Leclerc*. 2 vols. Paris: Charles-Antoine Jombert, 1774.

La Fontaine, Jean de. *Fables choisies: Mises en vers. . . .* Pt. 3. Paris: Denys Thierry and Claude Barbin, 1678.

————. *Fables, contes et nouvelles*. Edited by René Groos and Jacques Schiffrin. Paris: Gallimard, 1954.

————. *Œuvres diverses*. Edited by Pierre Clarac. Paris: Gallimard, 1958.

La Martinière, Pierre-Martin de. *A Monseigneur Colbert . . . sur le sujet de la transfusion du sang*. [Paris], n.d.

Lamy, Bernard. *Entretiens sur les sciences, dans lesquels on apprend comme l'en doit étudier les sciences*. Lyon: J. Certe, 1706.

Lamy, Guillaume. *Discours anatomiques de M. Lamy, . . . avec des Réflexions sur les objections qu'on luy a faites contre sa manière de raisonner de la nature de l'homme et de l'usage des parties qui le composent et cinq lettres du même autheur sur le sujet de son livre*. Rouen: J. Lucas, 1675.

————. *Explication mécanique et physique des fonctions de l'âme sensitive, ou Des sens, des passions et du mouvement volontaire*. 2nd ed. Paris: L. Roulland, 1681.

La Ramée, Pierre de. "Advertissements sur la reformation de l'Université de Paris." In *Archives curieuses de l'histoire de France depuis Louis XI jusqu'à Louis XVIII*, edited by L. Cimber and F. Danjou. Paris: Beauvais, 1834.

La Varenne, Pierre. *Le cuisinier François*. Paris: Pierre David, 1651.

Le Clerc, Daniel, and Jean-Jacques Manget. *Bibliotheca anatomica, medica, chirurgica, &c. Containing a Description of the Several Parts of the Body*. 2 vols. London: John Nutt for W. Lewis et al., 1711.

————. *Bibliotheca anatomica, sive Recens in anatomia inventorum thesaurus locupletissimus*. 2 vols. Geneva: J.-A. Chouët, 1685.

————. *Bibliotheca anatomica; Sive, Recens in anatomia inventorum thesaurus locupletissimus*. 2nd ed. 2 vols. Geneva: Jean-Antoine Chouët and David Ritter, 1699.

L'Écluse, Charles de. *Exoticorum libri decem*. [Antwerp]: Plantin-Raphelengius, 1605.

Lecoq, Anne-Marie, ed. *La querelle des anciens et des modernes: XVIIe.–XVIIIe. siècles*. Paris: Gallimard, 2001.

Le Gallois, Pierre, ed. *Conversations académiques, tirées de l'Académie de Monsieur l'Abbé Bourdelot*. 2 vols. Vol. 1, Paris: Claude Barbin, 1674; vol. 2, Louis Bilaine, 1674.

Le Laboureur, Louis. *Avantages de la langue Françoise sur la langue Latine*. Paris: Guillaume De Luyne, 1669.

Le Maire, Charles. *Paris ancien et nouveau*. 3 vols. 1685. Reprint, Paris: Nicolas LeClerc, 1697–98.

Lettres écrites sur le sujet d'une nouvelle découverte touchant la vue. Paris: Jean Cusson, 1682.

Lister, Martin. *A Journey to Paris in the Year 1698*. Edited by R. P. Stearns. Urbana: University of Illinois Press, 1967.

Lower, Richard. *Tractatus de corde*. London: J. Redmayne for James Allestry, 1669.

Lyser, Michael. *Culter anatomicus*. Copenhagen: Georg Lamprecht, 1653.

Magalotti, Lorenzo, ed. *Saggi di natvrali esperienze fatte nell'Accademia del Cimento*. Florence: Giuseppi Cocchini, 1667.

Malebranche, Nicolas. *The Search after Truth*, translated by Thomas M. Lennon and Paul J.

Olscamp; *Elucidations of "The Search after Truth,"* translated by Thomas M. Lennon. Columbus: Ohio State University Press, 1980.

Malpighi, Marcello. "An Account of Some Discoveries Concerning the Brain, and the Tongue, Made by Signior Malpighi, Professor of Physick in Sicily." *Philosophical Transactions* 2 (1666–67): 491–92.

Malpighi, Marcello, and Carlo Fracassati. *Tetras anatomicarum epistolarum de lingua, et cerebro.* Bologna: Vittorio Benati, 1665.

Mariotte, Edmé. *Œuvres.* 2 vols. Leiden, Netherlands: P. Van der Aa, 1717.

Mariotte, Edmé, and Jean Pecquet. *Nouvelle découverte touchant la veüe.* Paris: Frédéric Léonard, 1668.

———. *Seconde lettre de M. Mariotte à M. Pecquet pour montrer que la choroïde est le principal organe de la veüe.* Paris: Jean Cusson, 1671.

Martet, Jean. *Abbrégé des nouvelles expériences anatomiques.* Paris: C. de Sercy et J. Guignard, 1664.

Mémoires pour servir à l'histoire naturelle des animaux et des plantes per Messieurs de L'Academie Roiale des Sciences. The Hague: P. Gosse & I. Neaulme, 1731.

Mercier, Louis-Sébastien. *Tableau de Paris.* 12 vols. Amsterdam: n.p., 1782.

Mersenne, Marin. *Correspondance de P. Marin Mersenne.* Edited by P. Tannery et al. 17 vols. Paris: CNRS, 1933–88.

———. *Harmonie universelle contenant la théorie et la pratique de la musique.* 2 vols. Paris: S. Cramoisy, 1636.

Méry, Jean. *Observations de la manière de tailler . . . Nouveau système de la circulation du sang par le trou ovale.* Paris: Jean Boudot, 1700.

Molière. *Les femmes sçavantes.* Paris: Pierre Promé, 1672.

———. *Le malade imaginaire.* Paris: Christophe Ballard, 1673.

———. *Les précieuses ridicules.* Paris: Jean Ribou, 1660.

———. *Le tartuffe, ou l'imposteur.* Paris: Claude Barbin, 1673.

Oldenburg, Henry. *Correspondence.* Edited by A. Rupert Hall and Marie Boas Hall. 10 vols. Madison: University of Wisconsin Press, 1965–75.

Pardies, Ignace-Gaston. *Discours de la connoissance des bestes.* Paris: Sébastien Mabre-Cramoisy, 1672.

Paradis de Moncrif, François-Augustin de. *Les chats.* Paris: Gabriel-François Quillau, 1727.

Pascal, Blaise. *De l'esprit géométrique et l'art de persuader.* In *Pascal, l'art de persuader, précédé de l'art de conférer de Montaigne.* Paris: Payot et Rivages, 2001.

Pecquet, Jean. *Experimenta nova anatomica.* Paris: S. et G. Cramoisy, 1651.

———. *Experimenta nova anatomica.* 2nd ed. Paris: Officina Cramoisiana, 1654.

———. *New Anatomical Experiments.* Translated by Nicholas Culpepper. London: T.W. for Octavian Pulleyn, 1653.

———. "Observations sur un concrétion calculeuse de la matrice." *Mémoires de l'Académie Royale de Chirurgie* 2 (1753): 585–86.

Peiresc, Nicolas-Claude Fabri de. *Lettres à Cassiano dal Pozzo, 1626–1637.* Edited by Jean-François Lhote and Danielle Joyal. Clermont-Ferrand, France: Adosa, 1989.

———. *Lettres de Peiresc.* Edited by P. Tamizey de Larroque. 7 vols. Paris: Imprimerie nationale, 1888–98.

Perrault, Charles. *Contes.* Edited by Marc Soriano. Paris: Flammarion, 1989.

[———]. *Critique de l'opéra.* Paris: Louis Bilaine, 1674.

———. *Les hommes illustres qui ont paru en France pendant ce siècle.* The Hague: Mathieu Rogguet, 1698.

————. *Labyrinte de Versailles*. Paris: Imprimerie royale, 1677.

————. *Recueil de diverses ouvrages en prose et en vers*. Paris: Jean-Baptiste Coignard, Guillaume De Luynes, Jean Guignard fils, 1675.

————. *Le siècle de Louis le Grand*. Paris: Jean-Baptiste Coignard, 1687. In *Le querelle des anciens et des modernes*, edited by Lecoq, 256–73.

[Perrault, Charles, and Claude Perrault]. *Le cabinet des beaux-arts ou recueil des plus belles estampes, gravées d'après les tableaux originaux, où les beaux-arts sont représentez, avec l'explication de ces mêmes tableaux*. Paris: G. Edelinck, 1690.

————. *Mémoires de ma vie par Charles Perrault, Voyage à Bordeaux par Claude Perrault*. Edited by Paul Bonnefon. Paris: Renouard, 1909.

Perrault, Charles, Claude Perrault, and Pierre Perrault. *Les murs de Troye ou l'origine du burlesque*. Paris: Louis Chamhoudry, 1653.

[Perrault, Claude]. *Les dix livres d'architecture de Vitruve, corrigez et traduits nouvellement en François, avec des notes et des figures*. Paris: Jean-Baptiste Coignard, 1673.

————. *"Du bruit" et "De la musique des anciens."* Edited by François Lesure. Geneva: Editions Minkoff, 2003.

————. *Essais de physique*. 4 vols. Paris: Jean-Baptiste Coignard, 1680–88.

[————, ed.]. *Mémoires pour servir à l'histoire naturelle des animaux*. Paris: Imprimerie royale, 1671.

————, ed. *Mémoires pour servir à l'histoire naturelle des animaux*. Paris: Imprimerie royale, 1676.

[————, ed.]. *Memoir's [sic] for a Natural History of Animals. Containing the Anatomical Descriptions of Several Creatures Dissected by the Royal Academy of Sciences at Paris*. Translated by Alexander Pitfeild. London: Joseph Streater for T. Basset, 1688.

————. *Ordonnance des cinq espèces de colonnes selon des méthodes des anciens*. Paris: Jean Coignard, 1683.

————. *A Treatise of the Five Orders of Columns in Architecture*. Translated by John James. London: Benjamin Motte, 1708.

Physiologus. Translated by Michael J. Curley. 1979. Reprint, Chicago: University of Chicago Press, 2009.

Pineau, Sévérin. *Opusculum physiologicum & anatomicum in duos libros distinctum*. Paris: Steph. Prevosteau, 1597.

Platter, Felix. *Beloved Son Felix: The Journal of Felix Platter, a Medical Student in Montpellier in the Sixteenth Century*. Translated and with an introduction by Seán Jennett. London: Frederick Muller, 1961.

Quesnay, François. *Recherches critiques et historiques sur l'origine, sur les divers états et sur les progrès de la chirurgie en France*. Paris: Charles Osmont, 1744.

Quinault, Philippe. *Alceste, suivi de La Querelle d'Alceste*. Edited by William Brooks, Buford Norman, and Jeanne Morgan Zarucchi. Geneva: Droz, 1994.

Ray, John. *Historia plantarum generalis*. Vol. 1. London: Samuel Smith and Benjamin Walford, 1693.

————. *Synopsis methodica animalium quadrupedum et serpentini generis: Vulgarium notas characteristicas, rariorum descriptiones integras exhibens: Cum historiis & observationibus anatomicis perquam curiosis*. London: S. Smith and B. Walford, 1693.

Riolan, Jean. *Encheiridium anatomicum et pathologicum*. Leiden, Netherlands: Adriaan Wyngaerden, 1649.

————. *Oeuvres anatomiques*. Translated by Pierre Constant. 2 vols. Paris: Denys Moreau, 1628–29.

———. *Opuscula anatomica nova*. London: Miles Flesher, 1649.

Rondelet, Guillaume. *Libri de piscibus marinis: In quibus verae piscium effigies expressae sunt*. Lyon: Mathiam Bonhomme, 1554.

Rorario, Girolamo [Hieronymus Rorarius]. *Quod animalia bruta ratione utantur melius homine*. Paris: Sébastien and Gabriel Cramoisy, 1648.

Rudbeck, Olof. "A Translation of Olof Rudbeck's *Nova exercitatio anatomica* Announcing the Discovery of the Lymphatics (1653)." Translated by Aage E. Nielsen. Biographical note by Göran Liljestrand. *Bulletin of the History of Medicine* 11 (1942): 304–39.

Ruini, Carlo. *Anatomia del cavallo, infermita et suoi rimedii . . . del signor Carlo Ruini*. 1598. Reprint, Venice: Fioravanti Prati, 1618.

Saunders, J. B. D. M., and C. D. O'Malley. "The Preparation of the Human Skeleton by Andreas Vesalius of Brussels: An Annotated Translation of the 39th Chapter of the *De humani corporis fabrica*." *Bulletin of the History of Medicine* 20 (1946): 433–60.

Scudéry, Madeleine de. "Histoire de deux caméléons." In *Nouvelles conversations de morale*, vol. 2, 496–629. Paris: Veuve Sébastien Mabre-Cramoisy, 1688.

———. *La promenade de Versailles*. Paris: Claude Barbin, 1669.

Severino, Marco Aurelio. *Zootomia democritaea*. Nuremberg: Literis Endterianis, 1646.

Sévigné, Marie de. *Correspondance de Mme de Sévigné*. Edited by Roger Duchêne. 3 vols. Paris: Bibliothèque de la Pléiade, 1973–78.

Sibbald, Robert. *The Memoirs of Sir Robert Sibbald*. Edited by Francis Paget Hett. Oxford: Oxford University Press, 1932.

Sorbière, Samuel. *Discours sceptique sur le passage du chyle et sur le mouvement du cœur, où sont touchées quelques difficultés sur les opinions des veines lactées et de la circulation du sang*. Leiden, Netherlands: Jean Maire, 1648.

———. *Lettres et discours de M. de Sorbière sur divers matières curieuses*. Paris: François Clousier, 1660.

———. *Relations, lettres, et discours de Mr de Sorbière sur diverse matières curieuses*. Paris: Robert de Ninville, 1660.

Stensen, Niels. *De musculis et glandulis observationum specimen, cum epistolis duabus anatomicis*. Copenhagen: Matthias Godichenius, 1664.

———. *Discours de Monsieur Stenon, sur l'anatomie du cerveau*. Paris: Robert de Ninville, 1669.

Tardy, Claude. *Cours de médecine*. Paris: Chez l'Autheur, Jean du Bray, Claude Barbin, 1662.

———. *Traitté de la monarchie du cœur en l'homme*. Paris: Veuve du Puis, 1656.

———. *Traitté de l'ecoulement du sang d'un homme dans les venes d'un autre, & de ses utilitez*. Paris: Chez l'autheur, Jean du Bray, Claude Barbin, 1667.

Tauvry, Daniel. *Nouvelle anatomie raisonnée*. Paris: Étienne Michallet, 1690.

Topsell, Edward. *The History of Foure-Footed Beastes*. London: William Jaggard, 1607.

Tyson, Edward. *Phocaena, or the Anatomy of a Porpess*. London: Benjamin Tooke, 1680.

Valentin, Michael Bernhard. *Amphiteatrum zoocomicum*. Frankfurt am Main, Germany: Heirs of Zunnerianus and Johann Adam Jung, 1720.

Vitruvius. *Les dix livres de architecture*. Translated by Claude Perrault. Paris: Jean Coignard, 1673.

Willughby, Francis. *The Ornithology of Francis Willughby*. Edited by John Ray. London: A.C. for John Martyn, 1678.

Zipes, Jack, ed. and trans. *Beauties, Beasts, and Enchantment: Classic French Fairy Tales*. New York: New American Library, 1989.

PERIODICALS

Gazette de France.
Histoire de l'Académie royale des sciences, avec les mémoires.
Journal des sçavans.
Mercure galant.
Nouvelles de la république des lettres.
Philosophical Transactions.

REFERENCE WORKS

Collège de France. *Liste des professeurs depuis la fondation du collège de France en 1530.* Paris: Collège de France, affaires culturelles et relations extérieures, 2007.

Complete Dictionary of Scientific Biography, edited by Noretta Koertge. Gale, online.

Desmond, Ray. *Dictionary of British and Irish Botanists and Horticulturalists.* 2nd ed. London: Taylor and Francis and the Natural History Museum, 1994.

Dictionnaire de l'Académie française, 1694. http://www.lib.uchicago.edu/efts/ARTFL/projects /dicos/academie/premiere/artfl.

Eloy, Nicolas Francis Joseph. *Dictionnaire historique de la médecine ancienne et moderne, ou mémoires disposés en ordre alphabétique pour servir à l'histoire de cette science et à celle des médecins, anatomistes, botanistes, chirurgiens et chymistes de toutes nations.* 4 vols. Mons, Belgium: H. Hoyois, 1778.

Furetière, Antoine. *Dictionnaire universel.* The Hague: A. and R. Leers, 1690.

Guiffrey, Jules, ed. *Comptes des bâtiments du roi sous la règne de Louis XIV.* 5 vols. Paris: Imprimerie nationale, 1881–1901.

Halleux, Robert, James McClellan, Daniela Berariu, and Geneviève Xhayet. *Les Publications de l'Académie royale des sciences de Paris (1666–1793).* 2 vols. Turnhout, Belgium: Brepols, 2001.

Oxford Dictionary of National Biography. Edited by Colin Matthew and Brian Harrison. Oxford: Oxford University Press, 2004–, and online.

SECONDARY SOURCES

Ablondi, Fred. "Géraud de Cordemoy (1626–1684)." In *Stanford Encyclopedia of Philosophy.* http://plato.stanford.edu/entries/cordemoy/.

Acheson, Katherine. "Gesner, Topsell, and the Purposes of Pictures in Early Modern Natural Histories." In *Printed Images in Early Modern Britain: Essays in Interpretation,* edited by Michael Hunter, 127–44. Burlington, VT: Ashgate, 2010.

Adkins, G. Matthew. "The Montmor Discourse: Science and the Ideology of Stability in Old Regime France." *Journal of the Historical Society* 1 (2005): 1–28.

Albou, Philippe. "L'*Arrêt burlesque* de Boileau, Gui Patin et l'Académie Lamoignon." *Histoire des sciences médicales* 28 (1994): 25–32.

Ambrose, Charles T. "Immunology's First Priority Dispute: An Account of the 17th-Century Rudbeck-Bartholin Feud." *Cellular Immunology* 242 (2006): 1–8.

Apostolidès, Jean-Marie. *Le Roi-machine: Spectacle et politique au temps de Louis XIV.* Arguments. Paris: Éditions de Minuit, 1981.

Ariew, Roger. "Damned If You Do: Cartesians and Censorship, 1663–1706." *Perspectives on Science* 2 (1994): 255–74.

Armogathe, Jean-Robert. "Le groupe de Mersenne et la vie académique Parisienne." *XVIIe siècle* 44 (1992): 131–39.

Ashworth, William B., Jr. "Emblematic Natural History of the Renaissance." In *Cultures of Natural History*, edited by Nicholas Jardine, Anne Secord, and Emma Spary, 17–37. New York: Cambridge University Press, 1996.

———. "Marcus Gheeraerts and the Aesopic Connection in Seventeenth-Century Scientific Illustration." *Art Journal* 44 (1984): 132–38.

Asúa, Miguel de, and Roger French. *A New World of Animals: Early Modern Europeans on the Creatures of Iberian America*. Farnham, UK: Ashgate, 2005.

Atran, Scott. *Cognitive Foundations of Natural History: Towards an Anthropology of Science*. Cambridge: Cambridge University Press, 1990.

Atran, Scott, et al. *Histoire du concept d'espèce dans les sciences de la vie*. Paris: Editions de la Fondation Singer-Polignac, 1987.

Azouvi, François. "Entre Descartes et Leibniz: L'animisme dans les 'Essais de physique' de Claude Perrault." *Recherches sur le XVIIème siècle* 5 (1982): 9–19.

Balayé, Simone. *La Bibliothèque nationale des origines à 1800*. Geneva: Droz, 1988.

Baratay, Eric. "Des naturalistes dans les cages: L'exemple de Claude Perrault (1613–1688)." In *La Bête captive au Moyen Âge et à l'époque moderne*, edited by C. Beck and F. Guizard, 163–76. Amiens, France: Encrage, 2012.

Baratay, Eric, and Elisabeth Hardouin-Fugier. *Zoo: A History of Zoological Gardens in the West*. Translated by Oliver Welsh. London: Reaktion Books, 2002.

Baridon, Michel. *A History of the Gardens of Versailles*. Translated by Adrienne Mason. Philadelphia: University of Pennsylvania Press, 2008.

Barritault, Georges. *L'anatomie en France au XVIIIᵉ siècle. Les anatomistes du Jardin du roi*. Angers, France: L'imprimerie d'Anjou, 1940.

Bernard, Auguste. *Histoire de l'Imprimérie Royale du Louvre*. Paris: Imprimérie Impériale, 1867.

Berger, Robert W. *In the Garden of the Sun King: Studies on the Park of Versailles under Louis XIV*. Washington, DC: Dumbarton Oaks Research Library and Collection, 1985.

Bertoloni Meli, Domenico. "Authorship and Teamwork around the Cimento Academy: Mathematics, Anatomy, Experimental Philosophy." *Early Science and Medicine* 2 (2001): 65–95.

———. "The Collaboration between Anatomists and Mathematicians in the Mid-seventeenth Century with a Study of Images as Experiments and Galileo's Role in Steno's Myology." *Early Science and Medicine* 13 (2008): 665–709.

———. *Mechanism, Experiment, Disease: Marcello Malpighi and Seventeenth-Century Anatomy*. Baltimore: Johns Hopkins University Press, 2011.

———. *Thinking with Objects: The Transformation of Mechanics in the Seventeenth Century*. Baltimore: Johns Hopkins University Press, 2006.

Beullens, Pieter, and Allan Gotthelf. "Theodore Gaza's Translation of Aristotle's *De Animalibus*: Content, Influence, and Date." *Greek, Roman, and Byzantine Studies* 47 (2010): 469–513.

Biagioli, Mario. "Etiquette, Interdependence, and Sociability in Seventeenth-Century Science." *Critical Inquiry* 22 (1996): 193–238.

———. *Galileo's Instruments of Credit*. Chicago: University of Chicago Press, 2006.

Bigourdan, G. "Les premières sociétés scientifiques de Paris au XVIIᵉ siècle." *Comptes rendus hebdomadaires des séances de l'Académie des sciences* 163 (1916): 937–43.

———. "Les premières sociétés scientifiques de Paris au XVIIᵉ siècle—Les réunions du P. Mersenne et l'Académie de Montmor." *Comptes rendus hebdomadaires des séances de l'Académie des sciences* 164 (1917): 129–34, 159–62, 216–20.

Birn, Raymond. "Le Journal des savants sous l'ancien régime." *Journal des savants* (1965): 15–35.

Blair, Ann M. *Too Much to Know: Managing Scholarly Information before the Modern Age.* New Haven, CT: Yale University Press, 2011.

Boas, George. *The Happy Beast in French Thought of the Seventeenth Century.* Baltimore: Johns Hopkins University Press, 1933.

Bodson, Liliane. "Living Reptiles in Captivity: A Historical Survey from the Origins to the End of the XVIIIth Century." *Acta zoologica et pathologica Antwerpensia* 78 (1984): 15–32.

Brain, Peter. *Galen on Bloodletting.* Cambridge: Cambridge University Press, 1986.

Brantz, Dorothée, ed. *Beastly Natures: Animals, Humans, and the Study of History.* Charlottesville: University of Virginia Press, 2010.

Brian, Eric, and Christiane Demeulenaere-Douyère, eds. *Histoire et mémoire de l'Académie des sciences. Guide de recherches.* Paris: Lavoisier Tec-Doc, 1996.

Briggs, Robin. "The Académie royale des sciences and the Pursuit of Utility." *Past and Present* 131 (May 1991): 38–88.

———. *Early Modern France, 1560–1715.* New York: Oxford University Press, 1998.

Brockliss, L. W. B. *French Higher Education in the Seventeenth and Eighteenth Centuries: A Cultural History.* Oxford: Clarendon Press, 1987.

———. "Harvey, Torricelli, and the Institutionalization of New Ideas in the [*sic*] 17th Century France." In *Wissensideale und Wissenkulturen in der Frühen Neuzeit / Ideals and Cultures of Knowledge in Early Modern Europe,* edited by Wolfgang Detel and Claus Zittel, 115–34. Berlin: Academie Verlag, 2002.

———. "The Literary Image of the Médecins du Roi in the Literature of the Grand siècle." In *Medicine at the Courts of Europe, 1500–1837,* edited by Vivian Nutton, 117–154. London: Routledge, 1990.

———. "Medical Teaching at the University of Paris, 1600–1720." *Annals of Science* 35 (1978): 221–51.

———. "Seeing and Believing: Contrasting Attitudes towards Observational Autonomy among French Galenists in the First Half of the 17th Century." In *Medicine and the Five Senses,* edited by W. F. Bynum and Roy Porter, 69–84. Cambridge: Cambridge University Press, 1993.

Brockliss, L. W. B., and Colin Jones. *The Medical World of Early Modern France.* Oxford: Clarendon Press, 1997.

Brown, Harcourt. "Un cosmopolite du Grand Siècle: Henri Justel." *Bulletin—Société de l'histoire du protestantisme français,* 6th ser., 6 (1933): 187–201.

———. "History and the Learned Journal." *Journal of the History of Ideas* 22 (1972): 365–78.

———. "Jean Denis and Transfusion of Blood, Paris, 1667–1668." *Isis* 39 (1948): 15–29.

———. *Scientific Organizations in Seventeenth-Century France.* Baltimore: Williams and Wilkins, 1934.

Brunner, Bernd. *Bears: A Brief History.* Translated by Lori Lantz. New Haven, CT: Yale University Press, 2007.

Burke, Peter. *The Fabrication of Louis XIV.* New Haven, CT: Yale University Press, 1994.

———. "Images as Evidence in Seventeenth-Century Europe." *Journal of the History of Ideas* 64 (2003): 273–96.

———. "The Language of Gesture in Early Modern Italy." In *A Cultural History of Gesture,* edited by Jan Bremmer and Herman Roodenburg, 71–83. Ithaca, NY: Cornell University Press, 1992.

Bylebyl, Jerome J. "Cesalpino and Harvey on the Portal Circulation." In *Science, Medicine and Society in the Renaissance,* edited by A. G. Debus, 39–52. New York: Science History Publications, 1972.

———. "The Medical Side of Harvey's Discovery." In *William Harvey and His Age*, edited by Jerome J. Bylebyl, 28–102. Baltimore: Johns Hopkins University Press, 1979.

———. "Nutrition, Quantification, and Circulation." *Bulletin of the History of Medicine* 51 (1977): 369–85.

Campardon, Émile. *Les spectacles de la foire*. 2 vols. Paris: Berger Levrault, 1877.

Cap, Paul-Antoine. *Le muséum d'histoire naturelle*. Paris: L. Curmer, 1854.

Carlino, Andrea. *Books of the Body*. Chicago: University of Chicago Press, 1999.

———. "Paper Bodies: A Catalogue of Anatomical Fugitive Sheets, 1538–1687." Supplement 19, *Medical History*. London: Wellcome Institute for the History of Medicine, 1999.

Carlino, Andrea, and Hélène Cazès. "Plaisir de l'anatomie, plaisir du livre." *Cahiers de l'Association internationale des études françaises* 55 (2003): 251–74.

Castle, Terry. *Masquerade and Civilization*. Stanford, CA: Stanford University Press, 1986.

Cavallo, Sandra. *Artisans of the Body*. Manchester, UK: Manchester University Press, 2007.

Chardans, Jean-Louis. *Le Châtelet: De la prison au théâtre*. Paris: Pygmalion, W. Watelet, 1980.

Chartier, Roger, ed. *La correspondance: Les usages de la lettre aux XIXᵉ siècle*. Paris: Fayard, 1991.

———. *The Cultural Uses of Print in Early Modern France*. Translated by Lydia G. Cochrane. Princeton, NJ: Princeton University Press, 1987.

———. *The Order of Books*. Translated by Lydia G. Cochrane. Stanford, CA: Stanford University Press, 1994.

Chauvois, L. "Molière, Boileau, La Fontaine et la circulation du sang." *La presses médicale* 62, no. 59 (September 1954): 1219–20.

Clarke, Desmond. *Descartes' Philosophy of Science*. State College: Pennsylvania State University Press, 1982.

———. "Descartes' Philosophy of Science and the Scientific Revolution." In *The Cambridge Companion to Descartes*, edited by John Cottingham, 258–85. Cambridge: Cambridge University Press, 1992.

———. *Occult Powers and Hypotheses: Cartesian Natural Philosophy under Louis XIV*. Oxford: Clarendon Press, 1989.

Clarke, Jack A. "Abbé Jean-Paul Bignon, 'Moderator of the Academies' and Royal Librarian." *French Historical Studies* 8 (1973): 213–35.

Cole, Francis Joseph. "Harvey's Animals." *Journal of the History of Medicine and Allied Sciences* 12 (1957): 106–13.

———. *A History of Comparative Anatomy: From Aristotle to the Eighteenth Century*. 1949. Reprint. Mineola, NY: Dover, 1975.

Cook, Harold J. *Matters of Exchange: Commerce, Medicine and Science in the Age of Empire*. New Haven, CT: Yale University Press, 2007.

———. "Victories for Empiricism, Failures for Theory." In *The Body as Object and Instrument of Knowledge: Embodied Empiricism in Early Modern Science*, edited by Charles T. Wolfe and Ofer Gal, 9–32. Dordrecht, Netherlands: Springer, 2010.

Cooper, Alix. *Inventing the Indigenous: Local Knowledge and Natural History in Early Modern Europe*. Cambridge: Cambridge University Press, 2007.

Cordey, Jean. "Colbert, Le Vau, et la construction du château de Vincennes au XVIIe siècle." *Gazette des beaux-arts*, 6th ser., 9 (1933): 273–93.

Cordier, G. *Paris et les anatomistes au cours de l'histoire*. Paris: I.A.C., 1955.

Corlieu, August. *L'ancienne faculté de médecine de Paris*. Paris: Adrien Delahaye, 1877.

———. *Jacques Mentel, docteur-régent et professeur à la faculté de médecine de Paris (1599–1670)*. Paris: Adrien Delahaye and E. LeCrosnier, 1880.

Courtine, Jean-Jacques. "Le miroir de l'âme." In *Histoire du corps*. Vol. 1, *De la Renaissance aux Lumières*, edited by Georges Vigarello, 303–9. Paris: Seuil, 2005.

Cowart, Georgia. *The Triumph of Pleasure: Louis XIV and the Politics of Spectacle*. Chicago: University of Chicago Press, 2008.

Cunningham, Andrew. *The Anatomical Renaissance: The Resurrection of the Anatomical Projects of the Ancients*. Aldershot, UK: Scolar Press, 1997.

———. *The Anatomist Anatomis'd: An Experimental Discipline in Enlightenment Europe*. Farnham, UK: Ashgate, 2010.

Dagognet, François. *Le catalogue de la vie: Étude méthodologique sur la taxonomie*. Paris: Presses universitaires de France, 2004.

Darmon, Albert. *Les corps immatériels: Esprits et images dans l'œuvre de Marin Cureau de la Chambre*. Paris: Vrin, 1985.

Darnton, Robert. *The Great Cat Massacre: And Other Episodes in French Cultural History*. New York: Vintage, 1985.

Daston, Lorraine. "Description by Omission: Nature Enlightened and Obscured." In *Regimes of Description: In the Archive of the Eighteenth Century*, edited by John Bender and Michael Marrinan, 11–24. Stanford, CA: Stanford University Press, 2005.

———. "Marvelous Facts and Miraculous Evidence in Early Modern Europe." In *Questions of Evidence*, edited by James Chandler, Arnold I. Davidson, and Harry Harootunian, 243–74. Chicago: University of Chicago Press, 1994.

Daston, Lorraine, and Peter Galison. *Objectivity*. New York: Zone Books, 2007.

Daston, Lorraine, and Katharine Park, eds. *Early Modern Science*. The Cambridge History of Science, vol. 3. Cambridge: Cambridge University Press, 2006.

———. *Wonders and the Order of Nature, 1150–1750*. New York: Zone Books, 1998.

Dear, Peter. *Discipline and Experience: The Mathematical Way in the Scientific Revolution*. Chicago: University of Chicago Press, 1995.

———, ed. *The Literary Structure of Scientific Argument: Historical Studies*. Philadelphia: University of Pennsylvania Press, 1991.

———. "Marin Mersenne: Mechanics, Music and Harmony." In *Number to Sound: The Musical Way to the Scientific Revolution*, edited by Paolo Gozza, 267–88. Dordrecht, Netherlands: Springer, 2000.

———. *Mersenne and the Learning of the Schools*. Ithaca, NY: Cornell University Press, 1988.

———. "Method and the Study of Nature." In *The Cambridge History of Seventeenth-Century Philosophy*, edited by Daniel Garber and Michael Ayers, 147–77. Cambridge: Cambridge University Press, 1998.

———. *Revolutionizing the Sciences: European Knowledge and Its Ambitions, 1500–1700*. 2nd ed. Princeton, NJ: Princeton University Press, 2009.

DeJean, Joan. *Ancients against Moderns: Culture Wars and the Making of a Fin de Siècle*. Chicago: University of Chicago Press, 1997.

Delaporte, François. "Animal Blood." In *Figures of Medicine: Blood, Face Transplants, Parasites*, translated by Nils F. Schott, 1–24. New York: Fordham University Press, 2013.

Deleuze, J. P. F. *Histoire et description du Muséum Royal d'Histoire Naturelle*. 2 vols. Paris: M. A. Royer, au Jardin du Roi, 1823.

Delort, Robert. *Les animaux ont une histoire*. Paris: Seuil, 1993.

Demeulenaere-Douyère, Christiane. "Esquisse d'une histoire des archives de l'Académie des sciences." In *Histoire et mémoire de l'Académie des sciences: Guide de recherches*, edited by Christiane Demeulenaere-Douyère and Eric Brian, 45–54. Paris: Lavoisier Tec-Doc, 1996.

———. "Les sources documentaires conservées à l'Académie des sciences." In *Histoire et mé-*

moire de l'Académie des sciences: Guide de recherches, edited by Christiane Demeulenaere-Douyère and Eric Brian, 55–106. Paris: Lavoisier Tec-Doc, 1996.

Denonain, J. J. "Les problèmes de honnête homme vers 1635: *Religio medici* et les *Conférences du Bureau d'Adresse*." *Études anglaises* 18 (1965): 235–57.

Derrida, Jacques. *The Animal That Therefore I Am*. Translated by David Wills. New York: Fordham University Press, 2008.

Des Chene, Dennis. "*Animal* as Category: Bayle's 'Rorarius.'" In *The Problem of Animal Generation in Early Modern Philosophy*, edited by Justin E. H. Smith, 215–31. Cambridge: Cambridge University Press, 2006.

———. "Mechanisms of Life in the Seventeenth Century: Borelli, Perrault, Régis." *Studies in the History and Philosophy of Biological and Biomedical Sciences* 36 (2005): 245–60.

Diamond, Solomon. "Marin Cureau de la Chambre (1594–1669)." *Journal of the History of the Behavioral Sciences* 4 (1968): 40–50.

Di Renzi, Silvia. "Writing and Talking of Exotic Animals." In *Books and the Sciences in History*, edited by Marina Frasca-Spada and Nicholas Jardine, 151–67. Cambridge: Cambridge University Press, 2000.

Dupré, Sven, ed. *Optics, Instruments, and Painting, 1420–1720: Reflections on the Hockney-Falco Thesis*. Special issue, *Early Science and Medicine* 10 (2005).

Eales, Nellie B. "The History of the Lymphatic System, with Special Reference to the Hunter-Monro Controversy." *Journal of the History of Medicine and Allied Sciences* 29 (1974): 280–94.

Eamon, William. "Court, Academy, and Printing House: Patronage and Scientific Careers in Late Renaissance Italy." In *Patronage and Institutions: Science, Technology, and Medicine at the European Court, 1500–1750*, edited by Bruce T. Moran, 25–50. Rochester, NY: Boydell Press, 1991.

Elias, Norbert. *The Court Society*. Translated by Edmund Jephcott. New York: Pantheon Books, 1983.

Elkins, James. "Art History and Images That Are Not Art." *Art Bulletin* 77 (1995): 553–71.

Erlmann, Veit. "The Physiologist at the Opera: Claude Perrault's *Du bruit* (1680) and the Politics of Pleasure in the Ancien Régime." Unpublished manuscript, Microsoft Word file, 2012.

———. *Reason and Resonance: A History of Modern Aurality*. New York: Zone Books, 2010.

Farber, Paul L. "Buffon and Daubenton: Divergent Traditions within the *Histoire naturelle*." *Isis* 66 (1975): 63–74.

———. "Buffon and the Concept of Species." *Journal of the History of Biology* 5 (1972): 259–84.

Farge, Arlette. *Vivre dans la rue à Paris au XVIIIe siècle*. 1979. Reprint, Paris: Gallimard, 1992.

Farr, A. D. "The First Human Blood Transfusion." *Medical History* 24 (1980): 143–62.

Fauré-Fremiet, E. "Les origines de L'Académie des Sciences à Paris." *Notes and Records of the Royal Society* 21 (1966): 20–31.

Favier, Thierry, and Manuel Couvreur, eds. *Le plaisir musical en France au XVIIe siècle*. Sprimont, Belgium: Mardaga, 2006.

Ferrari, Giovanna. "Public Anatomy Lessons and the Carnival: The Anatomy Theater of Bologna." *Past and Present* 117 (November 1987): 50–106.

Fierro, Alfred, and Jean-Yves Sarrazin. *Le Paris des Lumières d'après le plan de Turgot (1734–1739)*. Paris: Réunion des musées nationaux, 2005.

Findlen, Paula. "The Economy of Scientific Exchange in Early Modern Italy." In *Patronage and Institutions: Science, Technology, and Medicine at the European Court, 1500–1750*, edited by Bruce T. Moran, 5–24. Rochester, NY: Boydell Press, 1991.

——. *Possessing Nature: Museums, Collecting, and Scientific Culture in Early Modern Italy.* Berkeley: University of California Press, 1994.

Flandrin, Jean-Louis. "Dietary Choices and Culinary Technique, 1500–1800." In *Food: A Culinary History,* edited by Jean-Louis Flandrin and Massimo Montanari, translated by Albert Sonnenfeld, 403–17. New York: Penguin, 2000.

Fosseyeux, Marcel. *L'Hôtel-Dieu de Paris au XVIIe et au XVIIIe Siècle.* Paris: Berger-Levrault, n.d. [ca. 1911].

——. "Le prix des cadavres à Paris, aux XVIIe et XVIIIe siècles." *Aesculape* 3 (1913): 52–56.

Foucart-Walter, Elisabeth. *Pieter Boel, 1622–1674: Peinture des animaux de Louis XIV.* Paris: Réunion des Musées Nationaux, 2001.

Foucault, Michel. *Discipline and Punish: The Birth of the Prison.* Translated by Alan Sheridan. New York: Pantheon, 1977.

Fournel, Victor. *Tableau du vieux Paris: Les spectacles populaires et les artistes des rues.* Paris: E. Dentu, 1863.

France, Peter. *Politeness and Its Discontents: Problems in French Classical Culture.* Cambridge: Cambridge University Press, 1992.

——. *Rhetoric and Truth in France: Descartes to Diderot.* Oxford: Clarendon Press, 1972.

Frank, Robert G., Jr. *Harvey and the Oxford Physiologists.* Berkeley: University of California Press, 1980.

Franklin, Alfred. *La vie privée d'autrefois: Les animaux.* 2 vols. Paris: Librairie Plon, 1897–99.

Franklin, K. J. "Jean Méry (1645–1722) and His Ideas on the Foetal Blood Flow." *Annals of Science* 5 (1945): 203–48.

Freedberg, David. *The Eye of the Lynx: Galileo, His Friends, and the Beginnings of Modern Natural History.* Chicago: University of Chicago Press, 2002.

French, Roger. *Dissection and Vivisection in the European Renaissance.* Burlington, VT: Ashgate, 1999.

——. *William Harvey's Natural Philosophy.* Cambridge: Cambridge University Press, 1994.

Friedman, Ann. "The Evolution of the Parterre d'eau." *Journal of Garden History* 8 (1988): 1–30.

——. "What John Locke Saw at Versailles." *Journal of Garden History* 9 (1989): 177–98.

Fumaroli, Marc. "Les abeilles et les araignées." In *La querelle des anciens et des modernes,* edited by Lecoq, 7–218. Paris: Gallimard, 2001.

——. *L'âge de l'éloquence: Rhétorique et "res literaria" de la Renaissance au seuil de l'époque classique.* New ed. Geneva: Droz, 2002.

——. "Nicolas-Claude Fabri De Peiresc: Prince de la république des lettres." *IVe centenaire de la naissance de Gassendi. Conférence organisée par l'association pro-Peyresq dans la maison d'Erasme à Anderlecht le mercredi 3 juin 1992.* Brussels: Association Pro-Peyresq, 1993.

——. "Nicolas Fouquet, the Favourite *manqué.*" In *The World of the Favourite,* edited by J. H. Elliot and L. W. B. Brockliss, 239–55. New Haven, CT: Yale University Press, 1999.

——. "Préface." In *Pascal, l'art de persuader, précédé de l'art de conférer de Montaigne.* Paris: Payot et Rivage, 2001.

Gabbey, Alan. "The Bourdelot Academy and the Mechanical Philosophy." *Seventeenth-Century French Studies* 6 (1984): 92–103.

Gaillard, Aurélia. "Bestiaire réel, bestiaire enchanté: Les animaux à Versailles sous Louis XIV." In *L'animal au XVIIe siècle,* edited by Charles Mazouer, 185–98. Tübingen, Germany: Günter Marr, 2003.

Gannal, F. "Cours d'anatomie au Jardin du Roi." *Bulletin de la Société de l'Histoire de Paris et de l'Ile-de-France* 20 (1893): 21–24.

Garber, Daniel. "Descartes and Experiment in the *Discourse* and *Essays.*" In Garber, *Descartes Embodied*, 85–110. Cambridge: Cambridge University Press, 2001.

———. "Descartes and Occasionalism." In Garber, *Descartes Embodied*, 203–20. Cambridge: Cambridge University Press, 2001.

———. "Experiment, Community, and the Constitution of Nature in the Seventeenth Century." In Garber, *Descartes Embodied*, 296–328. Cambridge: Cambridge University Press, 2001.

Garber, Daniel, and Michael Ayers, eds. *The Cambridge History of Seventeenth-Century Philosophy.* Vol. 1. Cambridge: Cambridge University Press, 2003.

Gauja, Pierre. "Les origines de l'Académie des Sciences de Paris." In *Institut de France: Troisième centenaire, 1666–1966.* 2 vols., 1:1–51. Paris: Gauthier-Villars, 1967.

Gaukroger, Stephen. "The Académie des Sciences and the Republic of Letters: Fontenelle's Role in the Shaping of a New Natural-Philosophical Persona, 1699–1734." *Intellectual History Review* 18 (2008): 385–402.

Gaukroger, Stephen, John A. Shuster, and John Sutton, eds. *Descartes' Natural Philosophy.* London: Routledge, 2000.

Gelfand, Toby. *Professionalizing Modern Medicine: Paris Surgeons and Medical Science and Institutions in the 18th Century.* Westport, CT: Greenwood Press, 1980.

George, Wilma. "Sources and Background to Discoveries of New Animals in the Sixteenth and Seventeenth Centuries." *History of Science* 18 (1980): 79–104.

Gilis, P. "Jean Pecquet." *Revue scientifique* 60 (1922): 221–27, 291–99.

Gillot, Hubert. *La querelle des anciens et des modernes en France.* Paris: Honoré Champion, 1914.

Glardon, Philippe. *L'histoire naturelle au XVIe siècle: Introduction, étude et édition critique de la nature et diversité des poissons de Pierre Belon (1555).* Geneva: Librairie Droz, 2011.

Godard de Donville, Louise, ed. *De la mort de Colbert à la révocation de l'Edit de Nantes: Un monde nouveau?* Marseille, France: Imprimerie A. Robert, 1984.

———, ed. *D'un siècle à l'autre: Anciens et modernes.* Marseille, France: Imprimerie A. Robert, 1987.

Goldberg, Benjamin. *The Mirror and Man.* Charlottesville: University of Virginia Press, 1985.

Goldgar, Anne. *Impolite Learning: Conduct and Community in the Republic of Letters, 1680–1750.* New Haven, CT: Yale University Press, 1995.

Goldstein, Carl. *Print Culture in Early Modern France: Abraham Bosse and the Purposes of Print.* New York: Cambridge University Press, 2012.

Goldstein, Claire. *Vaux and Versailles: The Appropriations, Erasures, and Accidents That Made Modern France.* Philadelphia: University of Pennsylvania Press, 2008.

Goodman, Dena. *The Republic of Letters.* Ithaca, NY: Cornell University Press, 1994.

Gordon, Daniel. "'Public Opinion' and the Civilizing Process in France: The Example of Morellet." *Eighteenth-Century Studies* 22 (1989): 302–28.

Gorham, Geoffrey. "Mind-Body Dualism and the Harvey-Descartes Controversy." *Journal of the History of Ideas* 55 (April 1994): 211–34.

Gouk, Penelope. *Music, Science, and Natural Magic in Seventeenth-Century England.* New Haven, CT: Yale University Press, 1999.

Grafton, Anthony. "Libraries and Lecture Halls." In *Cambridge History of Science.* Vol. 3, *Early Modern Science*, edited by Lorraine Daston and Katharine Park. Cambridge: Cambridge University Press, 2006.

Grafton, Anthony, and Lisa Jardine. *From Humanism to the Humanities.* Cambridge, MA: Harvard University Press, 1986.

Grande, Nathalie. "Une vedette des salons: Le caméléon." In *L'animal au XVIIe Siècle*, edited by Charles Mazouer, 89–102. Tübingen, Germany: Günter Marr, 2003.

Grant, Hugh. "The Revenge of the Paris Hat." *Beaver* 68, no. 6 (1988): 37–44.

Grell, Ole Peter. "Between Anatomy and Religion: The Conversions to Catholicism of Two Danish Anatomists, Nicolaus Steno and Jacob Winsløw." In *Medicine and Religion in Enlightenment Europe*, edited by Ole Peter Grell and Andrew Cunningham, 205–21. Burlington, VT: Ashgate, 2007.

Grivel, Marianne. "Le Cabinet Du Roi." *Revue de la Bibliothèque Nationale* 18 (1985): 36–57.

Grivel, Marianne, and Marc Fumaroli, eds. *Devises pour les tapisseries du roi*. Paris: Editions Herscher, 1988.

———. "Genèse d'un manuscrit." In *Devises pour les tapisseries du roi*, edited by Marianne Grivel and Marc Fumaroli. Paris: Editions Herscher, 1988.

Grmek, M. D. "Le débat sur le siège de la perception visuelle." In *La première révolution biologique: Réflexions sur la physiologie et la médecine du XVIIe siècle*, 189–229. Paris: Editions Payot, 1990.

———. *La première révolution biologique: Réflexions sur la physiologie et la médecine du XVIIe siècle*. Paris: Editions Payot, 1990.

Guénoun, Anne-Sylvie. "Les publications de l'Académie des sciences." In *Histoire et mémoire de l'Académie des sciences: Guide de recherches*, edited by Eric Brian and Christiane Demeulenaere-Douyère, 107–40. Paris: Lavoisier Tec-Doc, 1996.

Guerrini, Anita. "Anatomists and Entrepreneurs in Early Eighteenth-Century London." *Journal of the History of Medicine and Allied Sciences* 59, no. 2 (April 2004): 219–39.

———. "Duverney's Skeletons." *Isis* 94 (2003): 577–603.

———. "The Eloquence of the Body: Anatomy and Rhetoric at the Jardin du roi." In *Sustaining Literature*, edited by Greg Clingham, 271–87. Lewisburg, PA: Bucknell University Press, 2006.

———. "The Ethics of Animal Experimentation in Seventeenth-Century England." *Journal of the History of Ideas* 50, no. 3 (1989): 391–407.

———. *Experimenting with Humans and Animals: From Galen to Animal Rights*. Baltimore: Johns Hopkins University Press, 2003.

———. "Experiments, Causation, and the Uses of Vivisection in the First Half of the Seventeenth Century." *Journal of the History of Biology* 46 (2013): 227–54.

———. "Inside the Charnel House: The Display of Skeletons in Europe, 1500–1800." In *The Fate of Anatomical Collections*, edited by Rina Knoeff and Rob Zwijnenberg. Burlington, VT: Ashgate, forthcoming [2014].

———. "The King's Animals and the King's Books: The Illustrations for the Paris Academy's *Histoire des animaux*." *Annals of Science* 67, no. 3 (July 2010): 383–404.

———. *Natural History and the New World, 1524–1770: An Annotated Bibliography of Printed Materials in the Library of the American Philosophical Society*. Philadelphia: American Philosophical Society Library, 1986.

———. "Perrault, Buffon, and the Natural History of Animals." *Notes and Records of the Royal Society of London* 66 (2012): 393–409.

———. "The Value of a Dead Body." In *Vital Matters*, edited by Helen Deutsch and Mary Terrall, 246–64. Toronto: University of Toronto Press, 2012.

———. "The Varieties of Mechanical Medicine: Borelli, Malpighi, Bellini, and Pitcairne." *Nuncius* 27 (1997): 111–28.

Guivarc'h, Marcel. "Lieux des dissections et morgues dans Paris, de 1200 à nos jours." *Histoire des sciences médicales* 4 (2002): 432–50.

Hahn, Roger. *The Anatomy of a Scientific Institution: The Paris Academy of Sciences, 1666–1803*. Berkeley: University of California Press, 1971.

Hallays, André. *Les Perrault*. Paris: Perrin, 1926.

Hamy, Ernest Théodore. "Recherches sur les origines de l'enseignement de l'anatomie humaine et de l'anthropologie au Jardin des Plantes." *Nouvelles archives du Muséum*, 3rd ser., 7 (1895): 1–21.

———. "The Royal Menagerie of France and the National Menagerie Established on the 14th of Brumaire of the Year II (November 4, 1793)." *Annual Report of the Board of Regents of the Smithsonian Institution* (Washington, DC, 1898): 507–17.

Hanks, Lesley. *Buffon avant l'"Histoire naturelle."* Publications de la faculté des lettres et sciences humaines de Paris, vol. 24. Paris: Presses universitaires de France, 1966.

Harding, Vanessa. *The Dead and the Living in Paris and London, 1500–1670*. Cambridge: Cambridge University Press, 2007.

Harrison, Peter. "Reading Vital Signs: Animals and the Experimental Philosophy." In *Renaissance Beasts*, edited by Erica Fudge, 186–207. Urbana: University of Illinois Press, 2004.

———. "The Virtues of Animals in Seventeenth-Century Thought." *Journal of the History of Ideas* 59 (1998): 463–84.

Harth, Erica. *Cartesian Women: Versions and Subversions of Rational Discourse in the Old Regime*. Ithaca, NY: Cornell University Press, 1992.

Hazan, Eric. *The Invention of Paris*. Translated by David Fernbach. London: Verso, 2010.

Herrick, James A. *The History and Theory of Rhetoric: An Introduction*. 2nd ed. Needham Heights, MA: Allyn and Bacon, 2001.

Herrmann, Wolfgang. *The Theory of Claude Perrault*. London: A. Zwemmer, 1973.

Hillairet, Jacques. *Connaissance du vieux Paris*. 1956. Reprint, Paris: Payot & Rivages, 1993.

Hoff, Hebbel, and Roger Guillemin. "The First Experiments on Transfusion in France." *Journal of the History of Medicine and Allied Sciences* 18 (1963): 103–24.

Howard, Rio. *La bibliothèque et le laboratoire de Guy de la Brosse au Jardin des plantes à Paris*. Geneva: Droz, 1983.

———. "Guy de La Brosse and the Jardin des Plantes." In *The Analytic Spirit: Essays in the History of Science in Honor of Henry Guerlac*, edited by Harry Woolf, 195–344. Ithaca, NY: Cornell University Press, 1981.

———. "Medical Politics and the Founding of the Jardin des plantes in Paris." *Journal of the Society for the Bibliography of Natural History* 9 (1980): 395–402.

Hsia, Florence C. *Sojourners in a Strange Land: Jesuits and Their Scientific Missions in Late Imperial China*. Chicago: University of Chicago Press, 2009.

Huard, Pierre, and M. D. Grmek. "L'oeuvre de Charles Estienne et l'école anatomique parisienne." In *La dissection des parties du corps humain, 1546*. Paris: Au cercle du livre précieux, 1965.

Huppert, George. *The Style of Paris: Renaissance Origins of the French Enlightenment*. Bloomington: Indiana University Press, 1999.

Hyman, John. "The Cartesian Theory of Vision." *Ratio* 28 (1986): 149–67.

Innis, Harold A. *The Fur Trade in Canada: An Introduction to Canadian Economic History*. Toronto: University of Toronto Press, 1956.

Iriye, Masumi. "Le Vau's Menagerie and the Rise of the Animalier: Enclosing, Dissecting, and Representing the Animal in Early Modern France." PhD diss., University of Michigan, 1994.

Isherwood, Robert M. *Farce and Fantasy: Popular Entertainment in Eighteenth-Century Paris*. Oxford: Oxford University Press, 1986.

Jardine, Nicholas, J. A. Secord, and E. C. Spary, eds. *Cultures of Natural History*. Cambridge: Cambridge University Press, 1996.

Johns, Adrian. *The Nature of the Book*. Chicago: University of Chicago Press, 1998.

Jones, Colin. *Paris: The Biography of a City*. New York: Viking, 2004.

Jones, Matthew L. *The Good Life in the Scientific Revolution*. Chicago: University of Chicago Press, 2006.

———. "Writing and *Sentiment*: Blaise Pascal, the Vacuum, and the *Pensées*." *Studies in the History and Philosophy of Science* 32 (2001): 139–81.

Kaufmann, Thomas DaCosta. *The Mastery of Nature: Aspects of Art, Science, and Humanism in the Renaissance*. Princeton, NJ: Princeton University Press, 1991.

Keeble, K. Corey. "The Sincerest Form of Flattery: 17th-Century French Etchings of the Battles of Alexander the Great." *Rotunda* (Bulletin of the Royal Ontario Museum, Toronto) 16 (1983): 30–35.

Kennedy, George A. *Classical Rhetoric and Its Christian and Secular Tradition from Ancient to Modern Times*. 2nd ed. Chapel Hill: University of North Carolina Press, 1999.

Kettering, Sharon. *Patrons, Brokers, and Clients in Seventeenth-Century France*. Oxford: Oxford University Press, 1986.

Klestinec, Cynthia. "Civility, Comportment, and the Anatomy Theater: Girolamo Fabrici and His Medical Students in Renaissance Padua." *Renaissance Quarterly* 60 (2007): 434–63.

———. *Theaters of Anatomy: Students, Teachers, and Traditions of Dissection in Renaissance Venice*. Baltimore: Johns Hopkins University Press, 2011.

Konvitz, Josef. *Cartography in France, 1660–1848*. Chicago: University of Chicago Press, 1987.

Koyré, Alexandre. *Metaphysics and Measurement: Essays in Scientific Revolution*. Translated by R. E. W. Maddison. London: Chapman and Hall, 1968.

Kronick, David R. *"Devant le déluge" and Other Essays on Early Modern Scientific Communication*. Lanham, MD: Scarecrow Press, 2004.

———. *A History of Scientific and Technical Periodicals*. Metuchen, NJ: Scarecrow Press, 1976.

Kuriyama, Shigehisa. "Interpreting the History of Bloodletting." *Journal of the History of Medicine and Allied Sciences* 50 (1995): 11–46.

Kusukawa, S. *Picturing the Book of Nature: Image, Text, and Argument in Sixteenth-Century Human Anatomy and Medical Botany*. Chicago: University of Chicago Press, 2012.

———. "The Sources of Gessner's Pictures for the *Historia animalium*." *Annals of Science* 67 (2010): 303–28.

Lacroix, Jean-Bernard. "L'approvisionnement des ménageries et les transports des animaux sauvages par la Compagnie des Indes au XVIIIe siècle." *Revue française d'Histoire d'Outre-Mer* 65 (1978): 153–79.

Laissus, Yves. "Le Jardin du roi." In *Enseignement et diffusion des sciences en France au XVIIIe siècle*, edited by René Taton, 287–341. Paris: Hermann, 1964.

Le Blant, Robert. "Le commerce compliqué des fourrures Canadiennes au début du XVIIe siècle." *Revue d'histoire de l'Amérique française* 26 (1972): 53–66.

LeFanu, William R. "Jean Martet, a French Follower of Harvey." In *Science, Medicine and History*, edited by E. A. Underwood, 1:54–62. 2 vols. London: Oxford University Press, 1953.

Legée, Georgette. "Bio-bibliographie de Marin Cureau de la Chambre (1596–1669)." *Histoire et Nature* 8 (1976): 21–32.

LeMaguet, Paul. *Le monde médical parisien sous le Grand Roi*. Paris: Protat Frères, 1899.

Lévy-Valensi, J. *La médecine et les médecins français au XVIIe siècle.* Paris: Librairie J.-B. Baillière et Fils, 1933.

Lewis, Sarah Janvier. "Jean Pecquet (1622–1674) and the Thoracic Duct: The Controversy over the Circulation of the Blood and Lymph in Seventeenth-Century Europe." PhD diss., Yale University, 2003.

Licoppe, Christian. "The Crystallization of a New Narrative Form in Experimental Reports (1660–1690)." *Science in Context* 7 (1994): 205–44.

———. *La formation de la pratique scientifique: Le discours de l'expérience en France et en Angleterre (1630–1820).* Paris: La Découverte, 1996.

Lindberg, David C. *Theories of Vision from al-Kindi to Kepler.* Chicago: University of Chicago Press, 1976.

Loisel, Gustave. *Histoire des ménageries, de l'Antiquité de nos jours.* 3 vols. Paris: O. Doin, 1912.

Lokhorst, Gert-Jan. "Descartes and the Pineal Gland." *Stanford Encyclopedia of Philosophy,* 5 November 2008. http://plato.stanford.edu/entries/pineal-gland/.

Long, Pamela O. *Artisan/Practitioners and the Rise of the New Sciences, 1400–1600.* Corvallis: Oregon State University Press, 2011.

Lonie, I. M. "The 'Paris Hippocratics': Teaching and Research in Paris in the Second Half of the Sixteenth Century." In *The Medical Renaissance of the Sixteenth Century,* edited by Andrew Wear, Roger French, and Iain Lonie, 155–74. Cambridge: Cambridge University Press, 1985.

Loveland, Jeff. "Another Daubenton, Another *Histoire naturelle.*" *Journal of the History of Biology* 39 (2006): 457–91.

———. *Rhetoric and Natural History: Buffon in Polemical and Literary Context.* SVEC, no. 3. Oxford: Voltaire Foundation, 2001.

Lunel, Alexandre. *La maison médicale du roi: Le pouvoir royal et les professions de santé.* Seyssel, France: Champ Vallon, 2008.

Lux, David S. "Colbert's Plan for the *Grande Académie*: Royal Policy toward Science, 1663–67." *Seventeenth-Century French Studies* 12 (1990): 177–88.

Mabille, Gérard. "La ménagerie de Versailles." *Gazette des beaux-arts,* 6th ser., 83 (1974): 5–36.

Mabille, Gérard, and Joan Pieragnoli. *La Ménagerie de Versailles.* Arles, France: Éditions Honoré-Clair, 2010.

Maisonnier, Élisabeth, and Alexandre Maral, eds. *Le Labyrinthe de Versailles: Du mythe au jeu.* [Paris]: Magellan et cie, 2013.

Mallon, Adrian. "L'académie des sciences à Paris (1683–1685): Une crise de direction?" In *De la mort de Colbert à la révocation de l'Édit de Nantes,* edited by Louise Godard de Donville, 19–31. Marseille, France: Imprimerie A. Robert, 1984.

Mancosu, Paolo. "Acoustics and Optics." In *Early Modern Science,* vol. 3 of *The Cambridge History of Science,* edited by Lorraine Daston and Katharine Park, 596–631. Cambridge: Cambridge University Press, 2006.

Mandressi, Rafael. "Dissections et anatomie." In *Histoire du corps: De la Renaissance aux Lumières.* Vol. 1, edited by Georges Vigarello, 311–33. Paris: Seuil, 2005.

———. *Le regard de l'anatomiste: Dissections et invention du corps en occident.* Paris: Seuil, 2003.

Mandrou, Robert. *From Humanism to Science, 1480–1700.* Translated by Brian Pearce. Harmondsworth, UK: Penguin, 1978.

Marchand, Joseph. *L'Université d'Avignon aux XVIIe et XVIIIe siècles.* Paris: A. Picard et fils, 1900.

Marcialis, Maria Teresa. "Species: Percezioni e idee tra sei e settecento." *Rivista di storia della filosofia* 44 (1989): 647–76.

Margócsy, Daniel. "The Camel's Head: Representing Unseen Animals in Sixteenth-Century Europe." *Netherlands Yearbook of Art History* 61 (2011): 62–85.

Martin, H. T. *Castorologia.* London: Edward Stanford, 1892.

Mason, H. A. "Hommage à M. Despréaux: Some Reflections on the Possibility of Literary Study." *Cambridge Quarterly* 3 (1967–68): 51–71.

———. "The Miraculous Birth or the Founding of Modern European Literary Criticism." *Cambridge Quarterly* 11 (1982): 281–97.

Mazauric, Simone. *Fontenelle et l'invention de l'histoire des sciences à l'aube des Lumières.* Paris: Fayard, 2007.

McClaughlin, Trevor. "Censorship and Defenders of the Cartesian Faith in Mid-seventeenth Century France." *Journal of the History of Ideas* 40, no. 4 (1979): 563–81.

———. "Le concept de science chez Jacques Rohault." *Revue d'histoire des sciences* 30 (1977): 225–40.

———. "Descartes, Experiments, and a First Generation Cartesian: Jacques Rohault." In *Descartes' Natural Philosophy*, edited by Stephen Gaukroger, John Shuster, and John Sutton, 330–46. London: Routledge, 2000.

———. "Sur les rapports entre la Compagnie de Thévenot et l'Académie royale des sciences." *Revue d'histoire des sciences* 28 (1975): 235–42.

McHugh, Timothy J. "Establishing Medical Men at the Paris Hôtel-Dieu, 1500–1715." *Social History of Medicine* 19 (2006): 209–24.

Melchior-Bonnet, Sabine. *The Mirror: A History.* Translated by Katherine H. Jewett. New York: Routledge, 2001.

Meynell, Guy. "The Académie des sciences at the rue Vivienne, 1666–1699." *Archives internationales d'histoire de sciences* 44 (1994): 22–37.

———. "Surgical Teaching at the Jardin des plantes during the Seventeenth Century." *Gesnerus* 51 (1994): 101–8.

Miller, Elizabeth Maxfield. "Molière, l'Affaire Cressé, and le Médecin Fouetté et le Barbier Cocu." *PMLA* 72 (1957): 854–62.

Miller, Gordon L. "Beasts of the New Jerusalem: John Jonston's Natural History and the Launching of Millenarian Pedagogy in the Seventeenth Century." *History of Science* 46 (2008): 203–43.

Miller, Peter N. *Peiresc's Europe: Learning and Virtue in the Seventeenth Century.* New Haven, CT: Yale University Press, 2000.

Modlin, I. M. "Regnier de Graaf: Paris, Purging, and the Pancreas." *Journal of Clinical Gastroenterology* 30 (2000): 109–13.

Moore, Pete. *Blood and Justice.* Chichester, UK: Wiley, 2003.

Moran, Bruce, ed. *Patronage and Institutions: Science, Technology, and Medicine at the European Court, 1500–1750.* Woodbridge, Suffolk, UK: Boydell, 1991.

Motley, Mark Edward. *Becoming a French Aristocrat: The Education of the Court Nobility, 1580–1715.* Princeton, NJ: Princeton University Press, 1990.

Mousnier, Roland. *L'age d'or du mécénat (1598–1661).* Paris: CNRS, 1985.

Murphy, James Jerome. *Rhetoric in the Middle Ages.* Berkeley: University of California Press, 1974.

Naylor, Ronald. "Galileo: Real Experiment and Didactic Demonstration." *Isis* 67 (1976): 398–419.

———. "Galileo's Method of Analysis and Synthesis." *Isis* 81 (1990): 695–707.

Nonnoi, Giancarlo. "Against Emptiness: Descartes's Physics and Metaphysics of Pleni-tude." *Studies in History and Philosophy of Science* 25 (1994): 81–96.

Nordström, Johan. "Swammerdamiana: Excerpts from the Travel Journal of Olaus Bor-richius and Two Letters from Swammerdam to Thévenot." *Lychnos* 15 (1954): 21–65.

Ogilvie, Brian W. *The Science of Describing: Natural History in Renaissance Europe.* Chicago: University of Chicago Press, 2006.

Packard, Francis R. *Guy Patin and the Medical Profession in Paris in the XVIIth Century.* New York: Paul B. Hoeber, 1935.

Pagel, Walter. *New Light on William Harvey.* Basel, Switzerland: S. Karger, 1976.

Pastoureau, Michel. *The Bear: History of a Fallen King.* Translated by George Holoch. Cam-bridge, MA: Harvard University Press, 2011.

Pearl, J. L. "The Role of Personal Correspondence in the Exchange of Scientific Informa-tion in Early Modern France." *Renaissance and Reformation* 8 (1984): 106–13.

Pecker, Jean. "La chirurgie au xviie siècle." *XVIIe siècle* 40 (1988): 191–216.

Peltier, L. F. "Joseph Guichard Duverney (1648-1730), Champion of Applied Compara-tive Anatomy." *Clinical Orthopedics and Related Research* 187 (1984): 308–11.

Perfetti, Stefano. *Aristotle's Zoology and Its Renaissance Commentators (1521–1601).* Leuven, Belgium: Leuven University Press, 2000.

Perkins, Wendy. "Samuel Sorbière: Writings on Medicine." *Seventeenth-Century French Stud-ies* 8 (1986): 216–28.

———. "The Uses of Science: The Montmor Academy, Samuel Sorbière and Francis Ba-con." *Seventeenth-Century French Studies* 7 (1985): 155–62.

Perret, Jacques. *Le Jardin des plantes.* Paris: Julliard, 1984.

Petitfils, Jean-Christian. *Fouquet.* Paris: Perrin, 2009.

Peumery, Jean-Jacques. "Conversations médico-scientifiques de l'Académie de l'abbé Bourdelot (1610-1685)." *Histoire des sciences médicales* 12 (1978): 127–35.

———. *Les origines de la transfusion sanguine.* Amsterdam: B. M. Israël, 1975.

Pickstone, John V. *Ways of Knowing: A New History of Science, Technology, and Medicine.* Chicago: University of Chicago Press, 2001.

Picon, Antoine. *Claude Perrault ou la curiosité d'un classique.* Paris: Picard, 1988.

Pinon, Laurent. *Livres de zoologie de la Renaissance: Une anthologie, 1450–1700.* Corpus iconographique de l'histoire du livre 2. Paris: Klincksieck, 1995.

Pintard, René. "Autour de Pascal: L'Académie Bourdelot et le problème du vide." In *Mé-langes d'histoire littéraire offerts à Daniel Mornet*, 73–81. Paris: Nizet, 1951.

———. *Le libertinage érudit dans la première moitié du XVIIème siècle.* 1943. Reprint, Paris: Honoré Champion, 2000.

Pomata, Gianna. "*Praxis historialis*: The Uses of *Historia* in Early Modern Medicine." In *Historia: Empiricism and Erudition in Early Modern Europe*, edited by Gianna Pomata and Nancy G. Siraisi, 105–46. Cambridge, MA: MIT Press, 2005.

Pomata, Gianna, and Nancy G. Siraisi. *Historia: Empiricism and Erudition in Early Modern Europe.* Cambridge, MA: MIT Press, 2005.

Portal, Antoine. *Histoire de l'anatomie et de la chirurgie.* 6 vols. Paris: Chez P. Fr. Didot le jeune, 1770–73.

Posner, Donald. "Charles Lebrun's *Triumphs of Alexander.*" *Art Bulletin* 41 (1959): 237–48.

Poulsen, Jacob E., and Egill Snorrason. *Nicolaus Steno, 1638–1686: A Re-consideration by Danish Scientists.* Copenhagen: Nordisk Insulinlaboratorium, 1986.

Préaud, Maxime. "'L'Académie des sciences et des beaux-arts': Le testament graphique de Sébastien Leclerc." *Racar: Revue d'art canadienne* 10 (1983): 73–81.

———. *Inventaire du fonds français, Graveurs du XVIIᵉ siècle*. Vol. 9, *Sébastien Leclerc II*. Paris: Bibliothèque nationale, 1980.

Rabinovitch, Oded. "Chameleons between Science and Literature: Observation, Writing, and the Early Parisian Academy of Sciences in the Literary Field." *History of Science* 51 (2013): 33–62.

———. "Versailles as a Family Enterprise: The Perraults, 1660–1700." *French Historical Studies* 36 (2013): 385–416.

Radner, Daisie, and Michael Radner. *Animal Consciousness*. Buffalo, NY: Prometheus Books, 1989.

Ragland, Evan. "Experimenting with Chemical Bodies: Science, Medicine, and Philosophy in the Long History of Reinier de Graaf's Experiments on Digestion, from Harvey and Descartes to Claude Bernard." PhD diss., Indiana University, 2012.

———. "Making Trials in Sixteenth- and Seventeenth-Century Medicine: On the History of Experiment from Medicine to Science." Unpublished manuscript, PDF file, 2012.

Ranum, Orest. *Paris in the Age of Absolutism*. 1968. Reprint, Bloomington: Indiana University Press, 1979.

Rather, Lelland J. "The 'Six Things Non-natural': A Note on the Origins and Fate of a Doctrine and a Phrase." *Clio Medica* 3, no. 4 (1968): 337–47.

Reif, Patricia. "The Textbook Tradition in Natural Philosophy, 1600–1650." *Journal of the History of Ideas* 30 (1969): 17–32.

Renaud, Jean. *Les communautés de maîtres chirurgiens avant la Révolution de 1789 en Forez*. Saint-Étienne, France: Éditions du Chevalier, 1946.

Renouard, Philippe. *Répertoire des imprimeurs parisiens: Libraires et fondeurs de caractères en exercice à Paris au XVIIe siècle*. Nogent le roi, France: Librairie des arts et métiers-éditions, 1995.

Revel, Jacques. "The Uses of Civility." In *A History of Private Life*, vol. 3, edited by Roger Chartier, 167–206. Cambridge, MA: Belknap Press of Harvard University Press, 1989.

Robbins, Louise E. *Elephant Slaves and Pampered Parrots: Exotic Animals in Eighteenth-Century Paris*. Baltimore: Johns Hopkins University Press, 2002.

Rodis-Lewis, Geneviève. *Descartes: His Life and Thought*. Translated by Jane Marie Todd. Ithaca, NY: Cornell University Press, 1999.

———. "L'écrit de Desgabets sur la transfusion du sang et sa place dans les polémiques contemporaines." *Revue de synthèse* 95 (1974): 31–64.

Roger, Jacques. *Buffon: A Life in Natural History*. Translated by Sarah Bonnefoi. Ithaca, NY: Cornell University Press, 1997.

———. *The Life Sciences in Eighteenth-Century French Thought*. Edited by Keith R. Benson. Translated by Robert Ellrich. Stanford, CA: Stanford University Press, 1997.

Rosenfield, Leonora Cohen. *From Beast-Machine to Man-Machine*. New ed. New York: Octagon, 1968.

Roule, Lucie. *Les médecins du Jardin du roi aux XVIIᵉ et XVIIIᵉ siècles. Origine médicale du Muséum nationale d'histoire naturelle*. Paris: R. Foulon, 1942.

Ruhlmann, Henri Charles. *Sur l'anatomie de quelques grands chirurgiens du xviᵉ siècle*. Angers, France: H. Sirardeau, 1939.

Sabatier, Gérard. "La gloire du roi: Iconographie de Louis XIV de 1661 à 1672." *Histoire, economie et société* 19 (2000): 527–60.

Sahlins, Peter. "The Royal Menageries of Louis XIV and the Civilizing Process Revisited." *French Historical Studies* 35 (2012): 237–67.

———. "A Story of Three Chameleons: The Animal between Science and Literature in the

Age of Louis XIV." Paper delivered at the Western Society for French History, Banff, Canada, 11–13 October 2012.

Salomon-Bayet, Claire. L'institution de la science et l'expérience du vivant. Paris: Flammarion, 1978.

Sarasohn, Lisa T. "Nicolas-Claude Fabri de Peiresc and the Patronage of the New Science in the Seventeenth Century." Isis 84 (1993): 70–90.

Saunders, E. S. "Politics and Scholarship in Seventeenth-Century France: The Library of Nicolas Fouquet and the Collège Royal." Journal of Library History 20 (1985): 1–24.

Saville, Merilyn. "The Triple Portrait of Pierre Bernard, Gérard Edelinck and Nicolas De Largillière and the Debate in the French Academy in 1686 over the Status of Engravers." Melbourne Art Journal 5 (2001): 41–52.

Schiller, Joseph. "Les laboratoires d'anatomie et de botanique à l'Académie des Sciences au XVIIᵉ siècle." Revue d'histoire des sciences 17 (1964): 97–114.

———. "La transfusion sanguine et les débuts de l'Académie des sciences." Clio medica 1 (1965): 33–40.

Schmaltz, Tad M. "What Has Cartesianism to Do with Jansenism?" Journal of the History of Ideas 60 (1999): 37–56.

Schmitt, Charles. "Experience and Experiment: A Comparison of Zabarella's View with Galileo's in De motu." Studies in the Renaissance 16 (1969): 80–138.

Schmitt, Charles, and Charles Webster. "Harvey and M. A. Severino: A Neglected Medical Relationship." Bulletin of the History of Medicine 45 (1971): 49–75.

Schouten, J. "Johannes Walaeus (1604–1649) and His Experiments on the Circulation of the Blood." Journal of the History of Medicine and Allied Sciences 29 (1974): 259–79.

Schwartz, Vanessa R. Spectacular Realities: Early Mass Culture in Fin-de-Siècle Paris. Berkeley: University of California Press, 1999.

Sedgwick, Adam. Jansenism in Seventeenth-Century France. Charlottesville: University of Virginia Press, 1977.

Senior, Matthew. "The Ménagerie and the Labyrinth: Animals at Versailles, 1662–1792." In Renaissance Beasts, edited by Erica Fudge, 208–32. Urbana and Chicago: University of Illinois Press, 2004.

———. "Pierre Dionis and Joseph-Guichard Duverney: Teaching Anatomy at the Jardin du Roi, 1673–1730." Seventeenth-Century French Studies 26 (2004): 153–69.

Shank, J. B. The Newton Wars and the Beginning of the French Enlightenment. Chicago: University of Chicago Press, 2008.

Shapin, Steven. "Pump and Circumstance: Robert Boyle's Literary Technology." Social Studies of Science 14 (1984): 481–520.

———. The Scientific Revolution. Chicago: University of Chicago Press, 1996.

———. A Social History of Truth. Chicago: University of Chicago Press, 1994.

Shapin, Steven, and Simon Schaffer. Leviathan and the Air-Pump. Princeton, NJ: Princeton University Press, 1985.

Shelford, April G. Transforming the Republic of Letters: Pierre-Daniel Huet and European Intellectual Life, 1650–1720. Rochester, NY: University of Rochester Press, 2007.

Simon, Gérard. "On the Theory of Visual Perception of Kepler and Descartes: Reflections on the Role of Mechanism in the Birth of Modern Science." Vistas in Astronomy 18 (1975): 825–32.

Siraisi, Nancy G. Medieval and Early Renaissance Medicine: An Introduction to Knowledge and Practice. Chicago: University of Chicago Press, 1990.

———. "Vesalius and Human Diversity in De humani corporis fabrica." Journal of the Warburg and Courtauld Institutes 57 (1994): 60–88.

Sloan, Phillip R. "The Buffon-Linnaeus Controversy." *Isis* 67 (1976): 356–75.

———. "From Logical Universals to Historical Individuals: Buffon's Idea of Biological Species," 101–40 in Atran et al., *Histoire du concept d'espèce dans les sciences de la vie*.

Smith, Pamela H. *The Body of the Artisan: Art and Experience in the Scientific Revolution*. Chicago: University of Chicago Press, 2004.

Soll, Jacob. "From Note-Taking to Data Banks: Personal and Institutional Information Management in Early Modern Europe." *Intellectual History Review* 20 (2010): 355–75.

———. "Jean-Baptiste Colbert's Republic of Letters." *Republics of Letters: A Journal for the Study of Knowledge, Politics, and the Arts* 1, no. 1 (May 1, 2009). http://rofl.stanford.edu /node/28.

Solomon, Howard. *Public Welfare, Science, and Propaganda in Seventeenth-Century France: The Innovations of Théophraste Renaudot*. Princeton, NJ: Princeton University Press, 1972.

Sorell, Tom. *Descartes Reinvented*. Cambridge: Cambridge University Press, 2005.

Sørensen, Madeleine Pinault. "Les animaux du roi: De Pieter Boel aux dessinateurs de l'Académie Royale des Sciences." In *L'animal au XVIIe siècle*, edited by Charles Mazouer, 159–83. Tübingen, Germany: Günter Marr, 2003.

Soriano, Marc. *Le dossier Perrault*. Paris: Hachette, 1968.

Spary, E. C. *Utopia's Garden: French Natural History from Old Regime to Revolution*. Chicago: University of Chicago Press, 2000.

Star, Susan Leigh, and James R. Griesemer. "Institutional Ecology, 'Translations' and Boundary Objects: Amateurs and Professionals in Berkeley's Museum of Vertebrate Zoology, 1907–39." *Social Studies of Science* 19 (1989): 387–420.

Stroup, Alice. *A Company of Scientists: Botany, Patronage, and Community at the Seventeenth-Century Parisian Royal Academy of Sciences*. Berkeley: University of California Press, 1990.

———. *Royal Funding of the Parisian Académie Royale des sciences during the 1690s*. Transactions of the American Philosophical Society, vol. 77, pt. 4. Philadelphia: American Philosophical Society, 1987.

Sturdy, David J. *Richelieu and Mazarin: A Study in Statesmanship*. London: Palgrave Macmillan, 2003.

———. *Science and Social Status: The Members of the Académie des sciences, 1666–1750*. Woodbridge, UK: Boydell, 1995.

Sutton, Geoffrey V. *Science for a Polite Society: Gender, Culture, and the Demonstration of Enlightenment*. Boulder, CO: Westview Press, 1997.

Sutton, John. *Philosophy and Memory Traces: Descartes to Connectionism*. Cambridge: Cambridge University Press, 1998.

Sweetser, Marie-Odile. *La Fontaine*. Boston: Twayne, 1987.

Taton, René, ed. *Enseignement et diffusion des sciences en France au XVIIIe siècle*. Paris: Hermann, 1964.

———. "Mariotte et l'Académie Royale des Sciences." In *Mariotte, Savant et Philosophe: Analyse d'une renommé*, edited by Pierre Costabel, 13–32. Paris: J. Vrin, 1986.

———. *Les origines de l'Académie Royale des Sciences*. Alençon, France: Presses de l'imprimerie Alençonnaise, 1966.

———. "Présentation d'ensemble espoirs et incertitudes de la science française à la mort de Colbert." In *De la mort de Colbert à la révocation de l'Édit de Nantes: Un monde nouveau?*, edited by Louise Godard de Donville, 9–17. Marseille, France: C.M.R., 1984.

Tenenti, Alberto. "Claude Perrault et la pensée scientifique française dans la seconde moitié du XVIIe siècle." In *Éventail de l'histoire vivante: Hommage à Lucien Febvre*, 2:308–16. 2 vols. Paris: Armand Colin, 1953.

Thomas, Keith. *Man and the Natural World: A History of the Modern Sensibility.* New York: Pantheon, 1983.

Tits-Dieuaide, Marie-Jeanne. "Colbert et l'Académie royale des sciences." In *L'histoire grande ouverte: Hommages à Emmanuel Le Roy Ladurie,* edited by André Burguière, Joseph Goy, and Marie-Jeanne Tits-Dieuaide, 214–20. Paris: Fayard, 1997.

———. "Une institution sans statuts: L'Académie royale des sciences de 1666 à 1699." In *Histoire et mémoire de l'Académie des sciences: Guide de recherches,* edited by Eric Brian and Christiane Demeulenaere-Douyère, 3–13. Paris: Lavoisier Tec-Doc, 1996.

Truc, Gonzague. *Le quartier St-Victor et le Jardin des plantes.* Paris: Firmin-Didot, 1930.

Tucker, Holly. *Blood Work.* New York: Norton, 2011.

Ultee, Maarten. "The Republic of Letters: Learned Correspondence, 1680–1720." *Seventeenth Century* 2 (1987): 95–112.

Van Vree, Wilbert. *Meetings, Manners and Civilization: The Development of Modern Meeting Behaviour.* Translated by Kathleen Bell. Leicester, UK: Leicester University Press, 1999.

Van Wymeersch, Brigitte. "Descartes et le plaisir de l'émotion." In *Le plaisir musical en France au XVIIe siècle,* edited by Thierry Favier and Manuel Couvreur, 49–59. Sprimont, Belgium: Mardaga, 2006.

Vickers, Brian, and Nancy S. Struever. *Rhetoric and the Pursuit of Truth: Language Change in the Seventeenth and Eighteenth Centuries.* Los Angeles: Clark Library, 1985.

Wade, Nicholas J. *A Natural History of Vision.* Cambridge, MA: MIT Press, 2000.

Walmsley, Peter. *Locke's "Essay" and the Rhetoric of Science.* Lewisburg, PA: Bucknell University Press, 2003.

Warner, Marina. *No Go the Bogeyman: Scaring, Lulling, and Making Mock.* New York: Ferrar, Straus, and Giroux, 1998.

Watson, E. C. "The Early Days of the Académie des sciences as Portrayed in the Engravings of Sébastien Le Clerc." *Osiris* 7 (1939): 556–87.

Wear, Andrew. "William Harvey and the 'Way of the Anatomists.'" *History of Science* 21 (1983): 223–49.

Wear, Andrew, Roger French, and Iain Lonie, eds. *The Medical Renaissance of the Sixteenth Century.* Cambridge: Cambridge University Press, 1985.

Weber, Jacques. "L'Inde française de la compagnie de Colbert à la session des comptoirs." *Monde et culture* 56 (1996): 233–46.

Weil, E. "The Echo of Harvey's *De motu cordis,* 1628 to 1657." *Journal of the History of Medicine and Allied Sciences* (1957): 167–74.

Wellman, Kathleen. *Making Science Social: The Conferences of Théophraste Renaudot, 1633–1642.* Norman: University of Oklahoma Press, 2003.

West, William N. *Theatres and Encyclopedias in Early Modern Europe.* Cambridge: Cambridge University Press, 2002.

Westfall, Richard S. *Force in Newton's Physics.* New York: Science History Publications, 1971.

———. "Unpublished Boyle Papers Relating to Scientific Method." *Annals of Science* 12 (1956): 63–73, 103–17.

Whitteridge, Gweneth. *William Harvey and the Circulation of the Blood.* London: Macdonald, 1971.

Wright, John P. "The Embodied Soul in Seventeenth-Century French Medicine." *Canadian Bulletin of Medical History* 8 (1991): 21–42.

Yates, Frances. *The French Academies of the Sixteenth Century.* 1947. Reprint, London: Routledge, 1989.

Zouckermann, R. "Air Weight and Atmospheric Pressure from Galileo to Torricelli." *Fundamenta scientiae* 2 (1981): 185–204.

INDEX

Page numbers followed by the letter *f* indicate a figure.